BIODIVERSITY CONSERVATION AND ENVIRONMENTAL MANAGEMENT

BIODIVERSITY CONSERVATION AND ENVIRONMENTAL MANAGEMENT

Editors:
D.R. Khanna
R. Bhutiani
Gagan Matta
Vikas Singh

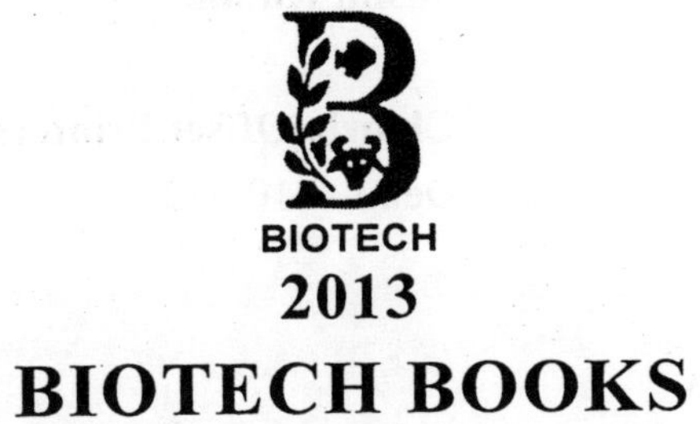

2013
BIOTECH BOOKS

ISBN: 978-81-7622-262-4

Published by : **BIOTECH BOOKS**
4762-63/23, Ansari Road, Darya Ganj
New Delhi - 110 002
Phone: +91-011-23262132
E-mail: biotechbooks@yahoo.co.in

Laser Typesetting : **Sushil Kumar**

Printed at : **Chawla Offset Printers**
Delhi - 110 052

PRINTED IN INDIA

PREFACE

Biodiversity Conservation and Environmental Management are on the agenda of almost all governments of developing and developed countries as conservationists stress on this issue from long time and again upon the fact that man can exist only if it has a symbiotic relationship with nature. Every day 150 species disappear and every hour 3 species bid good bye to the planet. While species have been appearing and disappearing on earth since its evolution, the accelerated rate of their disappearance has much to do with unethical exploitation of nature by man. By the end of this century, we *homosapiens* will wipe out more than 10 million species with desertification and climate change happening at such an alarming rate. The damage we cause to nature will boomerang is amply suggested a World Bank study that claims that climate change and farmer suicides in India are correlated. Thus it is also clear that neglecting the environment is having economic consequences as well.

Biodiversity Conservation and Environmental Management movement in India has always moved at an exciting tempo at the grass root level where it is directly concerned and linked with human survival. Now the time is ripe to address these issues holistically as very existence of life on earth is at stake. Environmental issues of the day like food security, livelihood vulnerability, poverty at the micro level and issues desertification, flooding, climate change, loss of biodiversity *etc.* at macro level need to be addressed collectively by stake holders to make a difference.

In pursuance of our goal of importing quality based education we have tried to provide real world situation about biodiversity conservation and environmental management in this book. The book mainly highlights the conservation, importance and other related issues of biodiversity. The present book features recent findings from leading academicians of national and international repute, researchers, scientists, policy makers, industrialists in their respective fields.

We trust this book will not only provide a viewful forum to readers but will also be professionally beneficial to them. We thank our colleagues, researchers who contributed their precious work, without whom it would not be possible for us to compile this book. We also thank State Biotechnology Department, Dehradun, Govt. of India, Uttarakhand for providing financial assistance for publishing this book. Last but not least we thank our publishers M/s Biotech Books for publishing this book.

Despite our best intentions, errors are sure to have slipped by us. Please let us know of any you find.

Authors

CONTENTS

Biodiversity Conservation and Envir. Management (2012)
Editors: D.R. Khanna et al.
Pub. by Biotech Books. *ISBN: 978-81-7622-262-4*

Pages: 1-13

BIODIVERSITY CONSERVATION AND WILD LIFE MANAGEMENT

G.N. Vankhede
Department of Zoology, SGB Amravati University,
Amravati, Maharashtra (INDIA)
E-mail: vganeshan2001@rediffmail.com

Biodiversity at gene, species or ecosystem level is a form of natural capital. India has a total of 89,451 animal species accounting for 7.31% of the faunal species in the world and the flora accounts for 10.78% of the global total. *The endemism of Indian biodiversity is high - about 33% of the country's recorded flora are endemic to the country and are concentrated mainly in the North-East, Western Ghats, North-West Himalayas and the Andaman and Nicobar islands. However, this rich biodiversity of India is under severe threat owing to habitat destruction, degradation, fragmentation and over-exploitation of resources.*

According to the Red List of Threatened species in India, 44 plant species are critically endangered, 113 endangered and 87 vulnerable. Amongst animals, 18 are critically endangered, 54 endangered and 143 are vulnerable. Ten species are Lower Risk conservation dependent, while 99 are Lower Risk near threatened. India ranks second in terms of the number of threatened mammals, while India is sixth in terms of countries with the most threatened birds.

The major proximate causes of species extinction are habitat loss and degradation affecting 89 percent of all threatened birds, 83 percent of mammals and 91 percent of all threatened plants assessed globally. The main causes of habitat loss are agricultural activities, extraction (including mining, fishing, logging and harvesting) and development (human settlements, industry and associated infrastructure). Habitat loss and fragmentation leads to the formation of isolated, small, scattered populations. These small populations are increasingly vulnerable to inbreeding depression, high infant mortality and susceptible to environmental stochasticity and consequently in the end, possible extinction. Changes in forest composition and quality and the resultant habitat type lead to declines in primary food species for wildlife.

The climate changes taking place in the world today are affecting not only humans, but also the wildlife. The natural habitat as well as migration patterns of the animals and birds is experiencing disturb patterns.

Exploitation such as hunting, collecting, fisheries and trade are a major threat to birds (37%), mammals (34%), plants (8% of those assessed), reptiles and marine fishes. In India, poaching is another insidious threat that has emerged in recent years as one of the primary reasons for the decline in numbers of species, such as the tiger. Poaching pressures, however, are unevenly distributed since certain selected species are more heavily targeted than others are. Alien invasive species are a significant threat affecting 350 (30% of all threatened) birds and 361 (15% of all threatened) plant species. Islands are particularly susceptible to invasions of alien species.

The underlying causes of biodiversity loss, however are poverty, macroeconomic policies, international trade factors, policy failures, poor environmental law/weak enforcement, unsustainable development projects and lack of local control over resources. Population pressures and concomitant increases in the collection of fuelwood and fodder, and grazing in forests by local communities too take their toll on the forests, and consequently its biodiversity.

Efforts for conservation and management of our biodiversity must derive from a set of clear objectives, mechanisms for action, and commitment from all stakeholders. Apart from this, halting the process of degradation and species loss requires specialized solutions and an understanding of ecological processes. Protecting biodiversity does not merely involve setting aside chunks of area as reserves. Instead, all the ecological processes that have maintained the area's biodiversity such as predation, pollination, parasitism, seed dispersal and herbivory, involving complex interactions between several species of plants and animal needs to be ensured. This, however, is possible only if reserves are large enough to maintain these processes and some of the other crucial links in the web of life.

There is also the need for greater involvement of communities and for models which decentralize the management and conservation roles and responsibilities. As of now, there are still major lacunae in information resources pertaining to forests, biodiversity - flora and fauna, causative factors for their degradation and major threats. The available data is alarmingly inadequate to provide a lucid picture of the current status and ongoing losses/gains. *More importantly, laws and policies governing natural resources are still not sufficient enough to tackle the scale of the problem, and these insufficiencies have not been addressed with a sense of urgency. In this section, we provide some indicators of the biological policy, technological, and institutional issues that will help mitigate the disturbingly accelerating biodiversity loss.*

CONSERVATION STRATEGIES

Expansion of the protected area network

Maintaining viable populations of species, whether plant or animal, is a crucial factor in biodiversity conservation and this requires the appropriate conservation of important ecosystems and habitats. Currently, India has 88 national parks, and 490 sanctuaries. In 1988, Rodgers and Panwar conducted a comprehensive study in which they drew up plans for a protected area network to cover the range of biological diversity in the country. They suggested that the percentage of the country's area under the protected area network, which was then only 3.3 %, be enhanced to 4.6 %. The recent update of the Rodgers and Panwar report suggests that the area under parks and sanctuaries be increased to 160 National Parks and 698 Sanctuaries accounting for 5.69 % of the country's geographical total. This coverage will provide a 'better distribution of protected areas with less gaps in the protection of biogeographic zones, biomes and species and fewer spatial or geographic gaps in the pattern of PA coverage'. Currently, the protected area network does not adequately cover some important biomes and mammal species of conservation significance.

Conservation strategies are urgently needed for the protection of species and ecosystems, involving a mix of *in situ* and *ex situ* strategies. Some of the steps in such a policy include the following.

Corridor concept

Biodiversity corridors are areas of habitat that provide functional linkages between protected areas to

(i) conserve habitat for species movement and for the maintenance of viable populations,
(ii) conserve and restore ecosystem services, and
(iii) enhance local community welfare through the conservation and sustainable use of natural resources.

Habitat loss and fragmentation are major threats to biodiversity conservation. One way of mitigating the negative effects of fragmentation is to improve habitat connectivity. Habitat corridors have been shown to be valuable for the conservation of various groups of wildlife and in various situations (*e.g.* urban, agricultural, production forest landscapes), although individual species vary in their use of corridors. Retained areas of native forest within plantations are beneficial for wildlife conservation, although their main benefit may be retained as habitat rather than dispersal corridors. It is in this context that corridors have become a significant factor in conservation management systems, in an attempt to reduce the isolation of spatially separated populations and to potentially increase the total area of habitat available.

Recent research in India has highlighted the value of wildlife habitat strips for conservation of vegetation and birds. Wildlife habitat strips form an important component of the conservation program in production forest landscapes. However, they cannot stand alone as a conservation measure; complementary measures such as reservation of extensive forested areas are also necessary.

The value of a corridor depends on its spatial configuration, landscape context, habitat type, scale, the nature of the connected areas and the species likely to use the corridor. These factors determine corridor 'quality', which also varies depending on the taxa concerned (Anderson and Danielson 1997). Conservation aims also determine how valuable a corridor is from a human perspective, and explicit goals are recommended in corridor design (Wilson and Lindenmayer 1995).

Population surveys, assessments and database creation

Adequate data on species diversity, populations, location and extent of habitat, major threats to different species, *etc.*, and changes in these aspects over time are not available to design a proper strategy for conservation. Given our extensive diversity, ecological surveys and taxonomic investigations need to be intensified, particularly for plants and insects. For conducting ecological studies, species for such studies can be prioritized including keystone species, large mammals, migratory species *etc.* The country's wildlife institutions need to network and coordinate their activities so that priority issues and areas are identified. The MoEF (Ministry of

Environment and Forests) through the Botanical Surveys of India and the Zoological Survey of India could play a guiding role by preparing a list of priority issues and areas for circulation to relevant institutions, based on a countrywide consultation of experts. Funding for these prioritized projects could be stepped up to ensure that research focuses on these issues. The MoEF must set up a database for the country as a whole.

Mapping of forest types, protected areas and natural forests

It is important to generate maps of the protected areas of the country showing their contiguity with the existing reserve and protected forests. This will provide a way for determining possible corridors, habitat contiguity, and buffer zones and facilitate biodiversity conservation. Further, vegetation mapping according to forest types needs to be done for the entire country. The Forest Survey of India data need to show the disaggregated changes in area according to forest types and natural forest areas. Current data on canopy densities make it difficult to estimate the changes in area under dense, natural or near-natural forests.

Improved protection efforts and a landscape approach to conservation

Owing to habitat fragmentation and consequent losses suffered by different populations, there is need for ensuring the safety of the biodiversity lying outside our protected areas. Population viability analyses for different species have revealed that the loss of even a single individual from a small isolated population could adversely affect the population structure and viability and push many species towards extinction. The need of the hour is a landscape approach to conservation where protected and non-protected areas are integrated through significant protection measures initiated at both the state and local community levels. This will include the following activities.

- Mapping the distribution of habitat types in the region as types of LSEs (landscape elements) with the help of satellite imagery along with field surveys.
- Establishing association of groups of species with different types of LSEs on the basis of field surveys.

- Assessing rates of transformations of LSE types with the help of satellite imagery of earlier years, official records and oral histories.
- Assessing the threats to different species as a result of ongoing landscape changes and other causes like commercial harvests.
- Assigning conservation priorities to species on the basis of threats to their populations, rarity, endemism and taxonomic distinctiveness.
- Assigning conservation priorities to different types of habitats or LSEs on the basis of richness and conservation significance of the threatened species they harbour.

In situ conservation

Good management of the extent populations in the wild is essential for their survival. This includes facilitating gene flow through the creation of corridors, introduction of new genetic stock and translocation of animals. While habitat protection and corridor creation may be possible, large-scale manipulations of the extant populations, *e.g.* through translocations, however, are yet to catch on in India because of financial and logistical problems. There are some exceptions such as the proposed transfer of some lions from Gir to Kuno Palpur in Madhya Pradesh.

Regular population-habitat viability and risk simulations

Computer models for regular population and habitat viability assessments should be generated on the basis of newly identified threats, population status and other relevant research. These models can provide estimates for the survival of these populations for the next 100 years while also indicating the various steps needed for improving the overall conservation status of the species. This would prove indispensable for saving many threatened and endangered species. Captive breeding and species reintroduction. Captive breeding aims at maintaining viable and healthy genetic captive stocks in conservation facilities and is meant to supplement *in situ* initiatives. Captive breeding can provide animals for possible reintroduction to the wild at a later stage or for supplementing current populations with new stock. *In situ* measures, however, are always preferred, since in most cases it is cheaper to protect populations in their natural habitat than to reintroduce captive-bred ones.

- establishing minimum target population goals to provide for maintenance of captive genetic diversity for at least the next 100 years.
- distributing founders through the various captive breeding programmes.
- compiling animal husbandry programmes for circulation to all breeding facilities.
- implement an overall plan that contributes to the objectives of maintaining viable captive populations across the globe - a strategy known as the GASP (Global Animal Survival Plan).

Because of limitations of space, finances, and facilities in the institutions that undertake captive breeding, species prioritization is a primary concern. An overwhelming 91 % of animals kept in Indian zoos are non-threatened, yet take up much of the space and resources of zoos. Zoos in India may need to restrict their efforts to a few species that can benefit from captive breeding initiatives, such as small-bodied species of Chiroptera, Rodentia, and Insectivora. These make up 60 % of mammalian taxa in the country and are easier to reintroduce than larger mammals. Instead, mini zoos and deer parks can act as a sink for the surplus, hybrid, aged and infirm animals while the larger zoos can focus on serious captive breeding.

Preservation plots

A number of states, *e.g.* Uttar Pradesh and Uttaranchal, have forest preservation plots. These are important means for conserving and protecting important floral species as well as for assessing ecological changes occurring in such areas over a period of time. These plots need to be demarcated and actively maintained.

Recommendations

Following recommendations can be put forth for active consideration and implementation to conserve the biodiversity in any sanctuary, protected area or national park in our country.

1. According to the Indian Philosophy, "whole is made of the part and part is made of the whole" and two understand the whole, understanding the part is must. The study of population dynamics and trends in MTR and its 4 constituents reveal that each has a different story of success and failure to tell and story of growth and decline in wild populations can not be well understood without focusing attention on its part constituents. In view of this, it is strongly recommended that management practices and monitoring techniques customized for each constituent should be evolved and implemented in a coherent, continuous and sustainable manner.
2. Wildlife populations in each constituent should be monitored periodically (preferably monthly at round level, quarterly at range level, six-monthly at division level and annually at Field Director level) without any breaks and published annually for public consumption.
3. NGOs and public participation should be encouraged at the time of annual monitoring and their views and approaches should be considered and addressed at Director level and not below, to avoid any confusion on policies and administrative instructions at field level.
4. The scope of monitoring should be enhanced to cover populations of major carnivores, herbivores and even predominant avifauna with special emphasis on pea-fowl, vultures, eagles, kites, kingfishers and owls as well as migratory and resident populations (ground as well as water birds).
5. Since recent studies have revealed role of smaller fauna like hares contributing largely towards diet of carnivores, techniques should be evolved to monitor populations of smaller fauna like hares, fowls, reptiles, quails and potridges. This can be effectively done at beat round and range level.
6. The monitoring process of carnivores and major herbivores should include differentiation in resident and transient populations as ecological disturbances are sometimes indicated by transient populations and aberrant behaviors can often be traced to such incidences.

7. As analysis of data on herbivores has revealed constant decline of populations in all constituents of the area and it forecasts a warning for decline of carnivorous population in future, it is very strongly recommended that monitoring process at division as well as director level must include the estimation of existing prey biomass/numbers and trends in their growth or decline before arriving at any figure for carnivores. The related parameters like prey-predator ratio, range, sex-ratio and ratio of existing to required prey biomass should also be closely monitored and a report should be generated on annual basis under close supervision of Field Director.
8. Several studies in protected areas in the past reveal high incidences of parasitic and other infections in carnivores and herbivores of the area, it is high time that appropriate measures are immediately taken to contain situation. For this following urgent steps are suggested:

    a. All deaths including natural deaths, of carnivores and as far as possible that of herbivores too, should be analyzed for the cause of death through post mortem and a systematic record generated.
    b. Waterholes in the entire Reserve/Sanctuary with more frequency and concentration on areas frequented by cattle for grazing should be subjected to periodic tests for contamination/poisoning and efforts should be made to make them as natural as possible.
    c. The process of immunization of cattle likely to come to area should be made more stringent and followed scrupulously.
    d. Entry /grazing of outside cattle should be banned in the area.
    e. Village gram panchayats/eco-development committees should be given incentives to enforce grazing and immunization regulations.

9. The periodic monitoring of changing patterns of habitats and their utilization like growth of shrubs/weeds in open areas/ meadows, drying of main waterholes, riverine flora, spread of water hyacinth/moss, fishing/camps by locals/labors should be

carried and converted to generation of monthly reports at range level.

10. In addition to methods of transect analysis and camera traps for monitoring prey and predators under central surveillance started recently, local level monitoring of carnivores through pugmarks coupled with other evidences and that of herbivores through waterhole (this may be restricted to 6 to 8 hours in the evening say 2 pm to 10 pm in April – May) must be continued failing which the local staff is likely to get detached and exhibit tendencies to disown the responsibility for any sudden or gradual decline in wild populations or even degradation of habitats.
11. Regular observations on Vegetation Monitoring Plots laid down in the area should be carried out and analyzed at an interval of 10 years to judge the impact of protection and anthropogenic factors on evolution, growth or decline of biodiversity lower flora.
12. The research related to wildlife behaviour and prey–predator relationships should be conducted through active involvement of staff and researchers from universities in the area.
13. The field staff should be trained regularly with a view to enhance their capabilities and motivation in order to ensure that monitoring process is meaningful and not remain obscured by ignorance in matters of wild animals or essential evidences. Organizing a regular training cum workshop at range level quaterly under close supervision of forest officers.
14. Villages from sanctuary should be relocated as early as possible to ensure full potential of the area.
15. The litigation on MUA needs to be brought to an end to give a meaningful direction to the valuable resources of the area. The steep decline in carnivore and herbivore population in the tract presents very sorry state of affairs. Present state of affairs is neither conducive to wildlife nor the people and may be it's only the unscrupulous people who have a field day here and therefore special request should be made to the H'ble High Court to decide the matter either in favor of wildlife or the people. As such both are sufferers here.

16. Eco-development works should be taken in large scale in MUA to ensure better participation of local people in wildlife conservation as shifting these villages is a distant possibility.
17. Tourism Zones should be strictly marked and visits should be allowed only in designated areas.
18. Celebrating Environment friendly days like:

    February 2 (World Wetland Day) - Wetlands are a very important part of our biodiversity and it is essential to see that they are well protected.

    March 21 (World Forestry Day) - Activities such as the planting of trees and highlighting the urgency to increase the green cover.

    March 22 (World Water Day) - The decision to celebrate this day has been taken recently as drinking water sources are fast depleting. The world must wake up to the problem and begin conserving it.

    March 23 (World Meteorological Day) – Everyone has to be reminded that weather is an integral part of the environment.

    April 18 (World Heritage Day) - Environment includes not just the natural surroundings but also the manmade ones.

    April 22 (Earth Day) - In 1970 a group of people in the United States of America got together to draw the attention of the world to the problems being caused to the earth due to modernisation. Since then this day has been celebrated all over the world as Earth Day.

    June 5 (World Environment Day) - On this day, in 1972, the Stockholm Conference on Human Environment was held in Sweden. There was a large gathering from all over the world and people expressed their concerns for the increasing environmental problems.

    July 11 (World Population Day) - Population has to be given special attention, as it is an ever-increasing problem especially in India.

    September 16 (World Ozone Day) - The United Nations declared this day as the International day for the preservation of the Ozone Layer. It is the day the Montreal Protocol was signed.

September 28 (Green Consumer Day) - The problems of consumerism and its impact on the environment is an area of major concern in today's world. Awareness building on the importance of recycling-reusing-reducing should be taken up seriously.

October 3 (World Habitat Day) - The earth is the habitat of not only human beings but also all living creatures. Increasing human activities is threatening the habitat of other living things.

October 1-7 (World Wildlife Week) - Celebrate this week by building awareness on the importance of preservation of our wildlife.

October 4 (World Animal Welfare Day) - The welfare of animals has to be looked into and given due importance.

October 13 (International Day for Natural Disaster Reduction)- Due to a change in the environment there has been an increase in the number of natural disasters. Efforts have to be taken to reduce these disasters.

December 2 (Bhopal Tragedy Day) - Mark this occasion by taking a pledge to put in your best efforts to prevent such a tragedy from occurring again.

References

Anderson, G.S. and Danielson, P.J. (1997). The effects of landscape (composition and physiognomy on metapopulation size: the role of corridors. Landscape Eccol. 12:261-271.

Radgers, W.A. and Panwar, H.S. (1988). Planning Wildlife protested area network in India. 2 vols. Project FO:IND/82/003. FAO, Dehradun.

Wilson, A.M. and Lindenmoyer, O.B. (1995). The role of wildlife corridors in the conservation of biodiversity: a review. A Report prepared for the National corridors of Green Program, Greening Australia.

Biodiversity Conservation and Envir. Management (2012)
Editors: D.R. Khanna et al.
Pub. by Biotech Books. *ISBN: 978-81-7622-262-4*

Pages: 15-22

TAXONOMY AND DIVERSITY OF THE FISH FAUNA OF THE GANGA RIVER IN GARHWAL HILLS

S.P. Badola[1] and Smita Badola Budakoti[2]✉

[1] Smiriti Bhawan, Upper Kalabarh, Kotdwara, Pauri- Garhwal, Uttrakhand (INDIA)

[2] Department of Zoology, Govt. Postgraduate College Kotdwara, Pauri-Garhwal, Uttarakhand (INDIA)

E-mail: drsmitabadola@gmail.com

River Ganga is very rich in fish diversity in Garhwal hills & they are tabulated clearely in this chapter.

Introduction

The credit for taxonomy of Indian freshwater fish goes to Francis Hamilton (Formerly known as Butchanan) who in 1822 gave pioneering an excellent illustrated "An account of the fish found in the river Ganga

✉ Corresponding author

and its branches". The greater important work "The Fishes of India being a Natural History of fishes known to inhabit the seas and freshwater of India, Burma and Ceylon" Francis Day (1878) contributed to the ichthyologists of India.

The fish fauna of Uttarakhand hills have been described by various workers. The fish fauna of Kumoan hills have been reported by Hora (1937). Menon (1949, 1950, 1954, 1962 and 1971) and Pant (1970). Similarly the fish fauna of the Garhwal hills have been described by Badola and Pant (1973), Badola (1975), Badola and Singh (1977a, 1977b) and Singh *et al.* (1987), Lal and Chatterjee (1962) and Singh (1964). Khanna and Badola (1990) reported the fish fauna of Haridwar.

Physical Features of the Ganga River

The prominent parental tributaries of Ganga river are Bhagirathi and Alaknanda. Both the rivers originate from the Garhwal Himalayan glaciers. The Bhagirathi originating at "Gomukh" (4255m.) of the Gangotri glacier, which after receiving the Jadganga river at Bhaironghati and Ashiganga at Gangotri, passing through Uttarkashi town and then meeting the river Bhilangana at Tehri. The Bhagirathi from "Gomukh" to Dev-prayag covers 205 km distance. The second river Alaknanda is the biggest tributary of Ganga, which originates from "Kamet" glacier (7766 m) of the Badrinath hills. The main tributaries are Vishanuganga, Birahi, Nandakini; Pindar and Mandakini which join the Alaknanda at Vishnuprayag, Birahi, Mandprayag, Karanprayag and Rudraprayag respectively. At Dev-prayag the Alaknanda meets the Bhagirathi and beyond this confluence the river is called the Ganga. The Alaknanda river flows about 195 km in Garhwal Himalaya. Downwards Dev-prayag, river Ganga separates Pauri and Tehri districts. The Gular river which originates from Tehri hills, mixes with Ganga river on its western side at Gular. After Laxmanjhoola, river Ganga separates the Pauri and Dehradun districts. *The Ganga river receives its three prominent tributaries namely Nayar at Viyasghat, Hinwal at Phool Chatti and Vindiyasani at Chilla on its eastern side, other hand in its western side the river Ganga receives two prominent Doon Valley rivers Song and Suswa near Satyanarayan. The Ganga river passes through Laxmanjoola, Rishikesh and Haridwar in Garhwal hill. In Garhwal region the Ganga river covers 96 km distance.*

Systematic Account of Fishes

S. No.	*Name of species*	*Local Name*	*Altitudinal Distribution (m) in Ganga River*	*Migratory Behaviour*
	Series: Pisces Class : Teleostomi Sub-class: Actinoptergyii Order : Cypriniformes Sub-order : Cyprinoidei Family : Cyprinidae Sub-family : Cyprinidae			
1.	*Tor tor* (Ham.)	Mahseer	340-472	Upward
2.	*Tor putitora* (Ham.)	Mahseer	-do-	-do-
3.	*Tor chilinoides* (Mc Clell.)	"Dansulu"	389-472	-
4.	*Tor hexastichus* (Mc Clell.)	"Dansulu"	-do-	-
5.	*Labeo dyocheilus* (Mc Clell.)	"Kharont"	340-472	Upward
6.	*Labeo dero* (Ham.)	"Kharont"	-do-	-do-
7.	*Labeo boga* (Ham.)	"Jabu"	250-300	-
8.	*Labeo bata* (Ham.)	"Bata"	-do-	-
9.	*Labeo gonius* (Ham.)	-	-do-	-
10.	*Chagunius chagunio* (Ham.)	"Kharont"	292-472	Upward
11.	*Puntius ticto* (Ham.)	"Damra"	292-340	-
12.	*Puntius conchonius* (Ham.)	"Damra"	-do-	
13.	*Puntius sophore* (Ham.)	"Damra"	-do-	-
14.	*Puntius sarana* (Ham.)	"Bara Damra"	-do-	
Sub-Family: Rasborinae				
15.	*Barilius bendelisis* (Ham.)	"Fulra"	300-472	Upward

contd...

S. No.	*Name of species*	*Local Name*	*Altitudinal Distribution (m) in Ganga River*	*Migratory Behaviour*
16.	*Barilius shacra* (Ham.)	"Fulra"	-do-	-do-
17.	*Barilius barna* (Ham.)	"Fulra"	-do-	-do-
18.	*Barilius barila* (Ham.)	"Fulra"	-do-	-do-
19.	*Barilius vagra* (Ham.)	"Fulra"	-do-	-do-
20.	*Raiamas bola* (Ham.)	"Trout"	250-340	-do-
21	*Danio (Brachydanio) rerio* (Ham.)	"Dharidar"	-do-	-
22.	*Danio aequipinnatus* (Mc Clell.)	"Fulra"	-do-	-
23.	*Danio devario* (Ham.)	"Patukari"	-do-	-
24.	*Esomus danricus* (Ham.)	"Damra"	-do-	-
25.	*Rasbora daniconius* (Ham.)	"Patukari"	-do-	-
Sub-family: Garrinae				
26.	*Garra gotyla gotyla* (Gray)	"Gunthala"	300-472	Upward
27.	*Garra lamta* (Ham.)	"Gunthala"	-do-	-do-
28.	*Crossocheilus latius latius* (Ham.)	"Sunhara"	-do-	-do-
Sub-Family: Schizothoracinae				
29	*Schizothorax richardsonii* (Gray)	"Maseen"	300-472	Downward
30.	*Schizothorax plagiostomus* (Heckel)	"Dhibrua"	-do-	-do-
31.	*Schizothorax sinuatus* (Heckel)	"Maseen"	-do-	-do-
32.	*Schizothoraichthys progastus* (Mc Clell.)	"Chongu"	-do-	-do-
33.	*Schizothoraichthys esocinus* (Heckel)	"Chongu"	-do-	-do-
34.	*Schizothoraichthys micropogon* (Heckel)	"Chongu"	-do-	-do-
35.	*Schizothoraichthys curvifrons* (Heckel)	"Chongu"	-do-	-do-
36.	*Schizothoraichthys niger* (Heckel)	"Chongu"	-do-	-do-

contd...

S. No.	*Name of species*	*Local Name*	*Altitudinal Distribution (m) in Ganga River*	*Migratory Behaviour*
Family: Psilorhynchidae				
37.	*Psilorhynchus balitora* (Ham.)	-	300-472	Upward
Family: Cobitidae Sub- Family: Cobitinae				
38.	*Lepidocephalus guntea* (Ham.)	"Gadiyal"	300-340	-
Sub-family: Botiinae				
39.	*Botia dario* (Ham.)	"Bara Gadiyal"	300-472	Upward
40.	*Botia geto* (Ham.)	"Bara Gadiyal"	-do-	Upward
41	*Botia almorhae* (Gray.)	"Bara Gadiyal"	-do-	Upward
Sub-family: Noemachilinae				
42.	*Noemacheilus botia* (Ham.)	"Gadiyal"	300-472	-
43	*Noemacheilus rupicola* (Mc Clell.)	"Gadiyal"	-do-	-
44.	*Noemacheilus montanus* (Mc Clell.)	"Gadiyal"	-do-	-
45.	*Noemacheilus bevani* (Gunther)	"Gadiyal"	-do-	-
46.	*Noemacheilus savona* (Ham.)	"Gadiyal"	-do-	-
47.	*Noemacheilus denisonii* (Ferdon)	"Gadiyal"	-do-	-
48.	*Noemacheilus zonatus* (Mc Clell.)	"Gadiyal"	-do-	-
49.	*Noemacheilus multifasciatus* (Day)	"Gadiyal"	-do-	-
Order: Siluriformes Family : Schilbeidae				
50	*Clupisoma garua* (Ham.)	"Garuwa"	300-472	Upward
Family : Bagridae				
51.	*Mystus tengara* (Ham.)	"Tengar"	300-340	-
52.	*Mystus vittatus* (Ham.)	"Tengar"	-do-	-
53.	*Rita rita* (Ham.)	"Reeta"	-do-	-
Family : Amblycepidae				
54.	*Amblyceps mangois* (Ham.)	"Nain"	300-340	-

contd...

S. No.	Name of species	Local Name	Altitudinal Distribution (m) in Ganga River	Migratory Behaviour
Family : Sisoridae				
55.	*Bagarius bagarius* (Ham.)	"Goonch"	300-340	upward
56.	*Euchiloglanis hodgarti* (Hora)	"Kthrua"	-do-	-
57.	*Glyptothorax madraspatanum* (Day)	"Naou"	-do-	-
58.	*Glyptothorax petinopterus* (Mc Clell.)	"Sepliya"	-do-	-
59.	*Glyptothorax telchitta* (Ham.)	"Sepliya"	-do-	-
60.	*Glyptothorax conirostris* (steind.)	"Noau"	-do-	-
61.	*Glyptothorax cavia* (Ham.)	"Noau"	-do-	-
62.	*Glyptothorax trilineatus* Blyth	"Noau"	-do-	-
63	*Glyptothorax kashmirensis* Hora	"Noau"	-do-	-
64.	*Glyptothorax brevipinnis* Hora	"Noau"	-do-	-
Family: Clariidae				
65.	*Clarias batrachus* (Linn.)	"Mangur"	292-340	-
Family: Heteropneustidae				
66.	*Heteropneustes fossilis* (Bloch)	"Singhi"	292-340	-
Order: Channiformes Family: Channidae				
67.	*Channa gachua* (Ham.)	"Soanl"	300-340	-
68.	*Channa striatus* (Bloch)	"Soanl"	-do-	-
69.	*Channa punctatus* (Bloch)	"Soanl"	-do-	-
Order: Perciformes Sub-order : Anabantoidei Family : Anabantidae				
70.	*Colisa fasciatus* (Sch.)	"Cohlisha"	-do-	-
71	*Mastacembelus armatus* (Lacep.)	"Gairee"	-do-	-
Order : Beloniformes Family: Belonidae				
72.	*Xenentodon cancila* (Ham.)	"Kuwa"	-do-	-

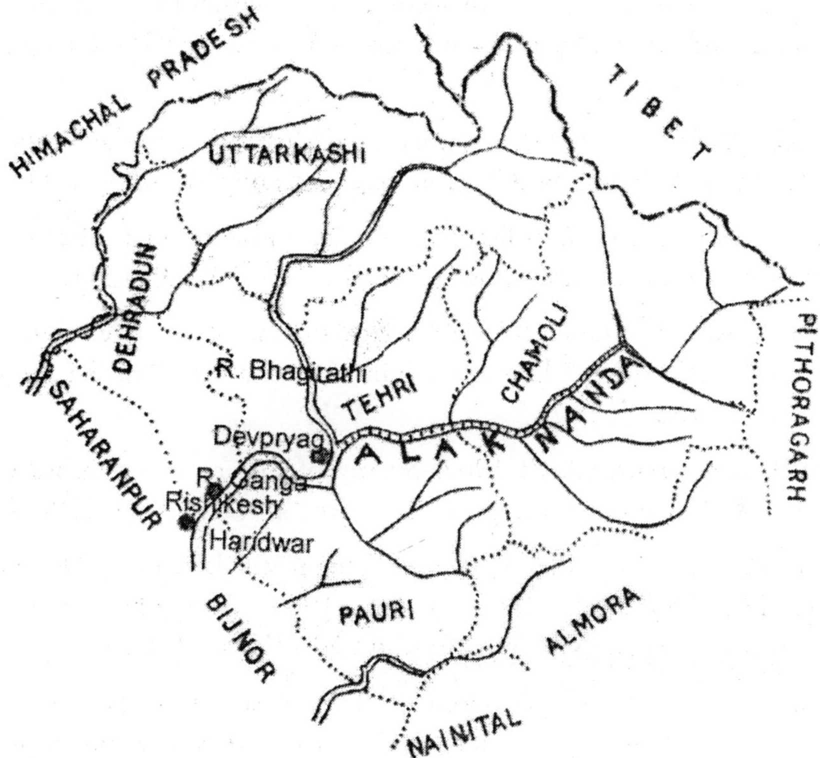

MAP OF GANGA RIVER IN GARHWAL HILLS

References

Badola, S.P. and Pant, M.C. 1973. Fish Fauna of the Garhwal Hills, Part I. Ind. 1. Zoot, 14(1): 37-44.

Badola, S.P. 1975. Fish Fauna of the Garhwal Hills, Part II (Pauri Garhwal-U.P.). Ind. J. Zoot., XVI(l): 57-70.

Badola, S.P. and Singh, H.R. I 977a. Fish Fauna of the Garhwal Hills Part III (Chamoli district). Ind. 1. Zoot, 18(2): 119-122.

Badola, S.P, and Singh, H.R. I 977b. Fish Fuana of the Garhwal hills. Part IV (Tehri district). Ind. 1. Zoot., 18(2): 115-118.

Day, F. 1878. The Fishes of India being a Natural History of the Fishes Known to Inhabit the Seas and Freshwaters of India, Burma and Ceylon, Vol. I & II: XX+778, pis. CXCV.

Hamilton-Buchanan, F. 1822. An account of the Fishes found in the river Ganges and its branches. Edinburgh and London. VII+405 pp., 39 pis.

Hora, S.L. 1923. Notes on fishes of the Indian Museum, V on the composite genus Glyptøstemum McClelland. Rec. Indian Mus., XXV: 1-44.

Hora, S.L. 1937. Notes on fishes in the Indian museum on a collection of fish from Kumaon Himalayas. Rec. Indian Mus. 39: 338-341.

Khanna, D.R. and Badola, S.P. 1990. Icthyofauna of the river Ganga at foot hills of Garhwal Himalaya. Journal of Natural Phy. Sci., Vol. 4 (1-2): 153-162.

Lal, M.B. and Chatterjee, P. 1962. Survey of eastern Doon fishes with certain notes on their biology. J. Zool. Soc. India, 14(1-2): 229-243.

Menon, A.G.K. 1949. Fishes of Kumaon Himalayas. J. Bombay Nat. Hist. Soc, 48(3): 535-542. Menon, A.G.K. 1950. Fishes from the Kosi Himalayas, Nepal. Rec. Indian Mus., XLVII(34): 231-237.

Menon, A.G.K. 1954. Fish geography of the Himalaya. Proc. Nat. Inst. Sci. India, 20(4): 467-493. Menon, A.G.K. 1962. A distributional list of fishes of the Himalayas. J. Zool. Soc. India, 14(1-2): 23-32.

Menon, A.G.K. 1962. Monograph of the cyprinoid fishes of the genus Garra Hamilton. Mem. Indian Mus., 14(4): 173-260.

Menon, A.G.K. 1971. Taxonomy of fishes of the genus Schizothorax Heckel with the description of a new species from Kumaon Himalayas. Rec. Zool. Surv. India, 68(1-4): 195-207.

Pant, M.C. 1970. Fish Fauna of the Kumaon hills. Rec. Zool. Surv. India, 64(1-4): 85-96.

Singh, H.R., Badola, S.P. and Dobriyal, A.K. 1987. Geographical distributional list of Ichthyofauna of the Garhwal Himalaya with some new records. Bombay Nat. Hist. Soc, 84(1): 126-132.

Singh, P.P. 1964. Fishes of the Doon valley. Ichthyologica, 111(1-2): 86-92.

Biodiversity Conservation and Envir. Management (2012)
Editors: D.R. Khanna et al.
Pub. by Biotech Books. *ISBN: 978-81-7622-262-4*

Pages: 23-35

3

UNSAFE WATER CONTACT: CAUSE FOR NEW AMOEBIC DISEASE

A.K. Sharma[✉] **and Tabrez Ahmad**
Department of Zoology, University of Lucknow,
Lucknow, Uttarpradesh (INDIA)
E-mail: sharma_ajayk@lkouniv.ac.in

Medical interest in amoebae developed when Gros (1849) described a parasitic amoeba (*Entamoeba gingivalis*) from oral mucosa responsible for Pyorrhea. In the year 1857 F. Losch a Russian Zoologist first discovered *Entamoeba histolytica* in the stool and intestinal ulcers of a patient suffering from chronic dysentery. Till 1958 it was generally believed that *E.histolytica* was the only pathogenic amoeba causing dysentery and other forms of amoebiasis. This notion became untenable with the discoveries made within the last 5 decades that the small free living amoebae belonging to

✉ Corresponding author

the genera *Naegleria* and *Acanthamoeba* were pathogenic and causing new serious and uncommon new amoebic diseases in human and other animals. The new amoebic diseases are: Primary Amoebic Meningeoencephalitis (PAM) caused by *N.fowleri;* Granulomatous Amoebic Encephalitis (GAE) caused by *A.culbersoni* and Amoebic Keratitis (AK) by *A.polyphaga*. After 1965, many cases of PAM have been reported worldwide. Butt (1966) named this new disease as Primary Amoebic Meningeoencephalitis to differentiate from the disease caused by the secondary infection of *E.histolytica* to brain. *N. fowleri* primarily infect brain taking entry through nasal route and occur in healthy young persons having recent history of swimming in contaminated water or had some contact with water or mud. *A.culbertsoni* is the etiological agent of sub-acute, chronic illness of immunologically weak individuals having no history of swimming and the disease is named Granulomatous Amoebic Encephalitis (GAE). *A.polyphaga* is known to cause Amoebic Keratitis (AK), which occurs mostly in people wearing contact lenses. No effective treatment is available for Amoebic Meningoencephalitis, except for Amphotericine-B alone or in combination with antibiotic. Since these diseases are new and understanding is limited, therefore public awareness about preventive and control measures are very essential. The impact of these new diseases is greater because of the ubiquitous presence and extreme virulence with absence of effective therapy.

Introduction

Water is a prime natural resource, a basic human need and a national asset. Pure and safe water is of paramount importance for the life of animals and plants and so termed as the "Elixir of life", whereas unsafe polluted water endangers the life. According to a WHO estimate, about 80% diseases globally are associated due to contamination of water and more than 15 million deaths worldwide occur annually from water borne infections (Chatterjee *et al.*, 2007). As per WHO (2004) estimate the causative organism are bacteria, viruses, protozoan, helminthes, nematodes, blue green algae *etc*; which find ways into water and cause Cholera. Gastroententis, Typhoid, Tuberculosis, Dysentery, Meningitis, Polio, Hepatitis, Schistonomiasis, Giardiases, Skin Rashes, Guinea worm, Diarrhoea *etc.* (Gopal, 2003).

Although the nature and diversity of protozoa in variously polluted waters has been studied, the existence of potential human pathogens in the water has not been adequately assessed on the basis of current knowledge; the most likely health risk to swimmer and divers is from Amphizoic protozoa. Amphizoic protozoa are those that are capable of being either endozoic (symbiotic or parasitic) or exozoic (free living). But exposure to obligate parasitic species in polluted waters can not be dismissed (Page, 1974).

Earlier interest in small free-living amoebae was largely academic, because they were known to play an important role in aquatic and soil ecosystem (Singh, 1975). But Gros (1849) was the first to discover a parasitic amoeba *Entamoeba gingivalis* from oral mucosa responsible for Pyorrhea. Later F. Losch (1875) first discovered *Entamoeba histolytica* in the stool and ulcers of a patient suffering from chronic dysentery. This anaerobic amoeba has its primary site of infection in colon region of large intestine. It may infect other body organs such as liver, brain, lungs *etc.* by haemotogenous route (secondary infection). Till 1958, it was generally believed that *E.histolytica* was the only pathogenic amoeba causing dysentery and other forms of amoebiasis. This notion became untenable with the discoveries made within the last 5 decades that these small free living amoebae belonging to the genera *Naegleria* and *Acanthamoeba* were pathogenic and causing new serious and uncommon new amoebic diseases in human and other animals.

Amphizoic amoebae are capable of existing both in free living and in parasitic form depending on the actual conditions. Two genera (*Naegleria* and *Acanthamoeba*) have become recognized as opportunist human parasites. Since the first description in 1965 of a lethal case of Primary Amoebic Meningeoencephalitis (PAM) caused by *Naegleria*, many more (mostly lethal) cases have been reported, while Granulomatous Amoebic Encephalitis (GAE), as well as eye (Keratitis, Conjunctivitis), ear, nose, skin and internal organ infection caused by *Acanthamoeba* species have also occurred in rapidly increasing numbers. Both pathogenic and nonpathogenic species of *Naegleria* and *Acanthamoeba* are found world wide in soil, water, air, dust, well *etc.* Small free living amoebae, mainly found in water and soil are known to cause fatal diseases in human being and domestic animals (Visvesvara, 2010).

Epidemiology

So far more than 500 cases of amoebic encephalitis and more than 3000 cases of amoebic keratitis have been reported world wide (John, 1993, Sharma, 2004 and Rai *et al.*, 2008). Amphizoic amoebae, *Naegleria* sp. and *Acanthamoeba* sp. causing diseases called PAM, GAE and AK respectively, reported world wide from Australia, England, India, Brazil, Japan, New Zealand, Mexico, Nigeria, Panama, South Africa, United Kingdom, Zambia, Pakisthan, Italy, Thialand, Korea, Peru, Germany *etc.* (John, 1993; Martinez, 1985 and Rezaeian *et al.*, 2008).

New Amoebic Diseases

Naegleria fowleri is the causal agent of a fulminating, rapidly fatal infection of the central nervous system called PAM (Primary Amoebic Meningeoencephalitis) (Plate-1, Fig. 1-6) but other species such as *N. australiensis* and *N. italica* also having pathogenic potential (Schuster, 2002). Mostly children, teenager and adults in a good health are infected by swimming or washing in contaminated water, where amoebae enter the nostrils, migrate along the olfactory nerves to the cribriform plate and gain access to the central nervous system. Amoebae proliferate rapidly and cause extensive damage to neural tissue. Symptom includes headache, fever, neck stiffness, nausea, vomiting, sign of meningeal irritation and disturbance in sense of smell or taste and visual disturbance (Martinez, 1985) followed by confusion, somnolence, seizures and coma (Jamil *et al.*, 2008).

Occurrence of amoebic meningitis has also been reported from India. Till date a total of ten cases have been reported between 1971 and 2008 from India. Pan and Ghosh (1971) from Calcutta first reported the occurrence of PAM in two children. Third case was reported by Bedi *et al.* (1972) from Udaipur, Rajasthan in a 45 years old women. Jain *et al.* (2002) from Chandigarh, India reported survival of a patient after the infection of *Naegleria*. Kaushal *et al.* (2008) and Rai *et al.* (2008) also reported a case infection with *Naegleria*.

PLATE 1

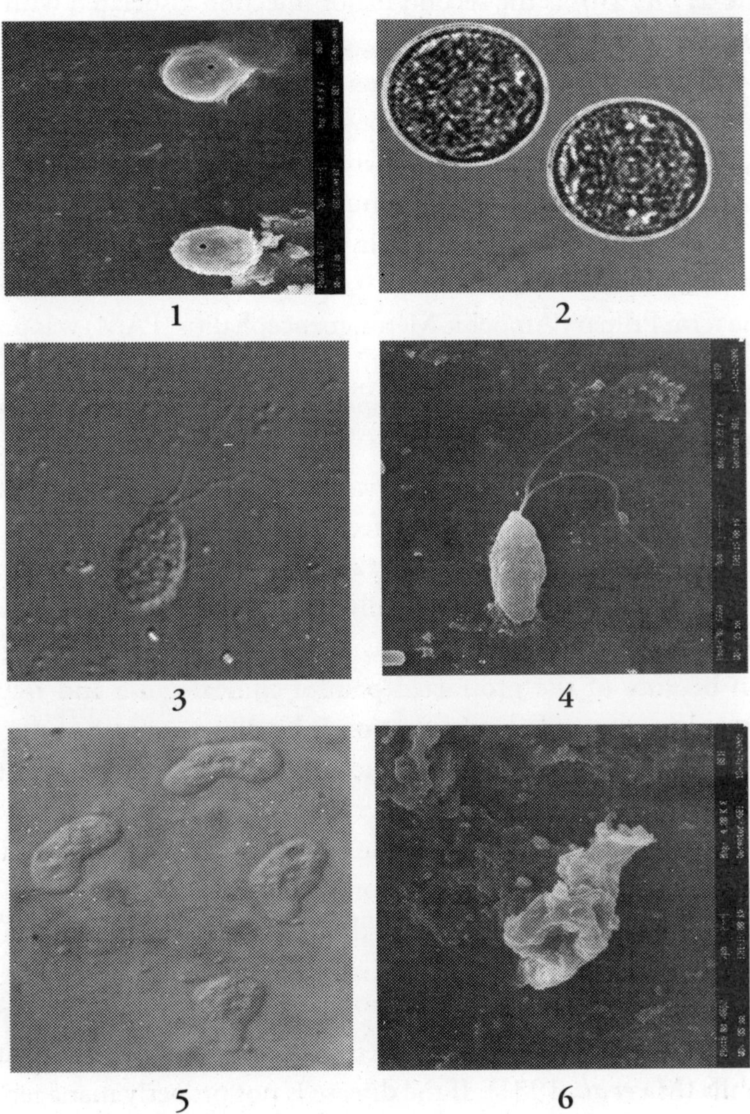

Fig 1: Shows SEM of Cyst of *Naegleria fowleri*

Fig 2: Shows photograph of Cyst of *Naegleria fowleri*

Fig 3: Shows photograph of Amoeba flagellate stage of *Naegleria fowleri*

Fig 4: Shows SEM of Amoeba flagellate stage of *Naegleria fowleri*

Fig 5: Shows photograph of Trophozoites of *Naegleria fowleri*

Fig 6: Shows SEM of Trophozoites of *Naegleria fowleri*

Granulomatous Amoebic Encephalitis (GAE) caused by *Acanthamoeba* sp. (Plate-2, Fig. 1-6) is the second major infection associated with central nervous system. GAE is chronic, progressive disease of the central nervous system occurring most often in persons with poor immune system or other debilitating health problems. Predisposing factors include chemotherapy, dialysis, diabetes mellitus, chronic alcoholism, smoking, bone marrow or renal transplantation, or acquired immunodeficiency syndrome (Marciano-Cabral *et al.*, 2000), chronic skin infections have been reported from patients with GAE.

The term Primary Amoebic Meningoencephalitis (PAM) was proposed for infection by *Naegleria* (Butt, 1966) where as the term Granulomatous Amoebic Encephalitis (GAE) was proposed for infection by *Acanthamoeba* (Martinez, 1985).

In addition of GAE, *Acanthamoeba* sp. also caused keratitis. *Acanthamoeba* keratitis is a chronic infection of the cornea by several species of *Acanthamoeba* (Marciano-Cabral *et al.*, 2003). *Acanthamoeba* keratitis is one of the most severe and potentially sight threatening ocular parasitic infectious diseases and is recognized as the most challenging among ocular infection because of the protracted painful clinical cause and frequently encountered treatment failure. Infection is by direct contact of the cornea with amoebae which may be introduced through minor corneal trauma or by exposure to contaminated water or contact lenses. The wearing of contact lenses and use of home made saline solution are important risk factors associated with the disease (Jeong and Yu, 2005). Amoebae invade the corneal stroma through a break in the epithelium or through the intact epithelium (John *et al.*, 1989). The disease characterized by severe ocular pain, inflammation (Mannis *et al.*, 1986), affected vision and a stromal infiltrate that frequently is ring shaped and composed predominantly of neutrophils (Ma *et al.*, 1981). If the disease is not properly managed, it can lead to loss of vision and even loss of eye (Khan, 2005).

Culture of amoebae

About two liters of water sample was collected from different water bodies and filtered through sterile filter paper (Whatman No.-1) in a conical funnel. Sediment was collected in the cone of filter paper. About 1 cm cone of filter paper was cut and placed on the thick suspension of *Escherichia coli* and incubated in BOD incubator at 28 °C-37 °C for 10-12 days. The culture was then observed under microscope for growth of amoebae.

PLATE 2

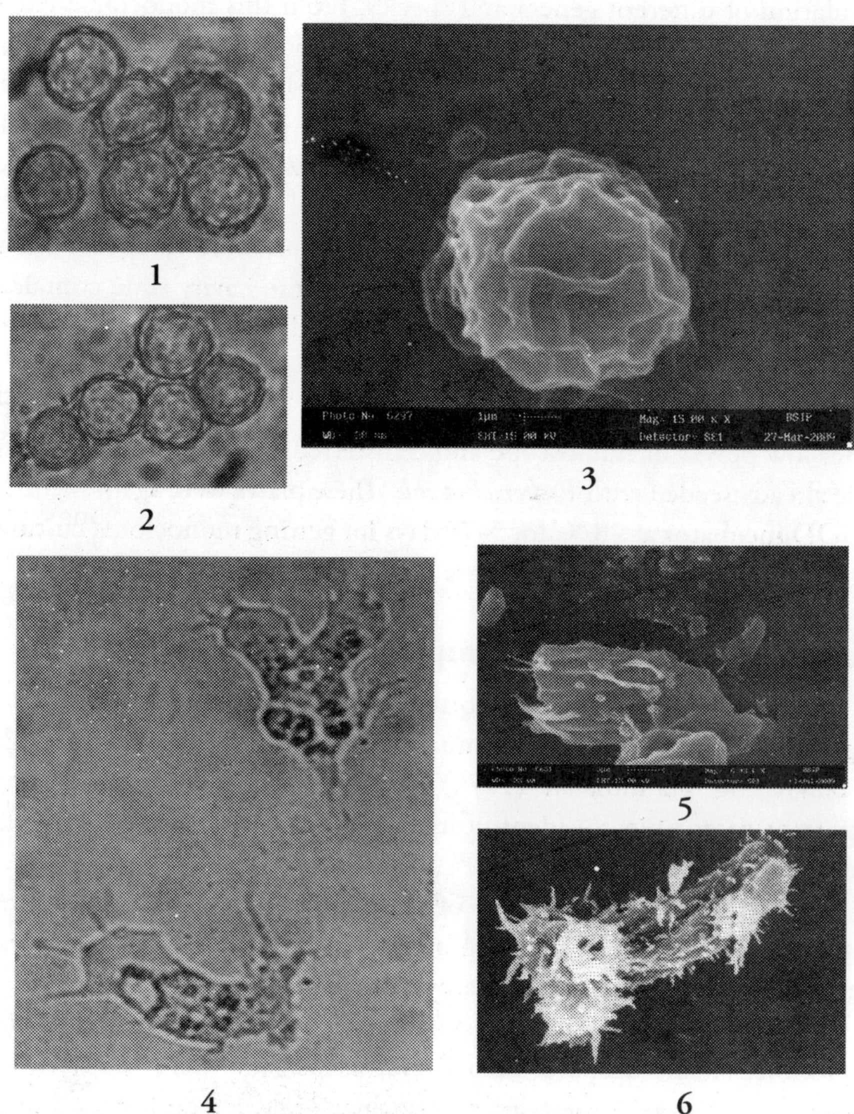

Fig 1-2: Shows photographs of Cysts of *Acanthamoeba culbertsoni*

Fig 3: Shows SEM of Cyst of *Acanthamoeba culbertsoni*

Fig 4: Shows Trophozoite of *Acanthamoeba culbertsoni*

Fig 5-6: Show SEM of Trophozoite of *Acanthamoeba culbertsoni*

Samples of Nasal swab, Eye tears and Cerebro spinal fluid (CSF) were taken for the isolation of amoebae using standard method of Singh and Hanumaiah (1979).

Isolated amoebae from different water samples contained a mixed population of different genera and species. From this monoclonal culture of isolated amoebae was prepared by the method of single cyst picking. Isolated amoebae were sub-cultured 3-4 times by cutting a small square piece of agar on which plenty of amoebae were present and placed facing downward on the fresh bacterial circle (*Escherichia coli*) in a non-nutrient agar plate. These plates were incubated in BOD incubator at 28 °C for 5-10 days. Mature cysts from mixed population were scraped with the help of sterilized microme wire loop and put into cavity slide containing distilled water). The cyst suspension was diluted 3-4 times so that the few cysts remained in the cavity slide.

Single cyst was picked up with the help of sterilized micropipette under low power of microscope and transferred on the non-nutrient agar plate already seeded with *Escherichia coli.* These plates were again incubated in BOD incubator at 28 °C for 5-10 days for getting monoclonal culture of different strains of isolated amoebae (Singh and Hanumaiah, 1979).

Identification of isolated Amphizoic amoebae

For the identification of isolated amoebae, studies on trophozoite stages, their locomotion, type of nuclear division were done. The method of Singh and Hanumaiah (1979) was used for identification. Other important parameters for identification were studied on amoebo-flagellate and cyst morphology.

Cyst morphology was one of the main bases of identification of amoebae. For this 10-15 days old monoclonal culture of different strains were taken and washed 5-6 times by centrifuging at low speed (400 rpm up to 20 minutes) in sterilized distilled water, to remove as much bacteria as possible. A few drops of settled cyst suspension were transferred on microslides to make camera lucida drawings of each strain from different water bodies. Photographs of cysts were also taken.

For the study of trophozoite stage, particular strain of isolate amoebae was sub-cultured 2-3 times after every 24 hours to get young and actively multiplying trophozoite. Amoebae were taken and centrifuged at low speed

(400 rpm up to 20 minutes) to get clean pallet of trophozoite in distilled water. After final washing amoebae were transferred to micro cavity slide to observe the trophozoite stage under microscope photographs were also taken. Camera Lucida drawings were traced for each strain isolated from different water bodies along with ocular micrometer under low power of magnification (12.5 $x \times 40.0\ x$).

Treatment

At present there exists no satisfactory treatment for PAM. The antibiotics used to treat bacterial meningitis are ineffective in *Naegleria* infection, as are the antiamoebic drugs. Amphotericin-B, a drug of considerable toxicity, is the antinaeglerial agent for which there is evidence of clinical effectiveness. The four known survivors of PAM were treated with amphotericin-B, given intravenously and intrathecally. Other compounds that have been shown to afford protection against naeglerial infection have been cyclophosphamide (Zhang *et al.*, 1988), lipopolysaccharide (Adams *et al.*, 1976), the immunomodulator muramyl dipeptide (Ferrante and Lederer, 1986), 9-tetra hydro cannabinol (Pringle *et al.*, 1979), and combinations of Amphotericin-B and rifampicin (Thong *et al.*, 1979b) and Amphotericin-B and tetracycline (Thong *et al.*, 1979c). Rifampicin and tetracycline shown to act synergistically with Amphotericin-B (Rai *et al.*, 2008). Among the drugs that have been used with success in treating GAE cases are pentamidine isethionate, imidazoles, flucytosine, amphotericin-B, sulpha containing antibiotics and macrolides (Bloch and Schuster, 2005).

Acanthamoeba keratitis can be managed by medical treatment if infection is identified soon. The choice of drugs used is topical use of miconazole, propamdine, and clotrimazole and systemic use of ketoconazole, neosprin (Schuster and Visvesvara, 2004).

Prevention and control

For control of *N.fowleri* infections, awareness in public and within medical community, adequate chlorination of public water supply, including swimming facilities are needed. Chlorination remains the most effective controlling measures for pathogenic *Naegleria fowleri* in public waters; with an emphasis on the level of residual free chlorine of about 0.5 mg/litre.

For keratitis controlling measures includes, use of sterile solutions, disinfections of lenses frequently, avoid using homemade saline solutions. Do not wear lenses while swimming.

Although infections by *N.fowleri* and *Acanthamoeba* sp. are rare circumstantial evidence which suggests that human meddling with water may aggravate the problem. Because of the relationship of swimming to *N.fowleri* infection, the swimming areas should be subjected to proper cleaning, disinfections and intensive investigations.

References

Adams, A.C., John, D.T. and Bradley, S.G. (1976). Modification of resistance of mice to Naegleria fowleri infections. ***Infect. Immun.*** **13**:1387-1391.

Bedi, H.K., Devapura, J.C. and Bomb, B.S. (1972). Primary amoebic meningoencephalitis. ***J. Indian Med. Assoc.*** **58**: 13-14.

Bloch, K.C. and Schuster, F.L. (2005). Inability to make a pre- mortem diagnosis of *Acanthamoeba* sp. Infection in a patient with fatal granulomatous amoebic encephalitis. ***J. Clin. Micriobiol.*** **43**(6): 3003-3006.

Butt, C.G. (1966). Primary amoebic meningo-encephalitis. New England ***J. Med.*** **274**: 1473-1476.

Chatterjee, P.R., Chatterjee, C. and Raziuddin, M. (2007): Impact of human activity on water quality of a lentic water body in Asansol. ***Nat. Environ. & Pollut. Tech.*, 6 (1): 77-84.**

Ferrante, A., and Lederer, E. (1986). Curative properties of muramyl dipeptide in experimental Naegleria meningoencephalitis. ***Trans. Roy. Soc. Trop Med. Hyg.*** **80**: 323-326.

Gopal, K. (2003). Fundamentals of water and waste water. APH. Pub. Corp. New Delhi.

Gros, G. (1849). Fragments d helminthologie et de physiologie microscopique. ***Bull. Soc. Imp. Nat. (Moscow)***, **25**:*549.*

Jain, R., Prabhakar, S., Modi, M., Bhatia, R.and Sehgal, R. (2002).*Naegleria* Meningitis: A rare survival. 470 ***Neurology*** India: 50.

Jamil, B., Ilyas, A. and Zaman, V. (2008). Primary Amoebic Meningoencephalitis. ***Infectious Diseases journal of Pakistan.*** **Vol. 17, issue 02**, 66-68.

Jeong, H.J. and Yu, H.S. (2005). The role of domestic tap water in *Acanthamoeba* containing in contact lens storage cases in Korea. ***Korean J. Parasitol.*** **43**: 47-50.

John, D.T. (1993). Opportunistically pathogenic free-living amoebae. Parasitic Protozoa, 2nd Edition, volume 3, Edited by Julius P. Kreier and John R. Baker Academic Press, Inc. San Diego.

John, D.T., Desai, D. and Sahm, D. (1989). Adherence of *Acanthamoeba castellanii* cysts and trophozoites to unknown contact lenses. ***Am. J. ophthmol.*** **10**: 579-584.

Kaushal, V., Chhina, D.K., Ram, S., Singh, G., Kaushal, R.K. and Kumar, R. (2008). Primary Amoebic Meningoencephalitis due to *Naegleria fowleri*. ***JAPI*, Vol. 56**.

Khan, N.A. (2005). The immunological aspects of *Acanthamoeba* infections. ***American Journal of Immunology*** **1**(1): 24-30.

Ma, P., Willaert, E., Juechter, K.B. and Stevens, A.R. (1981). A case of Keratitis due to *Acanthamoeba* in New York and features of 10 cases. ***J. Infect. Dis.*** **143**: 662-667.

Mannis, M.J., Tamaru, R., Rath, A.M., Burns, M. and Thirkill, C. (1986). *Acanthamoeba* sclerokeratitis. ***Arch. Ophthalmal.*** **104**: 1313-1317.

Maraciano-cabral, F.R., Puffenbarger, R. and Carbal, G.A. (2000). The increasing importance of *Acanthamoeba* infection. ***J. Eukaryot. Microbiol.*** **47**: 29-36.

Marciano-Cabral, F., MacLean, R., Mensah, A. and LaPat-Polasko, L. (2003). Identification of *Naegleria fowleri* in domestic water sources by nested PCR. ***Appl. Environ. Microbiol.*** **69:** 5864-5869.

Martinez, A.J. (1985). "Free-living amoebas: natural history, prevention, diagnosis, pathology and treatment of disease". CRC Press, Boca Ratan, Florida.

Page, F.C. (1974). A further study of taxonomic criteria for limax amoebae with description of new species and a key to the genera. ***Arch. Protistenk.*****116**:149-184.

Pan, N.R. and Ghosh, T.N. (1971). Primary Amoebic Meningoencephalitis in two Indian children. ***J. Indian Med. Assoc.*** **56**: 134-137.

Pringle, H.L. Bradley, S.G., and Harris, AL.S. (1979). Susceptibility of Naegleria fowleri to Δ9 tetrahydrocannabinol. ***Antimicrob. Agents Chemother.*** **16:** 674-679.

Rai R., Singh, D.K., Srivastava, A.K. and Bhatgava, A. (2008). A case report of primary amoebic meningoencephalitis. ***Indian Pediatrics.*** **1004**, Vol. 45.

Rezaeian, M., Niyyati, M., Farnia, S.N. and Haghi Motevalli, A. (2008). Isolation of *Acanthamoeba* sp. From different environment sources. ***Iranian J. Parasitol.*** **Vol. 3**, No.1, 44-47.

Schuster, F.L. (2002). Cultivation of pathogenic and opportunistic free-living amoebas. ***Clinical Microbiology Review.*** **15**(3): 342-354.

Schuster, F.L. and Visvesvara, G.S. (2004). Amoebae and ciliated protozoa as causal agents of waterborne zoonotic disease. ***Vet. Parasitol.*** **126**: 91-120.

Sharma, A.K. (2004). The increasing importance of 'free-living' Amoebae casing human disease. ***Journal of Science.*** **1 (2):** 29-35.

Singh, B.N. and Hanumaiah, V. (1979). Studies on pathogenic and nonpathogenic amoebae and the bearing of nuclear division and locomotive form and behaviour on the classification of order Amoebida. Monograph No.1 of the Association of Microbiologist of India. ***Published by Indian J. Microbiol.*** **1**-80.

Singh, B.N. (1975). Pathogenic and Non Pathogenic amoebae. The Macmillan Press Ltd.,London and Basingstoke.

Thong, Y.H., Rowan-kelly, B., and Ferrante, A. (1979c). Delayed treatment of experimental amoebic meningo-encephalitis with amphotericin-b and tetracycline. ***Trans. Roy. Soc. Trop. Med. Hyg.*** **73**: 336-337.

Thong, Y.H., Rowan-Kelly, B., Shephered, C. and Ferrante, A. (1979b). Delayed treatment of experimental meningoencephalitis with Amphotericin-B and tetracycline. ***Trans. Roy. Soc. Trop. Med. Hyg.*** **73**: 336-337.

Visvesvara, G.S (2010). Free-living Amoebae as opportunistic Agents of Human Disease. ***Journal of Neuroparasitology.*** **1,** Article IDN 100802, 1-13.

WHO (2004). Guidelines for drinking water, 3rd ed. Geneva; WHO.www.iwha.net\a_abstract.htm.

Zhang, L., Marciano-Cebral. F., and Bradley, S.G. (1988). Effects of cyclophosphamide and metabolite, acrolein, on Naegleria fowleri n vitro and in vivo. ***Antimicrob. Agents Chemother*** **32**: 495-502.

Biodiversity Conservation and Envir. Management (2012)
Editors: D.R. Khanna et al.
Pub. by Biotech Books. *ISBN: 978-81-7622-262-4*

Pages: 37-42

EVALUATION OF CONSTRUCTION INDUSTRY ROLE AND ITS RELATED ACTIVITY IN THE CHANGING CLIMATE CIRCUMSTANCES

D.R. Khanna[1], Dheeraj Kumar[2], R. Bhutiani[1], Gagan Matta[1] and Vikas Singh[1]✉

[1] Department of Zoology and Environmental Science, G.K. V. Haridwar, Uttarakhand (INDIA)

[2] LANCO Constructions Ltd., Udyog Vihar, Gurgaon, Haryana

E-mail: vikassinghenv@gmail.com

On a global scale, construction sector is related to many of the environmental problems we are facing today which we have perceived from the repeated signals of serious impending changes in our global environment just in last decades. The construction process and building use not only consume the most energy of all sectors and create the most CO_2 emissions, they also create the most waste,

✉ Corresponding author

use most non-energy related resources, and are responsible for the most pollution.The present chapter evaluates the contribution of the construction sector to climate change in terms of resource use, destruction of habitat and contribution to pollution.

Introduction

Warming of the climate system is clear, as is 'now evident from observations of increase in global average air and ocean temperatures, widespread melting of snow and ice, and rising global average sea level' (Climate change, 2007). Greenhouse gases (GHGs) released due to human activities are the main cause of global warming and climate change, which is the most serious threat that human civilization has ever faced. Carbon dioxide produced from burning of fossil fuels, is the principle GHG (Sengupta, 2008). We have realized, and even experienced first hand, that the changes can and will affect the built environment and therefore our daily lives (Holm, 2003).

Construction industry input in climate change

Modern buildings consume energy in a number of ways. Energy consumption in buildings occurs in five phases. The first phase corresponds to the manufacturing of building materials and components, which is termed as embodied energy. The second and third phases correspond to the energy used to transport materials from production plants to the building site and the energy used in the actual construction of the building, which is respectively referred to as grey energy and induced energy. Fourthly, energy is consumed at the operational phase, which corresponds to the running of the building when it is occupied. Finally, energy is consumed in the demolition process of buildings as well as in the recycling of their parts, when this is promoted (Building climate change). Productivity in construction may be high on constituent processes that are of repetitive nature, however overall construction has low productivity and low quality (Fernandez-Solis, 2007; Butler, 2002) because of diversification and design uniqueness. Impact of construction industry on environment is because of:

Resource Use

The construction industry is the major consumer of resources, it accounts for 90% of all non-fuel mineral use. A distinction needs to be made between sustainable and non-sustainable resources. Sustainable resources can be divided into renewable resources (those which can be renewed – particularly those that are grown in short time cycles such as food and certain kinds of timber) and plentiful resources (such as clay, chalk, and sand). In addition materials which can be indefinitely re-used (or recycled easily) are to some extent sustainable. Non-sustainable resources are those of which there a known limited supply are and which cannot be replaced or easily reused or recycled with minimal extra energy input. These non-sustainable resources include many minerals, oil and some timber (which is very slow growing or where the extraction causes the extinction of the habitat and therefore of the resource) at our current levels and forms of use. Construction industry is the main consumer of non-renewable resources, as well as a huge consumer of renewable resources and this means it bears greatest responsibility for addressing this situation.

A study has revealed that construction industry to be a wasteful sector (Ekanayake and Ofori 2000). An estimate of the resources consumed by the construction industry is given in Table 1 and Fig. 1 in comparison of the total consumption.

Habitat Destruction

While the three greatest and most imminent threats to the survival of our civilization are global warming, peak oil (the growing energy gap between supply and demand) and resource depletion, habitat destruction can have a more immediate and disastrous effect on certain localized areas and species. Sometimes these can also have a global impact. It is hard to keep track of the number of species becoming extinct every year and of the further erosion of biodiverse and rare habitats. It is equally hard to relate this destruction to construction. However, the fact that the construction industry is such a huge consumer of materials, particularly of chemicals, minerals, metals and organic materials, inevitably means it has a huge impact on habitat erosion and destruction globally. At the present rate of resource consumption we will lose huge areas of unique habitat forever in the coming years unless we change the way we consume such materials. This is particularly as regards how we build with less waste.

Table1: Resource consumption made by the construction industry.

Resource	*Percentage of total Consumption*
Water consumption	16
Wood	25
Energy	40
Other material	40

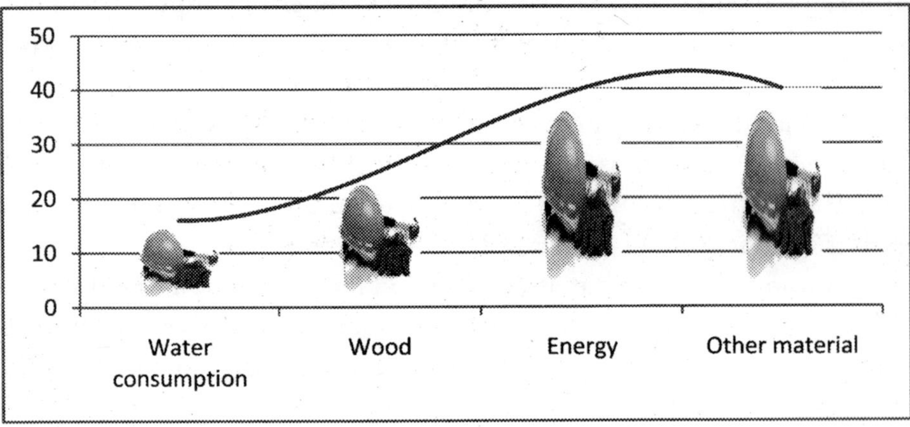

Fig.1: Percentage of resource consumed by the construction industry relative to overall consumption made.

Pollution

Finally the environmental impact of construction is also felt in terms of pollution. This is not in the extraction but in the processing of materials for construction. And again, not surprisingly, the construction industry has the biggest effect of all sectors because of the quantity of materials used in construction. In the past there was a simple general equation between the amount of pollution and the amount of energy in a process. On the whole the more energy required, and the more processes, the more waste and the more pollution was generated. The loss of control of manufacturing processes therefore has a considerable environmental impact. Input of pollutants made by construction industry is tabulated in Table 2 and Fig. 2.

Table 12: Addition of pollutants made by construction industry into the environment.

Pollution cause	*Percentage contribution by construction sector to pollution cause*
GHG	30
Waste in stream	40
Landfill	30
Other	15

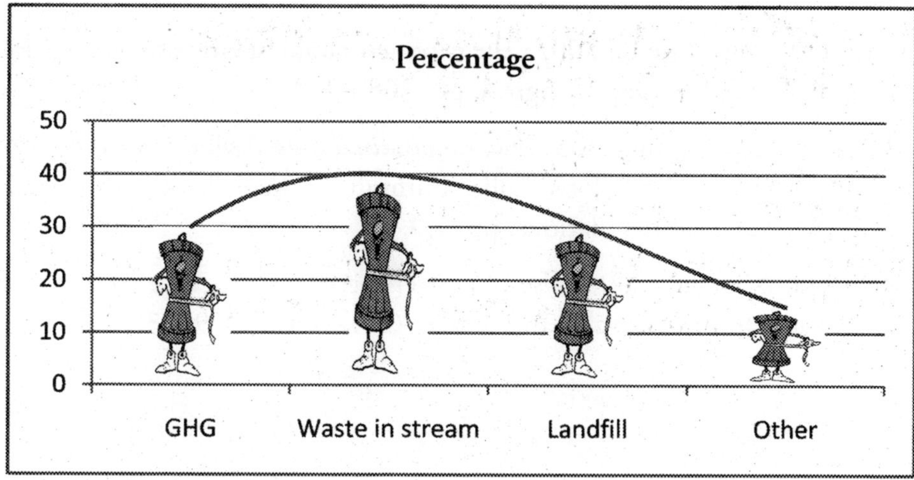

Fig. 2: Pollutant output hold by construction

Every developmental activity has some side effect and humanity has to pay for it in long run in term of environmental degradation and as a side effect seems to be slow but continuous changing climatic condition. What a construction industry can do to reduce its input value in causing climate change is to reduce high energy material use, use local and low energy materials as much as possible. We have to stick to what we are sure of and also what is inherently non-polluting along with reducing our dependency on fresh material, alternative material requirement for construction need, developing a new economic way for recovery, repair, recycle and reuse and creating new consumption pattern. Until there is proper global control of polluting processes or a clear legislation/ incentives along with proper assessment lifecycle assessment of all materials and manufacturers.

Reference

Butler, J.R. Jr., 2002. Construction quality stinks. Engineering news record.

Climate Change, 2007. The Physical Science Basis – Summary for Policymakers, IPCC. pp: 5.

Ekanayake, L.L. and Ofori, G., 2000. Construction material waste source evaluation, Proceedings of the 2nd Southern African Conference on Sustainable Development in the Built Environment, Pretoria, South Africa.

Fernandez-Solis, Jose L., 2007. The exponentialoid of emission generation. CIB World Building Congress. pp: 2685-2698.

Holm, Frank Henning, 2003. *Towards sustainable built environment prepared for climate change*? Global Policy Summit on the Role of Performance-Based Building Regulations. pp: 1:9.

Sengupta, Nilanjan, 2008. Use of cost-effective construction technologies in India to mitigate climate change. *Current science*, 94: 38-43.

Biodiversity Conservation and Envir. Management (2012)
Editors: D.R. Khanna et al.
Pub. by Biotech Books. *ISBN: 978-81-7622-262-4*

Pages: 43-54

5

A CASE STUDY OF AMBIENT NOISE POLLUTION IN AGRA CITY (U.P.) INDIA

Shakun Singh[1] and J.P. Singh[2] ✉
Department of Applied Science,
[1]BMAS Engineering College (SGI Group), Agra
[2]UP Pollution Control Board, Aligarh

Agra city is a district in the northern Indian state Uttar Pradesh having world famous historical monuments such as Taj Mahal, Agra Fort, Sikandra, Fathepur Sikri *etc*. Due to Taj Mahal, this city is also known as Taj Nagari, where lakhs of national and international tourists visit these historical monuments every year. It is located at about 200 km distance from New Delhi. A national highway passes through the city and has a population of more than half a million. Sound level monitoring was conducted to study the ambient noise level at different places in the

✉ Corresponding author

city. Monitoring places were categorized as commercial, residential, silence zone or sensitive area. Silence zone has been mainly defined as hospitals, courts, educational institutions, historical monuments and eminent religious places where no sources of sound generating equipments, loud speakers, musical instruments on the occasion of functions/parties, burning of crackers, diesel generators, blowing of horns of automobiles *etc* are allowed in the radius of 100 meter of the silence zone. It has been found that during the day time sound level monitoring, the noise level average data range between 52.6-93.2 dB(A) which was beyond the standards limits prescribed for different categorized area than the noise level average data range 36.6-63.1 dB(A) during night time, except the noise level average data range 59.6-77.4 dB(A) observed at national highway road. Higher noise level pollution can disturb our work, study of students, rest, sleep, communications, hearing power, nervous system of human beings, disturbance to animals and its impact may cause uneasy, stress, tension, heart problem, blood pressure, loss of hearing *etc* to the inhabitants of that area. Noise pollution is mainly due to automobiles, D.J. sets, musical instruments during the occasion of functions/parties, loud speakers, burning of crackers, industrial activities *etc*. Residential areas even silence zones are not exceptional from the exposure to high noise levels. Hence, it should be the duty of every citizen to minimize or control the noise pollution and people must be aware about its impact on the health of living beings. Local authorities must take a necessary action to prevent this important city from the menace noise by following some important steps against the creators of noise pollution:

1. Strict enforcement to ban the use of loud speakers, musical instruments and bursting of crackers on the various occasions of functions and parties.

2. DG sets should not be allowed without noise pollution control system like acoustic enclosures / canopy in the city.
3. Restrict the movement of vehicles and blowing of horns.

Introduction

Agra city is a district in the North Indian state Uttar Pradesh having world famous historical monuments such as Taj Mahal, Agra Fort, Sikandra, Fathepur Sikri *etc.* Due to Taj Mahal, this city is also known as Taj Nagari, where lakhs of national and international tourists visit these historical monuments every year. It is located about 200 km in the south-east of New Delhi from where a national highway road passes through the city, and has a population of more than half a million. Dayal Bagh is most densed residential area of this city, where a famous Radha Swami temple and a Dayal Deemed University is situated. In this zone there is no industrial and commercial activity. As per the Radha Swami trust temple is being constructed since more than 50 years by absolutely marble stones having a great architecture work where lakhs of visitors also come to visit this amazing temple. There is Yamuna River which touches main city as well as Taj Mahal. National Highway passes over the bridge of Yamuna river and this highway crosses mainly four chaurahas named Khandari chauraha, Bhagwan Talkies chauraha, Water works chauraha and Rambagh chauraha. Bhagwn Talkies chauraha is a place where National highway NH2 passes and below this highway MG road crosses to main city. Khandari chauraha has internal road to city where light vehicles passes and Water works chauraha and Rambagh chauraha are the crowded places of the transport and there generally traffic Jam.

In the centre of city there is a S.N. Medical College as a main government health care facilities for the treatment of habitants of the district Agra around which commercial and residential areas are situated.

Taj Mahal is a world famous historical monument of this city. In view of safety and security of the Taj Mahal as well as the people of the city, there is a Military cant area where no commercial and industrial activities take place.

Keeping all these things in mind, it has been decided to assess the noise level pollution of this city. Eight places have been selected for ambient noise level monitoring at different categorized areas of the city in the month of October -2010. Number of studies have been performed on noise level pollution at various cities of India (Edison *et al.*, 1999; Singhal, 2000; Bhatt *et al.*, 2004 and Khanna *et al.*, 2004).

Materials and Methods

Regarding materials no more equipment was required during noise level monitoring. Only a sound level monitor was used to observe the values of sound level at different places. Methodology of sound level monitoring was adopted as per the manual of the instruments and CPCB guidelines. Sound pressure level was taken in "A" weightage. The values of noise level have been recorded in the form of minimum, maximum and average. Sound level monitoring was conducted at different categorized areas.

Monitoring stations

Eight monitoring stations were selected to conduct the ambient noise level monitoring at different categorized areas in the city which are given below:

S. No.	*Name of monitoring stations*	*Categorised area*	*Monitoring point*
1.	Bhagwan Talkies Chauraha/NH2-Crossing	-do-	A
2.	Dayalbagh (Adanbagh colony)	- Residential	B
3.	S.N. Medical College	- Silence	C
4.	Military Cantt Area (Canteen)	- Silence	D
5.	Taj Mahal	- Silence/Sensitive	E
6.	Water Works Chauraha/NH2-Crossing	- Residential	F
7.	Khandari Chauraha/NH2-Crossing	- Commercial	G
8.	Rambagh Chauraha/NH2-Crossing	- Commercial	H

Brief details about monitorig stations

Bhagwan Talkies Chauraha/NH2-Crossing is generally a busy and crowded crossing/chauraha from where generally light vehicles and city buses pass below the national highway at MG road to the city. This crossing is known as Bhagwan Talkies Chauraha as there is a Bhagwan Talkies picture hall. This chauraha is having commercial area. This monitoring station is represented as –A.

Dayalbagh (Adanbagh colony) has been selected as reference point of view as residential area where no commercial and industrial activities take place. Dayalbagh is a famous place due to its renowned Radha Swami temple and an educational institute named Dayalbagh Deemed University where light vehicular movement passes. This monitoing station is represented as –B.

S.N. Medical College is a main government health care facility in the heart of city around which mixed areas of residential and commercial purposes are situated. This point was selected as a silence zone to study the noise pollution and is represented as monitoring station-C.

Military Cantt Area (Canteen) is situated far away from the dense city of Agra, however road passes to Agra Cantt Railway station and Idgah road ways bus stand. This point was choosen as a silence zone which is represented as monitoring station –D.

Taj Mahal is situated in Tajganj on the river bank of Yamuna around which several activities of hotels and parkings held. Taj Mahal is a world famous historical monument. So this point was also selected as silence/sensitive point to assess the intensity of sound level. This monitoring station is represented as –E.

Water works Chauraha/NH2-Crossing is a bussiest and crowded chauraha at National highway due to generally occurring traffic Jam. This crossing is known as water works chauraha as there is a water works situated on the river bank of Yamuna river. Here, National highway road passes over the bridge of Yamuna river. There is mostly residential activities around this crossing.This monitoing station is represented as –F.

Khandari Chauraha/NH2-Crossing is a bussy crossing/chauraha which crosses NH2 from where generally light vehicle passes. This crossing is having light commercial activities.This monitoing station is represented as –G.

Rambagh Chauraha/NH2-Crossing is also a bussiest and crowded chauraha below the National highway due to generally occurring traffic Jam. This chauraha is situated on the other river bank of Yamuna around which mainly commercial and industrial activities held and is considered as commercial area to study the sound level pollution at this point. This monitoring station represented as –H.

Sound level monitoring was conducted for at least 10 minutes at each sampling point during day and night time to study the ambient noise level at different places in the city. Monitoring places were categorized as Commercial, Residential, Silence Zone or Sensitive area. Silence Zone has been mainly defined as Hospitals, Courts, Educational institutions, historical monuments and eminent religious places where no sources of sound generating equipments, loud speakers, musical instruments on the occasion of functions/ parties, burning of crackers, diesel generators, blowing of horns of automobiles *etc.* are allowed in the radius of 100 meter from the Silence Zone.

Results and Discussion

Values of noise level data are given in Tables 1 and 2. Ambient noise level standards are mentioned in Table 3.

It is evident from the Table 1 that during day time minimum sound level values varied between 41.4 dB(A) to 62.2 dB(A) and maximum values 66.4 dB(A) to 98.6 dB(A) where as during night time the values of minimum sound level found between 26.3 dB(A) to 50.3 dB(A) and its maximum values varied from 48.4 dB(A) to 88.9 dB(A).The average values of minimum and maximum of day time monitoring recorded 52.63 dB(A) and 83.20 dB(A) and during night time monitoring average values of minimum, maximum observed 36.63 dB(A) to 63.10 dB(A) of all sampling stations shown in Table 1. In Table 2 sound level data of minimum, maximum and their average values of National highway have computed for only day time which only two monitoring stations have been observed. The average values were found between 59.6 dB (A) to 77.35 dB(A) of minimum, maximum values of both sampling stations (G & H).

The values of minimum and maximum were observed 60.6 dB(A) and 96.2 dB(A) and their average value observed 78.4 dB(A) during day time while during night time minimum, maximum and its average values

Table 1: Results of sound level data of different monitoring stations of the Agra city.

Sl.	*Monitoring place*	*Station Denoted as*	*Date of monitoring*	*Categorised Area*	*Values in dB(A) Leq.*					
					During Day time			During Night time		
					Min.	Max.	Avg.	Min.	Max.	Avg.
1	Bhagwan Talkies Chauraha	A	3.10.2010	Commercial	60.60	96.20	78.40	48.02	84.30	66.15
2	Dayalbagh (Adanbagh colony)	B	3.10.2010	Residential	53.30	81.50	67.40	35.10	50.50	42.80
3	S.N.Medical college	C	3.10.2010	Silence Zone	51.20	87.60	69.40	32.00	55.80	43.90
4	Military Cantt area (Canteen)	D	3.10.2010	Silence Zone	41.40	66.40	53.90	26.30	48.40	37.35
5	Taj Mahal	E	3.10.2010	Silence/ Sensitive Zone	47.10	68.90	58.00	28.10	50.60	39.35
6	Water works Chauraha	F	3.10.2010	Residential	62.20	98.60	80.40	50.30	88.90	69.60
			Average		52.63	83.20	67.91	36.63	63.08	49.86

were found as 48.02 dB(A), 84.30 dB(A) and 66.15 dB(A) respectively at the monitoring point-**A**. The average values of day and night time both are found beyond the prescribed standards for a commercial area at this point which may due to blowing of horns by heavy vehicular movements at national highway and crowdyness of this place.

Table 2: Results of sound level data at National Highway in the city of Agra.

7	Khandari Chauraha	G	3.10.2010	Commercial	53.3	75.4	64.35
8	Rambagh Chauraha	H	3.10.2010	Commercial	65.9	79.3	72.6
			Average		59.6	77.35	68.475

The values of sound pressure level are found between 53.30 dB(A) to 81.50 dB(A) and its average value was found 67.40 dB(A) during day time and values during night time was recorded 35.10 dB(A) to 50.50 dB(A) at monitoring point-**B**. Average value of day time found beyond the prescribed standards for a residential area which may due to blowing of horns by vehicular movements and the range found of sound pressure level during night time as mentioned above which is obviously within the prescribed standards.

The values of sound pressure level at monitoring point-**C** ranges 51.20 dB(A) to 87.60 dB(A) and its average 69.40 dB(A) obtained during day time and during night time the values of noise level found between 32.00 dB(A) to 55.80 dB(A) and its average value observed 43.9 dB(A). The average values of day and night time both were obtained beyond the prescribed standards limit for silence zone but average value of night time is slightly higher than the limit which may be due to random vehicular movement of emergency cases to Health care facilities.

At monitoring station-**D**, the range of sound pressure level observed 41.40 dB(A) to 66.40 dB(A) and its average value noted 53.90 dB(A) during day time while during night time it was 26.30 dB(A) to 48.40 dB(A) and its average value found 37.35 dB(A).The average value of day time found slightly beyond the limit of prescribed standards for silence zone which may be due to light vehicular movements while on the other hand during night time average value of noise level was obtained well within the limit and it may be due to restriction of entering private vehicular movement of outsider in this prohibited Military cantt area.

The minimum, maximum and its average of noise level were observed as 47.10dB(A), 68.90dB(A) and 58.00dB(A) respectively during day time and 28.10 dB(A), 50.60 dB(A) and 39.35 dB(A) respectively during night time at sampling point-**E.** The average value of day time found more than the standards limit prescribed for silence/sensitive zone [50.0dB(A)] which may be due to vehicular movements around the Taj Mahal, blowing of horns, hotels activities parkings and visitors chitchating during visiting Taj in day time period while during night time the average value of noise level observed slightly within the prescribed limit [40.0 dB(A)].

Water works chauraha/crossing is a place from where some heavy traffics move to the city and it is generally known as Traffic Jammed Area. During day time monitoring the values of sound level found between 62.20 dB(A) to 98.60 dB(A) and its average value observed 80.40 dB(A) while during night time noise level were values found 50.30 dB(A) to 88.90 and its average value found 69.60 dB(A). Average sound level values of day and night time of this monitoring point-**F** were found beyond the prescribed standards for residential as well as commercial area which may due to blowing of horns by vehicular movements and being a Traffic Jammed Area.

The values of noise level varied between 53.30 dB(A) to 75.40 dB(A) and its average value obtained 64.35 dB(A) during day time at sampling point-**G.** The average value of this point found slightly within the prescribed standards limit for commercial area [65 dB(A)] which may be due to light vehicular movements and not having more traffic problems at this point.

Rambagh chauraha/crossing is a place from where some heavy traffics moves and it is generally known as Traffic Jammed Area. During day time monitoring, the values of sound level found between 65.90 dB(A) to 79.30 dB(A) and its average value observed 72.60dB(A). Average sound level value of day time of this monitoring point-**H** were found beyond the prescribed standards for commercial area which may due to blowing of horns by vehicular movements, industrial activities and being a Traffic Jammed area. Bhatt *et al.* (2004); Khanna *et al.* (2004) reported 65.0 dB(A) to 81.0 dB(A); and Babu (2003) reported noise level 30.0 to 90.00 dB(A) and Ingle *et al.* (2001) by traffic.

Highest average values are found at monitoring point-F during day and night time as compared to monitoring point-A, H, G which may be due to heavy traffics and traffic jam, blowing of horns, sirens of ambulance and VIP vehicles at the national highway roads. At other monitoring points average values of day time monitoring periods were not found within the prescribed standard limits of their categorised areas, where as during night time the average values of the noise level were observed within the prescribed standards limit at only three monitoring point-B, D and E and this may be due to negligence of vehicular movements at these points.

Table 3: Ambient Noise Level Standards.

Sl.No.	*Category of area*	*Limits in dB(A) Leq.*	
		Day time	Night Time
1	Industrial Area	75	70
2	Commercial Area	65	55
3	Residential Area	55	45
4	Silence Zone	50	40

Note:

1. Day time is reckoned in between 6:00 a.m.-10:00 p.m.
2. Night time is reckoned in between 10:00 p.m.-6:00 a.m.
3. Silence is defined as areas upto 100 metres around such premises as hospitals, educational institutions, courts and eminents temples/ monuments .
4. Use of vehicular horns, loudspeakers and bursting of crackers shall be banned in those zones (Notification, 2000).

As it is evident from the Tables 1 and 2 that a trend of higher noise pollution at sampling points-F > A > H > G were observed in respect of noise pollution. At these points the main factors of higher noise level pollution may be due to heavy traffic jam, use of pressure horns, operations of Diesel generators during power failure and other human activities *etc.*

Blowing of pressure horns, vehicular movements, bursting of crackers, playing of musical instruments, loudspeakers, operations of DG sets and industrial activities *etc* are responsible for higher level noise pollution and persons exposed to this level of noise pollution for long time may suffer the menace of noise pollution such as hearing loss can be temporary or permanent. Consequently, students in colleges, office bearers, visitors, tourists, shop owners, patients in hospitals, birds and other animals *etc* are exposed to very high noise level pollution.

Higher noise level pollution can disturb our work, study of students, rest, sleep, communications, hearing power, nervous system of human beings, disturbance to animals and its impact may cause uneasy, stress, tension, headache, heart problem, blood pressure, loss of hearing *etc* to the inhabitants of that area. Noise pollution is mainly due to automobiles, D G sets, musical instruments during the occasion of functions/parties, loud speakers, burning of crackers, industrial activities *etc* Residential areas even silence zones are not exceptional from the exposure to high noise levels. Hence, it should be the duty of every citizen to minimize or control the noise pollution and people must be aware about its impact on the health of living beings. Local authorities must take a necessary action to prevent this important city from the menace noise by following some main steps against the creators of noise pollution. Some suggestions may be adopted for control of noise pollution as given below:

1. Strict enforcement to ban the use of loud speakers, musical instruments and bursting of crackers on the various occasions of functions and parties.
2. DG sets should not be allowed without noise pollution control system like acoustic enclosures/canopy in the city.
3. Restrict the movement of unmaintained vehicles and blowing of horns.
4. Specific legislation and regulations should be proposed for designing and operation of machines to include vibration control, sound proof cabins and sound-absorbing materials.

References

Babu, S. Sarvana 2003 Noise Pollution health hazard. Noise Pollution. Environmental Training Institute, DANIDA(Denmark),Tamil Nadu Pollution Control Board Chennai. P.9.

Bhatt,C.S.; Sikander, Mohd; Singh, JP; Khanna, DR; Gautam, ashutosh; and Singh,Shakun;2004: Astudyof ambient air quality and noise pollution in Nainital city. Environment Conservation Journal 5 (13) 7-13.

Edison. R. Raja; C. Ravi Chandran and J. Christal Sagila, 1999. An assessment of Noise Pollution due to automobiles in Cuddalore, Tamil Nadu. India. J. Env. Health.41(4):312-316.

Ingle, ST; Attarde, SB; Dhake, R.B; and Panchpande, B.G. 2001:Noise Pollution- An insidious Hazardous in Urban Environment-"A case study of Jalgoon City-" Status of Indian Environment-(Pre conference Proceedings), ASEA, Rishikesh, P-31.

Khanna,DR; Singh, J.P.; and Singh, Shakun; 2004: Assessment of noise pollution in the city of Haridwar (Uttaranchal) India. Environment Conservation Journal 5(1-3)61-66,2004.

Singhal, S.P. 2000: Noise pollution and control in Urban and Industrial environment National Physical Laboratory, New Delhi. A short course on Ambient Air monitoring and management.

Biodiversity Conservation and Envir. Management (2012)
Editors: D.R. Khanna et al.
Pub. by Biotech Books. *ISBN: 978-81-7622-262-4*

Pages: 55-69

6

BRASSINOSTEROIDS: BIOCHEMICAL AND PHYSIOLOGICAL PERSPECTIVE(S)

H.Punetha[1], Shivom Singh[2✉], Kajal Srivastava[2] and A.K. Gaur[1]
[1]Department of Biochemistry
[2]Department of Biological Sciences
College of Basic Sciences and Humanities, G.B. Pant University of Agriculture & Technology, Pantnagar, Uttarakhand (INDIA)
✉E-mail: shivom101@rediffmail.com

Brassinosteroids (BRs) are phytohormones, having growth promotory role in plants, which show physiological action similar to auxin, gibberlin and cytokinins. These are polyhydroxy sterols which are denoted by sequential suffix from BR1 to BRn. BR1 denoted first isolated brassinosteroid, brassinoloide. Two biosynthetic pathways, an early C6 oxidation and late C6 oxidation are operated. Brassinosteroids are considered as hormones with

✉ Corresponding author

pleiotropic effects, as they influence varied processes such as growth, germination of seeds, rhizogenesis, flowering and senescence. They also confer resistance to plants against various abiotic stresses. There presence have been reported in 44 plant species including 37 angiosperms, one algae and one fern. Such a wide occurrence of BRs in lower and higher plants indicate their wide distribution and it seems to be an important growth factor. New discovery on physiological properties of these compounds is being utilized for plant protection and yield promotion in agriculture sector during recent past.

Keywords: Brassinosteroids, Phytohormone, Growth factor, Signal transduction, *In vitro*.

Phytochemicals are of interest from the early 20th century with isolation of Auxin and until 1979, five classes of phytohormones *viz*., auxin, gibberellin, cytokinin, abscisic acid and ethylene were reported. A sixth class of phytohormone, brassinosteroid which support plant growth and development was observed later. Brassinosteroids were first isolated and characterized from the pollen of Rape plant, *Brassica napus* L. On the basis of structure determination these should uniquely steroids supporting growth in plant systems, however.

Discovery and distribution of Brassinosteroids

In 1970, USDA first time published a paper entitled Brassin: A new class of plant hormone from Rape pollen. Today there is sufficient report on its presence by various scientists (Abe *et al.*, 1984a; Abe *et al.* 1984b; Takatsuto *et al.*, 1990; Arima *et al.*, 1984), who reported the presence of brassinosteroid in tea, rice, buckwheat and chestnut respectively. Later on brassinosteroids were also isolated from seed of different plant like *Raphanus sativus* (Schmidt *et al.*, 1993), Foxtail millet (Takatsuto *et al.,* 1999) and Pumpkin (Jang *et al.*, 2000). Brassinosteroids have been characterized from 44 plant species, which include 37 angiosperms (nine monocots and 28 dicots), five gymnosperms, one pteridophyte and one algae.

Structure and Classification of Brassinosteroids

Brassinosteroids are group of polyhydroxy sterols. Natural brassinosteroid classification system includes denotion by a sequential suffix for example BR1 denotes first isolated brassinosteroid namely brassinolide, others follow sequence like BR2 to BRn. Natural brassinosteroids characterized from certain plants are 5-α cholestan•derivatives but these are denoted as C-26, C-27 and C-29 steroids (Mandava and Bhushan 1988; Fujioka and Sakurai 1997 and Adams and Marquardt 1986). These variation of brassinosteroids are due to substitution either in A or B rings, otherwise side chains are created upon oxidation or reduction reactions found to occur during their biosynthesis. Brassinolide, 24-epibrassinolide and 28-homobrassinolide are three biologicaly active brassinosteroids.

Brassinosteroids are phytohormone

Brassinosteroids are a group of naturally occurring polyhydroxy steroids. Natural brassinosteroids have common skeleton 5α-cholestan skeleton. Brassinosteroids have shown biological activity like auxin, gibberellins and cytokinin. Although it became increasingly confusing initially but today these are recognized as a separate class of phytohormones. Slowly their functions are being clearly elucidated. Brassinosteroids (BR_s) have been identified as plant hormones with significant growth promotory activities. At present Brassinosteroids are regarded probably most ubiquitous in plant kingdom. Brassinosteroids are present in extremely low concentration, the pollen and immature seeds contain 1.0-100 ng/g fresh weight of brassinosteroids, while leaves and shoots contain lower amount ranging from 0.01 to 0.1 ng/g fresh weight (Takatsuto, 1994). These are highly mobile in the plant system and satisfy the translocation property originally attributed to plant hormones as proved by the fact that when exogenous brassinolide were applied to the roots of young tomato plants it affected hypocotyls and petioles (Takatsuto *et al.*, 1983).

Biosynthetic Pathway of Brassinosteroids

Biosynthetic pathway of brassinosteroids, have been studied by using culture cell through developing various BR deficient mutant which upon analysis have confirmed the existence of two parallel pathways *viz.*, early and late C-6 oxidation pathways which show possible branching at the time of conversion of campesterol. The scheme of the two pathways are described in Table 1 (Sakurai, 1999 and Noguchi *et al.*, 2000)

Table 1: Early and late C_6 oxidation biosynthetic pathway of brassinolide.

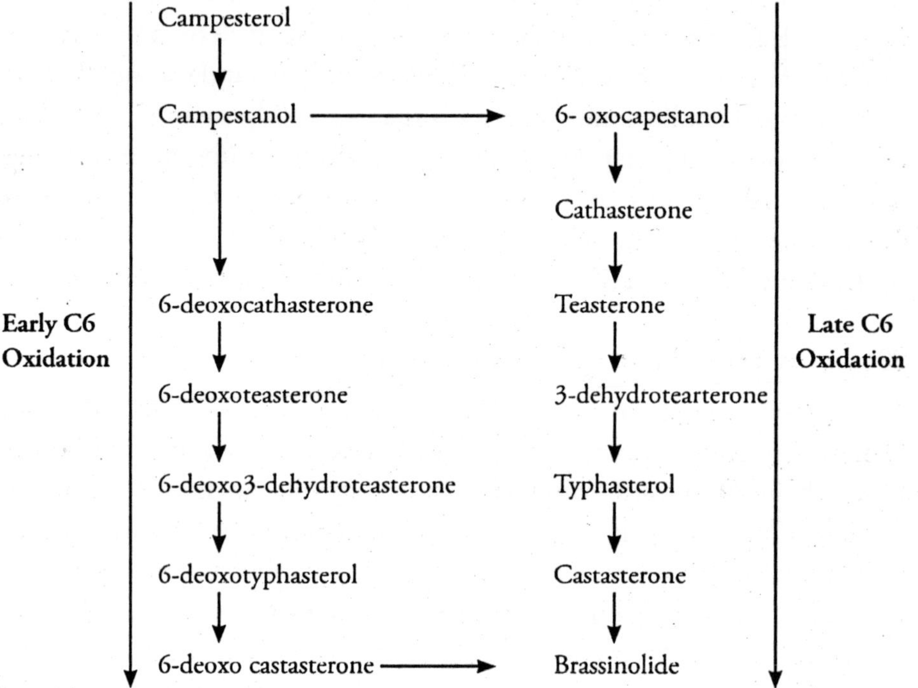

These two biosynthetic pathways are ubiquitous in plant system (Fujioka and Sakurai 1997). Experiment was also conducted with dwarf mutant of *Arabidopsis thaliana* and *Pisum sativum* showing defect in brassinosteroids biosynthesis (Yokota, 1997). In this experiment *det2* and *cpd* encode enzymes involved in the synthesis of brassinosteroids. The dark and light-grown phenotypes *det2* and *cpd* mutants can be completely reversed by the action of brassinolides to the growth medium. DET-2 catalyzes the conversion of 3-dehydro-Δ^{4-5}-campesterol to 3-dehydrocampestanol and CPD converts cathasterone to teasteron. In addition, characterization of many other brassinosteroid deficient mutants, such as *sax1*, *dwarf4*, *dim1/dwarf1/cbb1/lkb* and *dwf5/dwarf7* resulted in the identification of the other enzymes in the proposed biosynthetic pathway (Friedrichsen and Chory, 2001). Both DWF5 and DWF7 are desatuarase required for the formation of 24-methylenecholesterol. *DIM1/DWARF1/CBB1/LKB* are homologous to FAD-dependent oxidoreductase and is involved in converting 24-methylenecholesterol to campesterol. SAX-1 converts campesterol to 3-dehydro- Δ^{4-5}-campesterol, the substrate for DET2. Like CPD, DWF4, DDWF1 and DWARF are cytochrome P450 proteins that convert campestanol to cathasterone, typhasterol to catasterone and 6-deoxocatasterone to catasterone, respectively.

Signal Transduction Pathway of Brassinosteroids

Attempts were made to determine role of Brassinosteroids in protein and nucleic acid metabolism (Mandava and Bhushan 1988). The first cloning and characterization of gene, regulated primarily by brassinosteroids was found in soyabean epicotyl elongation system (Zurek, 1994). Brassinosteroids promotes elongation independently of auxins in this system and gene *bru1* was found to be specifically regulated by brassinosteroids. Sequence analysis of BRU1 showed the presence of signal peptide and homology with xylo-glucan endo trans glycosylase and bacterial lichenases. In classical model of steroid hormone action the steroid permeates the cell membrane and bind with the cell surface receptor rather than intracellular receptor and this finally operates as signaling cascade of protein kinase that activate certain transcription factor which interact with different DNA sequences in hormone responsive gene. Brassinosteroids are also shown to regulate gene expression at post transcriptional level (Clouse, 1997).

Brassinosteroids signal transduction pathway is modulated by light, perhaps by regulating the metabolism of brassinosteroids or by altering the cell's responsiveness to brassinosteroids (Kang *et al.*, 2001). Over expression of the *bas1* gene encoding a steroid 26-hydroxylase leads to decreased brassinosteroid levels and a BR-deficient dwarf phenotype and their is a evidence that *bas1* gene is regulated by multiple photoreceptor systems (Neff *et al.*, 1999b). A light regulated gene *pra2* encodes a small G protein that directly regulates the activity of a P450 protein DDWF1 involved in brassinosteroids biosynthesis (Kang *et al.*, 2001). In order to identify the components of brassinosteroid signal transduction pathway 20 BR-insensitive mutants have been identified in *Arabidopsis*. Most of the mutations have been found in the same locus named BRI1 which encode the protein that has a sequence homology to transmembrane leucine-rich repeat (LRR) receptor kinase (Li and Chory, 1997). Recent studies demonstrate that BRI1 functions as a brassinosteroid receptor. BRI1 is composed of an extracellular domain containing 25 LRRs interrupted by a 70 amino acid island domain, a single transmembrane domain and an intracellular serine/threonine kinase domain (Li and Chory, 1997). A chimeric receptor composed of BRI1 extracellular domain and kinase domain of Xa21, a receptor kinase for pathogen response in rice, confers brassinolide dependent pathogen responses to heterologous rice cells (He *et al.*, 2000), suggesting that BRI1's extracellular domain is responsible for brassinosteroid perception. The endogenous levels of many brassinosteroids such as castasterone, typhasterol and brassinolide increased in *bri1* mutants, hence brassinosteroid signaling through BRI1 inhibits hormone levels by a feedback inhibition mechanism (Friedrichsen and Chory, 2001). This feedback inhibition functions through brassinosteroid-biosynthetic genes such as *cpd* and *DWARF4* because these genes are down-regulated by brassinosteroids and up-regulated in *bri1* and other BR-deficient mutants.

Physiological role of Brassinosteroids on plant growth and development

Brassinosteroids are growth promoting hormones which play a role in promoting cell elongation in a wide range of plant species. It has been examined that the effect of brassinolide on the kinetics and final cell division frequencies by using regenerating leaf mesophyll protoplast of *Petunia hybrida* under optimal auxin and cytokinin condition, where it was noticed that 10-100 nm brassinolide accelerated the time of first cell divisions by 12 hours (Oh and Clouse, 1998). Salinity was found to increase accumulation of lectin and ABA after 14 hours of incubation in the roots of 4 day old plant of *Triticum aestivum* in association with presence of 24-epibrassionlide (Shakirova and Bezrukova 1998). Brassinolide was found to increase the germination percentage of corn seed (Kong and Zhang, 1998) and *Phaseolus vulgaris* (Takahashi *et al.* 1999). Brassinosteroids were found to increase the yield of IR-50 indica variety of rice (Krishnan *et al.*, 1999). In case of tomatoes and onions differences were found in the analogues as far as growth promotion and yield were concerned, whereas analogue BB-6 cause growth promotion in tomato, in case of onion the same responses were observed with BB-16 analogue (Nunez *et al.*, 2000). Inhibitory effect of ABA on seed germination and seedling growth of *Trigonella foenum graeoune* was found to reverse upon treatment with either 28-homobrassionolide or 24-epibrassinolide (Vardhini and Rao, 1997).

Brassinosteroids like other plant hormones regulate the composition of protein in wheat grain affecting the grain quality (Novikov and Vaiersa, 1995). 28-homobrassinolide treatment significantly increased seed/grain yield in wheat, rice, cotton, mustard and groundnut (Ramraj *et al.*, 1997). Homobrassinosteroids (HBRs) treatment influenced plant growth at particular stage of application and frequency of applications. In the green house trial, strawberry plant of the day neutral cultivar *viz.*, Miyorhi and Enrai, when sprayed with 0.01 ppm TS 303 (a brassinosteroid) at different time intervals showed increased total leaf area by 150-180% along with number of leaves, petiole length and number of crowns by 110-140% as compared with untreated control plant (Fujishige *et al.*, 1996). Wheat and potato also responded differently to epibrassinolide treatment by producing extra IAA and gibberellin, with an additional observation of increased transport of glucose, resulting increase in final yield upto 25% over control (Kurapov *et al.*, 1996).

It has been observed that when brassinosteroids were applied in 30 mg/l quant is 50-65 days after sowing of *Vicia faba* it showed significant growth and increased yield as well as significant increase in endogenous hormone levels (Helmy *et al.*, 1997). Occurrence of dwarfism due to lack of cell elongation could be rescued from dwarf plant by exogenous supply of brassinolide (Azpiroz *et al.*, 1998).

Recent studies on BR-insensitive and BR-deficient mutant confirmed that these compounds (BRs) regulate normal plant growth and development (Clouse and Sasse, 1998). Besides, these are also known to regulate the functions like cell elongation and cell division similar to cytokinine. Highly purified brassinolide caused not only cell elongation in the second internode of bean but also showed profound effect on curvature, swelling and increased length of internode due to increased cell division (Grove *et al.*, 1979). Brassinolide in the presence of auxin and cytokinin caused a minimum increase of 50% with respect to total number of cells in cultured explants of *Helianthus tuberosus* after 24h suggesting a strong promotive effect of brassinosteroid on cell division. Foliar application of brassinostreroids resulted in the increase in the number of flowers in strawberry (Pipattanawong *et al.*, 1996).

Brassinosteroids and Stress tolerance

Brassinosteroids increase the resistance of plants against various biotic and abiotic stresses. Roth *et al.*, (2000) reported that Homobrassinosteroids containing extract of *Lychniss viscarcia* are found to induce resistance in plants against biotic stress. Brassinosteroids have found to confer resistance against phytopathogen (Khripach *et al.*, 2000). The antiviral activity of brassinosteroids against Herpes and Arena virus was also reported (Wachsman *et al.*, 2000). Similarly the ability to confer tolerance/resistance against wide range of abiotic stresses has been established by various workers (Kamoro and Takatsuto, 1999). Improvement of the growth and vigour in rice and tomato (Kamuro and Takatsuto, 1991), maize (Hand and Wang) and cucumber (Katsumi) against low temperature stress was established. Brassinosteroids increased tolerance to high temperature in wheat (Kulaeva *et al.*, 1991) and bromegrass (Wilen *et al.*, 1995), increased tolerance to drought stress in sugar beet (Schilling *et al.*, 1991), resistance to moisture stress in wheat (Sairam, 1994), tolerance to saline media in rice (Kamuro and Takatsuto, 1999) and resistance against chilling stress (1-5°C) in rice (Wang and Zang, 1993). The role of Brassinosteroids in the improvement of plant growth against adverse environmental stress conditions will be a major thrust in further research work with this new class of phytohormone.

Effect of Brassinosteroids on *in vitro* culture

Rooting and adaptation in micro propagated plants of sugarcane were found to be enhanced by synthetic brassinosteroids (Fe and Ortiz, 1998). Brassinosteroids proved to be a hormonal substitute for callus induction in coffee (Garcia and Sasse, 1997). Brassinolide altered the abundance in *in vitro* specific translatable mRNA from peduncles of whole plant of *Arabidopsis thaliana* (Clouse *et al.*, 1998). Explants derived from plantlets of *Dianthus, Caryophyllus* grown *in vitro* and upon further transfer into MS medium containing brassinosteroid at concentration of 0.5 or 1.0 mg/l have shown that it is most effective substitute of IAA (Montes *et al.*, 1997). Since brassinosteroids act as a general growth enhancer, 24 epibrassinolide was found to affect specifically as a stem elongator in *in vitro* grown sweet pepper seedlings (Franck-Duchenne *et al.*, 1998). The cell division promotion by epibrassinolide was found to be influenced through Cyc D3 and thus can also substitute cytokinin in culturing calluses or cells in suspension of *Arabidopsis* (Hu *et al.*, 2000). With a variety of brassinosteroids in existence, BR-6 analogues were found to effect growth of plantlets and BB-16 analogues were found to increase callus growth in tissue culture of potatoes (Hernandez *et al.*, 1999). It has been observed that an increase upto 50% in the frequency of conversion of somatic embryo of *Thea sinensisin*, to plantlets when the medium was supplemented with brassin (Ponsamuel *et al.* 1996).When normal plantlets were treated with brassin, profuse rooting was observed in these plantlets.

Commercial usages in Agriculture

The discovery of brassinosteroids in plant system paves the way to explore the potential of these new compounds for improving the yield of economically useful crop plants. Satisfactory progress has been made by Meundt and Thompson (1998) in the improvement of vegetable crops *viz.*, radish, beans and lettuce.

Application of 28-homobrassinolide significantly increased fiber yield in tobacco (Ramraj *et al.*, 1997). Improvement in the yield parameters among wheat (Braun and Wild, 1984), mustard (Braun *et al.*, 1984), corn and tobacco (Yokota and Takahashi, 1986) through the foliar application of brassinolides has been established and further research on various other crops *viz.*, sugar beet (Schilling *et al.*, 1991), legumes (Kamuro and Takatsuto, 1992), french bean (Laksminarayan, 2002) and barley (Khripach *et al.*, 1997) are under progress.

Future prospects

With an immense role in performing several functions in the plant growth and development, brassinosteroids can be considered as one of the vital phytohormones. New discoveries of the physiological properties of brassinosteroids allowed us to consider them as highly promising environmental friendly natural substances suitable for wide spectrum application for plant protection and yield promotion in Agriculture (Khripach *et al.*, 2000). One of the major constraints in the application of brassinosteroids are their high cost. However, the recent progress in the chemical synthesis of brassinosteroids and their analogues leads to economical feasible products having good market in agriculture sector. In near future the greater role of brassinosteroids will be established for bettar crop yield including the food production during the regime of organic farming.

References

Abe, H.; Morishita, T.; Uchiyama, M.; Takatsuto, S. and Ikikawa, N. **1984a.** A new brassinolide- related steroid in the leaves of Thea sinensis. Agricultural and Biological Chemistry. **48:** 2172-2172.

Abe, H.; Nakamura, K.; Morishita, T.; Uchiyama, M.; Takatsuto, S. and Ikikawa, N. **1984**b. Endogenous brassinosteroids of the rice plant: catasterone and dolichosterone: *Agricultural and Biological chemistry,* **48**: 1103-1104.

Adams, G. and Marquartdt, V. **1986**. Brassinosteroid. *Phytochemistry* **25**: 1787 – 1799.

Arima, M.; Yokota, T. and Takashi, N. **1984**. Identification and quantification of brassinolide related steroids in the insect gall and healthy tissues of the chestnut plant, *Phytochemistry*, **23** (8): 1587-1592.

Azpiroz, R.; Wu, Y. Locascio, J.C. and Feldman, K.A. **1998**. An Arabidopsis brassinosteroid dependent mutant in blocked in cell elongation. *The Plant Cell.* **10**(2) 219-230

Braun, P. and Wild, A. **1984**, in Proc. 6th Congress Photosynthesis (ed. Syberma, C.) *The Hague Nijhoff* P. 11.

Clouse Steven, D. **1997**. Molecular genetic analysis of brassinosteroid action. *Physiologia plantanum.* **100**: 702-709

Clouse, S.D. and Sasse, J.M. **1998.** Brassinosteroid; essential regulators of plant growth and development. *Annu Rev. Plant Physio. Plant Mol. Biol*: 427-451.

Fe, C. and Ortig, R. **1998**. Some contribution to sugarcane micro propagation technology applied in Cuba. *Cultivos Tropicales* (Cuba). **19**(3): 45-48

Franck-Duchenne, M.; Wang, Y.W.; ben Tahar, S. and Beachy, R.N. **1998**. In vitro stem elongation of sweet pepper in media containing 24 epibrassinolide. *Plant Cell Tissue and Organ Culture*. **53** (2): 79-84

Friedrichsen, D., and Chory, J. **2001**. Steroid signaling in Plants: from the cell surface to nucleus, *Bio Essays*. **23**: 1028-1036

Fugishige, N; Pipatharawong-N; Yamane-K, Ogata, R. **1996**. Effects of brassinosteroide on vegetative and reproductive growth in two day neutral strawberries. *Journal of the Japanese society for Horticultural Science*, **65**:3, 651-654.

Fujioka, S. and Sakurai, A. **1997**. Biosynthesis and metabolism of brassinosteroids. *Physiologia plantarum* **100**: 710-715.

Garcia, D.; Marrero, M.T.; Cuba, M. and Nunez, M. **1997**. Qualitative effects of synthetic brassinosteroids as hormonal substitutes in callus formation in coffee. *Cultivos Tropicales*. **18**(2): 44-46.

He, Z., Wang; Z.Y., Li; J., Zhu;; Q., Lamb; C., Onald; P., and Chory, J.2000. Perception of brassinosteroids by the extracellular domain of the receptor kinase BR11.*Science* **288**:2360-2363.

Helmy, Y.I.; Sawan, O.M.M. and Abdel Halim, S.M. **1997**. Growth yield and endogenous hormone of broad bean plants as affected by brassinosteroids. *Egyptian Journal of Horticulture* **24** (1): 109-115

Hernandez, M.; Moore, O. and Nunez, M. **1999**. Use of brassinosteroid analogues in in vitro culture of potatoes *(Solonum tuberosum). Cultivos Tropicales*. **20** (4): 41-44.

Hu,Yu Xin; Bao Fang; Li Jia Yang; Hu, Y.X.; Bao, F. and Li, J.Y. **2000**. Promotive effect of brassinosteroids on cell division involves a distinct Cye D3 induction pathway in Arabidopsis. *Plant Journal*. **24**(5): 693-701.

Jang, M.S.; Han, K.S. and Kim, S.K. **2000.** Identification of brassinosteroids and their biosynthetic precursors from seeds of pumpkin. *Bulletin of the Korean Chemical Society*. **21** (2): 161-164.

Kamuro, Y. and Takatsuto, S., **1999.** In Brassinosteroids-Steroidal Plant Hormones (eds Sakurai., Yokota, T. and Clouse, S.D.), Springer, Tokyo, 223-241.

Kamuro,Y. and Takatsuto, S. **1991.** In Brassinosteroids–Chemistry, Bioactivity and Application, ACS Symp. Ser. (eds. Cutler, H.G., Yokota, T. and Adam,G.), *Am. Chem. Soc.*, Washington DC, 292-297.

Kamuro,Y. and Takatsuto, S. **1992** *Proc. Plant Growth Regul. Soc. Am.*, **19**, 275-277.

Kang, J.G.,Yun, J., Kim, D.H., Chung, K.S., Fujioka, S., Kin, J.I., Dae, H.W., Yoshida, S., Takatsuto, S., Song, P.S., and Park, C.M. **2001**. Light and brassinosteroid signals are integrated via a dark-induced small G protein in etiolated seedling groeth, *Cell.***105**: 625-636.

Khripach, V.A., Zhabinskii, V.N. and Malevannaya, N.N.**1997**. *Proc. Plant Growth Regul. Soc.*

Khripach, V.; Zhabinskii, V.; De Groot, A. and Groot, De, A. **2000**. Twenty years of brassinosteroids: steroidal plant hormones warrant better crops for XXI century, *Annuals of Botany*. **86**(3): 441-447.

Kong-Xiangsheng and Zhang Miaoxia. **1998**. Effect of brassinolids and multi effect triazol on seed germination and seedling growth in corn. *Chinese Agricultural Science bulletin* **14**(2): 21-23.

Krishnan, S.; Azhakanandam, K.; Ebenezr, E.A.I.; Samson, N.P. and Dayanandan. P. **1999**. Brassinosteroids and benzylaminopurine increased yield of IR-50. Indica rice. *Current Science.* **76**(2): 145-147.

Kulava, O.N.; Burkhanova, E.A.; Fedina, A.B.; Khokhlova, V.A., Bokebayeva, G.A., Vorbrodt, H.M. and Adam, G. **1991**. In Brassinosteroids Chemistry, Bioactivity and Applications ACS, Symp. Ser.(eds Culter, H.G., Yakota,T. and Adam, G.), *Am.Chem.Soc.*, Washington DC. 141-155.

Kurapov, P.B.; Skorobogatova, I.V.; Siusheva, A.G. and Kozik, T.a. **1996.** Hormonal balance and productivity of wheat, barley and potato plants under the influence of epibrassinolide treatment. *Sel- skokhoyaistvennaya Biologiya,* **5**: 99-104.

Laksminaraynan, A. **2002**. MSc thesis, G.B.P.U. Ag. and Tech. Pantnagar.

Li,J., and Chory, J. **1997**. A putative leucine-rich repeat receptor kinase involved in brassinosteroid signal transduction. *Cell* **90**: 929-938.

Mandava, and Bhushan, N. **1988**. Plant growth-promoting brassinosteroids. *Ann. Rev. Plant Mol. Biol.* **39**:23-52.

Meundt, W.J. and Thompson, M.J. **1998**. *Annu. Rev. Plant Physiol. Mol. Biol.* **49**, 427-551

Montes S.; Mesa, O.; Hernadez, M.M.; Santana, N.; Nunez, M. and Varela, M. **1997**. Use of the synthetic brassinosteroid BB-6 in carnation micropropatation. *Cutlivos Tropicales.* **18** (2): 51-55.

Neff, M.M., Nguyen, S.M., Malancharuvil, E.J., Fujioka, S., Naguchi, T., Seto, H., Tsubuki, M., Honda, T., Takatsuto, S., Yoshida, S., and Chory, J.**1999b.** *BAS1*: A gene regulating brassinosteroid levels and light responsiveness in *Arabidopsis. Proc Natl Acad Sci* U.S.A. **96**: 15316-15323.

Noguchi, T.; Fujioka, S.; Choe, S.; takatsuto S., Tax, F.E.; Yushida, S. and Feldmann, K.A. **2000.** Biosynthetic pathways of brassinolide in Arabidopsis. *Plant Physiology.* **124** (1): 201-209.

Novikov, NM, and Vaiersa, B.V. **1995**. Effect of growth regulator on composition of protein and quality of wheat grain. *Izvertiya Timiryazevskoi Sellskokhyzyais tvennoi.* Akademii No. **1**, 65-75.

Nunez, M.; Alfonso, J.L.; Arzuaga, J.; Hernadez, A. and Coll, F. **2000**. Influence of new Cuban bioregulators on vegetable production under tropical conditions. Proceedings of the inter *American society for Tropical Horticulture*, Barquisimeto, Venezuela. **42**:335-343.

Oh, M.H. and Clouse, S.D. **1998**. Brassenosteroids affect the rate of cell division in isolated leaf protoplast of petunia hybrida. *Plant cell reports.* **17**:12 921-924.

Pipattanawong, N., Fujishige, N., Yamane, K., J. **1996** *Jpn. Soc.Hortic. Sci.* **65**, 651-654.

Ponsamuel, J.; Samson, N.P. and Ganeshan, P.S. **1996**. Somatic embryogenesis and plant regeneration from the immature cotyledonary tissues of cultivated tea. *Plant cell reports*b. **16** (3-4): 405-214.

Ramraj, V.M.; Vyas, B.N.; Godrej, N.B.; Mistry, K.B.; Swami, B.N. and Singh, N. B. 1997. Effects of 28-homobrassinolide on yields of rice, wheat groundnut, mustard, potato and cotton. *Journal of Agricultural Science.* **128** (4) 405-413.

Roth, U.; Friebe, A. and Schnabl, H. **2000**. Resistance induction in plants by a brassinosteroid containing extract of Lychnis viscaria. Zcitschrift fur Naturforschung C.*A Journal of Biosciences.* **55**(7-8): 552-559.

Sairam, R.K. **1994**. *Plant Growth Regul,* **14**,173-181.

Sakurai, A. **1999**. Brassinosteroid biosynthesis. *Plant Physiology and Biochemistry.* 36 (5): 351-361.

Schilling, G., Schiller, Cand Otto, S.**1991**. In Brassinosteroids Chemistry, Bioactivity and Applications, ACS Symp. Ser. (eds Cutler, H.G., Yokota, T. and Adam, G.), *Am. Chem. Soc.*, Washington DC, 208-219.

Shakirova, F. M. and Begrukova, M.V. **1998**. Effect of 24-epibrasii olide and salinity on the levels of ABA and lectin. *Russian Journal of Plant Physiology.* **45** (3): 388-391.

Takahashi, H,; Musuoka, S.; Lee-Youngja and Lee, J.J. **1999**. Potential practical use of Abscisic acid, brassinosteroids and Jasmonic acid in seed germination and some characteristics of the mechanisms their effects. *Chemical Regulation of Plant.* **34** (1): 97-105.

Takatsuto, S.; Kosuga, N.; Abe, B.; Noguchi, T.; Fujioka, S. and Yokota, T. **1999**. Occurrence of potential brassinosteroid precursor seeds of wheat and foxtail millet. *Journal of Plant Research.* **112** (1105): 27-33.

Takatsuto, S.; Omote, K.; Gamoh, K. and Ishibashi, M. **1990**. Identification of brassinolide and caststerone in buck wheat (Fagopyrum esculentum moench) pollen. *Agricultural Biological Chemistry,* **54** (3): 757-762.

Takatsuto, S.,Yazawa, N.,Ikekawa, N., Takamatsu, T., Takeuchu,Y and Koguchi, M., **1983**, *Phytochemistry.* **22**, 2437-2441.

Takatsuto, S. **1994**. *J. Chratogr.*, **658**, 3-15.

Vardhini, B.V. and Rao, S.S.R. **1999**. Effect of brassinosteroids on nodulation and nitrogen activity in groundnut. *Plant Growth Regulation.* **28** (3) : 165-167.

Wachsman, M.B.; Lopez, E.M.F; Ramirez, J.A.; Galagovasky, L/r. and Coto, C.E. **2000**. Antiviral effect of brassinosteroids against herpes virus and arena virus. *Antiviral chemistry and Chemotherapy.* **11**(1):71-77.

Wang, B.K. and Zang, G.W. **1993**. *Acta Phtophysiol. Sin.*, **19**, 38-42.

Wilen, R.W., Sacco, M., Lawrence,V.G. and Krishna, P. **1995**., *Physiol. Plant.*, **95**, 195-202.

Yokota, T. **1997**. Topic on brassinosteroids mutants. *Chemical Regulation of Plants.* **32** (2). pp. 144-149.

Yokota, T. and Takahashi, N.**1986**. In plant growth substances (ed.Bopp, M.), Springer-Verlag, Berlin.129-138.

Biodiversity Conservation and Envir. Management (2012)
Editors: D.R. Khanna et al.
Pub. by Biotech Books. *ISBN: 978-81-7622-262-4*

Pages: 71-80

HERBACEOUS STRUCTURE AND SPECIES DIVERSITY PATTERN IN *QUERCUS LEUCHOTRICHOPHORA* FOREST OF PHAKOT WATERSHED (UTTARAKHAND HIMALAYA), INDIA

P. Pokriyal, G.K. Dhingra[✉], Ramdas and Sanjeev Lal
Department of Botany, Government P.G. College,
Uttarkashi, Uttarakhand (INDIA)

An investigation was undertaken in the Sub Himalayan stretch of Phakot in Tehri Garhwal District of Uttarakhand to understand the vegetation structure and diversity pattern. An altitudinal gradient from 1400-1900 m asl in the upper reaches was selected. Five transects were spatially distributed so as to minimize the auto correction among the vegetation.

✉ Corresponding author

Introduction

Indian Himalayan region which covers approximately 10% of the India's total land area is one of the largest and youngest mountain chains in the world. Indian Himalayan region represents the unique biological diversity (Samant *et al.*, 1993; Dhar *et al.*, 1994; Maikhuri *et al.*, 2001). The Himalayan moist temperate forest extent from 1500-3000 m asl. Within one altitudinal range co-factors like topography, aspect and inclination of slope and soil type effect the forest composition (Shank and Noorie, 1950). The phytosociological study incorporates mainly the composition of vegetation over any terrain and provides detailed information about composition of tree, shrub and herb communities and also their functional aspects. The present study aims to investigate the structure and diversity of herbaceous flora of Phakot watershed.

Materials and Method

Study Area: Phakot watershed is a part of Phakot beat of Saklana range of Garhwal Himalaya. It lies between 78° 19' 53" to 78° 22' 16" East and 30° 14' 29" to 30° 13' 17" North with elevation ranging from 1500 m to1900 m. The area experiences three distinct summer, winter and rainy seasons. Soils vary from loam to clay. Soils of the region have been formed either through pedogenetic processes or are transported soils. In the study area upper zone forest constituted the Himalayan moist temperate forest started from 1400-1600 m and 1700-1900 m asl and *Quercus leuchotrichophora* was found dominating in both of the altitudinal zone.

Ecological surveys were conducted periodically in summer, winter and rainy season. Stratified random sampling method was used for collecting the vegetation data. Baseline vegetation survey was conducted for whole watershed by using transects and quadrat methods. Species area curve was used to determine minimal sample area which is based on quantitative variations of the vegetation in terms of species number (Mishra 1968; Muller- Dombois and Ellenberg, 1974). Transects were spatially distributed so as to minimize the autocorrelation among the vegetation. In each plot four 1m² (1m × 1m) sample plots which were nested within 100 m² were used to enumerate herbs. About 5% of total forest area has been sampled of which 0.1% was enumerated.

Results

A total of 66 herb species were recorded in the altitudinal zone of 1400-1600 m asl (Table 1) whereas 82 herb species were found in the altitudinal zone of 1700-1900 m asl (Table 2). *Dioscorea belophylla* and *Apluda mutica* were observed dominant and co-dominant herb species in altitudinal zone of 1400-1600 m asl whereas *Apluda mutica* and *Mariscus paniceus* were observed dominant and co-dominant herb species respectively in the altitudinal zone of 1700-1900 m asl. Maximum frequency percentage (71.9%) was recorded for *Dioscorea belophylla* followed by (67.2%) for *Chrysopogon fulvus* in the altitudinal zone of 1400-1600m asl (Table 1) while maximum frequency percentage (51.88%) was recorded for *Boenninghausenia albiflora* followed by (51.56%) for *Apluda mutica* in the altitudinal zone of 1700-1900 m asl. Highest density (3.06 plants/m^2) was recorded for *Dioscorea belophylla* followed by (2.99 plants/ m^2) for *Apluda mutica* in the altitudinal zone of 1400-1600 m asl whereas highest density (2.92 plants/m^2) was recorded for *Apluda mutica* followed by (2.26 plants/ m^2) for *Mariscus paniceus* species in the altitudinal zone of 1700-1900 m asl (Table 2). Maximum IVI (9.17) was recorded for *Apluda mutica* at altitudinal zone of 1700-1900 m asl while all other the species were distributed contagiously (Table 2).

Diversity index

Diversity index (H$^-$) and Evenness index (J') showed similar pattern like species richness, at both of the altitudinal zones of the watershed. In the lower strata of the vegetation, Shannon diversity was highest (3.99) for 1700-1900 m asl and evenness was highest (0.92) for 1400-1600 m asl. Concentration of dominance showed nearly reverse trend to the species richness. Thus, the highest concentration of dominance for herbs was 0.028 at 1400-1600 m asl (Fig.1). Margalef index (8.3) was highest for 1700-1900 m asl and Berger Parker value (0.06) was highest for 1400-1600 m asl (Fig.1).

Table 1: Phytosociological analysis of herb species in altitudinal zone (1400-1600 m) of *Quercus leuchotrichophora* forest.

Species	*Family*	*F%*	*D*	*IVI*	*A/F*
Abelmoschus crinitus	Malvaceae	18.23	0.72	2.96	0.216
Achyranthes aspera	Amaranthaceae	2.60	0.11	0.44	1.613
Ageratum conyzoides	Asteraceae	7.29	0.14	0.87	0.255
Ageratum houstonianum	Asteraceae	22.92	0.68	3.26	0.130
Ajuga bracteosa	Lamiaceae	19.79	0.48	2.59	0.124
Andropogon munroi	Poaceae	26.56	1.36	4.96	0.193
Apluda aristata	Poaceae	27.08	1.33	4.95	0.182
Apluda mutica	Poaceae	61.98	2.99	11.20	0.078
Artemisia capillaris	Asteraceae	16.15	0.60	2.54	0.230
Bergenia ciliata	Saxifragaceae	4.17	0.06	0.45	0.330
Bidens bipinnata	Asteraceae	10.42	0.45	1.77	0.413
Bidens pilosa	Asteraceae	34.38	1.52	5.92	0.128
Bidens tripartita	Asteraceae	29.69	0.95	4.35	0.108
Blumea fistulosa	Asteraceae	10.94	0.33	1.56	0.274
Boenninghausenia albiflora	Rutaceae	46.88	1.49	6.87	0.068
Capillipedium parviflorum	Poaceae	7.81	0.21	1.06	0.341
Chrysopogon fulvus	Poaceae	67.19	2.38	10.33	0.053
Chrysopogon gryllus	Poaceae	18.23	0.65	2.81	0.194
Crotalaria calycina	Fabaceae	10.42	0.34	1.55	0.317
Curcuma aromatica	Zingiberaceae	37.50	1.13	5.35	0.080
Cynoglossum glochidiatum	Boraginaceae	4.17	0.13	0.59	0.720
Cyperus niveus	Poaceae	14.06	0.68	2.54	0.342
Desmodium trifloum	Fabaceae	18.75	0.68	2.91	0.193
Dicliptera bupleuroides	Acanthaceae	14.58	0.26	1.71	0.122
Digitaria ciliaris	Poaceae	7.29	0.38	1.38	0.715
Dioscorea belophylla	Dioscoreaceae	71.88	3.06	12.14	0.059
Dipsacus inermis	Dipsacaceae	10.42	0.52	1.92	0.480
Echinochloa colona	Poaceae	20.31	0.73	3.16	0.178
Epipactis helleborine	Orchidaceae	6.25	0.17	0.86	0.440
Eriophorum comosum	Cyperaceae	28.13	1.53	5.45	0.194
Eulaliopsis binata	Poaceae	4.69	0.24	0.89	1.114
Euphorbia dracunculoides	Euphorbiaceae	9.90	0.32	1.45	0.324
Euphorbia geniculata	Euphorbiaceae	13.54	0.63	2.39	0.341
Euphorbia hirata	Euphorbiaceae	24.48	0.73	3.48	0.122

contd.

Species	*Family*	*F%*	*D*	*IVI*	*A/F*
Euphorbia pilosa	Euphorbiaceae	32.29	1.48	5.68	0.142
Fragaria nubicola	Rosaceae	27.08	1.03	4.32	0.141
Gallium asperifolium	Rubiaceae	20.31	0.61	2.90	0.148
Gerbera maxima	Asteraceae	2.60	0.10	0.41	1.459
Gonatanthus pumilus	Araceae	25.52	0.78	3.66	0.119
Heteropogon contortus	Poaceae	18.23	0.78	3.09	0.235
Imperata cylindrica	Poaceae	3.65	0.11	0.52	0.823
Leucas lanata	Lamiaceae	40.10	1.72	6.81	0.107
Mariscus paniceus	Cyperaceae	27.60	0.95	4.19	0.124
Micromeria biflora	Lamiaceae	8.33	0.24	1.18	0.353
Ocimum americanum	Lamiaceae	17.19	0.56	2.54	0.189
Oxalis corniculata	Oxalidaceae	29.69	0.95	4.36	0.108
Panicum sumatrense	Poaceae	20.31	0.74	3.17	0.179
Perilla frutescens	Laminaceae	8.33	0.35	1.40	0.503
Pimpinella diversifolia	Apiaceae	3.65	0.10	0.50	0.744
Pogonatherum crinitum	Poaceae	4.69	0.14	0.67	0.640
Pouzolzia zeylanica	Urticaceae	14.06	0.47	2.10	0.237
Ranunculus diffusus	Ranunculaceae	3.13	0.10	0.46	1.013
Reinwardtia indica	Linaceae	29.69	1.09	4.65	0.124
Rosularia rosulata	Crassulaceae	38.02	2.46	8.19	0.170
Salvia lanata	Lamiaceae	4.17	0.10	0.54	0.570
Salvia nubicola	Lamiaceae	13.02	0.32	1.72	0.190
Salvia plevia	Lamiaceae	2.08	0.14	0.46	3.240
Scutellaria grossa	Lamiaceae	15.63	0.63	2.57	0.258
Shuteria vestita	Fabaceae	15.10	0.36	1.97	0.160
Smilax aspera	Smilacaceae	14.58	0.48	2.18	0.228
Taraxacum officinale	Asteraceae	28.65	1.17	4.73	0.142
Thalictrum secundum	Ranunculaceae	20.83	0.67	3.06	0.154
Thalictum foliolosum	Ranunculaceae	17.71	0.66	2.80	0.211
Tridax procumbens	Asteraceae	1.04	0.05	0.19	4.800
Vernonia cinera	Astercaceae	6.77	0.20	0.97	0.443
Vicia tenera	Fabaceae	9.38	0.30	1.37	0.338
Total		1252.1	47.8	200	

F% =Frequency percentage; D=Density/100m^2; TBC=Total Basal Cover; IVI= Importance Value Index; A/F= Abundance-Frequency Ratio

Table 2: Phytosociological analysis of herb species in altitudinal zone (1700-1900 m) of *Quercus leuchotrichophora* forest

Species	*Family*	*F%*	*D*	*IVI*	*A/F*
Abelmoschus crinitus	Malvaceae	3.13	0.20	0.59	2.016
Achyranthes aspera	Amaranthaceae	14.38	0.58	2.12	0.281
Aerva sanguinolenta	Amaranthaceae	1.88	0.04	0.21	1.067
Ageratum houstonianum	Asteraceae	0.94	0.04	0.15	4.978
Ainsliaea latifolia	Asteraceae	4.69	0.16	0.64	0.725
Alopecurus borii	Poaceae	1.88	0.08	0.29	2.400
Anaphalis contorta	Asteraceae	18.13	0.60	2.43	0.183
Andropogon spp.	Poaceae	15.31	0.78	2.57	0.335
Androsace lanuginosa	Primulaceae	5.31	0.14	0.64	0.487
Apluda aristata	Poaceae	31.25	1.42	4.91	0.145
Apluda mutica	Poaceae	51.56	2.92	9.17	0.110
Arenaria serphllifolia	Caryophyllaceae	1.56	0.07	0.24	2.816
Arisaema concinnum	Araceae	13.44	0.45	1.81	0.247
Arisaema denticulata	Araceae	13.44	0.33	1.59	0.183
Artemisia capillaris	Asteraceae	23.75	0.64	2.91	0.113
Bergenia ciliata	Saxifragaceae	6.25	0.10	0.64	0.264
Bidens bippinata	Asteraceae	10.94	0.25	1.26	0.212
Bidens pilosa	Asteraceae	21.56	0.96	3.35	0.206
Boenninghausenia albiflora	Rutaceae	51.88	1.74	7.00	0.065
Chloris dolichostachya	Zingiberaceae	3.44	0.13	0.49	1.111
Chrysopogon fulvus	Poaceae	26.25	1.30	4.32	0.189
Chrysopogon gryllus	Poaceae	25.94	0.99	3.72	0.147
Circium wallichii	Poaceae	2.81	0.09	0.37	1.146
Cissampelos pareira	Asteraceae	6.56	0.17	0.78	0.385
Corallodiscus lanuginosus	Menispermaceae	2.81	0.07	0.34	0.909
Conyza bonariensis	Asteraceae	38.13	1.23	5.05	0.084
Curcuma aromatica	Zingiberaceae	27.19	0.77	3.39	0.104

contd...

Species	*Family*	*F%*	*D*	*IVI*	*A/F*
Cyathula tomentosa	Amarantahceae	9.38	0.24	1.13	0.274
Cyperus niveus	Poaceae	25.00	1.62	4.82	0.259
Desmodium dichotomum	Fabaceae	5.63	0.15	0.69	0.474
Desmodium microphyllum	Fabaceae	29.06	1.45	4.81	0.172
Desmodium trifloum	Fabaceae	24.69	1.16	3.94	0.190
Dicliptera bupleuroides	Acanthaceae	8.44	0.28	1.14	0.399
Digitaria ciliaris	Poaceae	11.88	0.45	1.70	0.319
Dioscorea belophylla	Dioscoreaceae	28.13	0.92	3.75	0.117
Dipsacus inermis	Dipsacaceae	20.00	0.57	2.51	0.142
Epipactis helleborine	Orchidaceae	10.00	0.36	1.40	0.363
Eriophorum comosum	Cyperaceae	36.56	1.90	6.19	0.142
Eulaliopsis binata	Poaceae	5.00	0.10	0.55	0.413
Euphorbia hirata	Euphorbiaceae	4.69	0.19	0.70	0.882
Euphorbia pilosa	Euphorbiaceae	38.75	1.52	5.63	0.101
Fragaria nubicola	Rosaceae	10.63	0.31	1.35	0.277
Gallium asperifolium	Rubiaceae	35.63	1.32	5.04	0.104
Geranium ocellatum	Geraniaceae	3.13	0.10	0.42	1.056
Gerbera maxima	Asteraceae	25.63	0.83	3.39	0.126
Gonatanthus pumilus	Araceae	30.63	1.02	4.11	0.108
Heteropogon contortus	Poaceae	35.94	1.62	5.62	0.126
Hypericum dyeri	Hypericaceae	8.75	0.18	0.97	0.233
Hypericum oblongifolium	Hypericaceae	3.13	0.15	0.51	1.536
Imperata cylindrica	Poaceae	16.25	0.61	2.31	0.230
Impatiens racemosa	Poaceae	5.00	0.16	0.65	0.625
Leucas lanata	Laminaceae	42.19	2.00	6.79	0.113
Lindenbergia indica	Scrophullariaceae	10.94	0.38	1.50	0.316
Malva parviflora	Malvaceae	7.19	0.19	0.88	0.375
Mariscus paniceus	Cyperaceae	46.88	2.26	7.60	0.103
Micromeria biflora	Lamiaceae	7.19	0.29	1.07	0.569

contd...

Species	*Family*	*F%*	*D*	*IVI*	*A/F*
Origanum vulgare	Lamiaceae	40.94	1.49	5.74	0.089
Oxalis corniculata	Oxalidaceae	13.44	0.49	1.89	0.273
Panicum sumatrense	Poaceae	33.75	1.43	5.10	0.125
Pentanema indicum	Asteraceae	7.50	0.17	0.85	0.294
Pimpinella diversifolia	Apiaceae	8.13	0.23	1.01	0.341
Pogonatherum crinitum	Poaceae	9.38	0.28	1.19	0.313
Portulaca grandiflora	Portulaceae	4.06	0.19	0.64	1.136
Potentilla fulgens	Rosaceae	5.31	0.22	0.79	0.775
Potentilla geradiana	Rosaceae	2.81	0.17	0.52	2.133
Pouzalzia zeylanica	Urticaceae	19.06	0.58	2.45	0.158
Primula denticulata	Primulaceae	11.25	0.28	1.34	0.225
Ranunculus diffusus	Ranunculaceae	7.81	0.21	0.96	0.343
Reinwardtia indica	Linaceae	25.63	0.74	3.24	0.113
Rosularia rosulata	Crassulaceae	37.81	1.96	6.38	0.137
Saccharum rufipilum	Poaceae	4.38	0.18	0.65	0.931
Salvia plevia	Lamiaceae	7.81	0.20	0.94	0.333
Scutellaria grossa	Lamiaceae	23.13	1.41	4.29	0.263
Selinum candollii	Apiaceae	8.44	0.22	1.02	0.312
Shuteria vestita	Fabaceae	33.13	1.21	4.65	0.110
Smilax aspera	Smilacaceae	6.25	0.16	0.76	0.416
Strobilanthes atropurpureus	Acanthaceae	1.25	0.04	0.16	2.400
Taraxacum officinale	Asteraceae	36.25	1.19	4.84	0.091
Thalictrum secundum	Ranunculaceae	21.88	0.70	2.89	0.147
Thalictum foliolosum	Ranunculaceae	19.38	0.54	2.41	0.144
Vicia tenera	Fabaceae	20.00	0.73	2.81	0.183
Viola biflora	Violaceae	2.19	0.08	0.32	1.763
Total		1381.6	53.7	200	

F% = Frequency percentage; D=Density/100m^2; TBC=Total Basal Cover; IVI= Importance Value Index; A/F= Abundance-Frequency Ratio

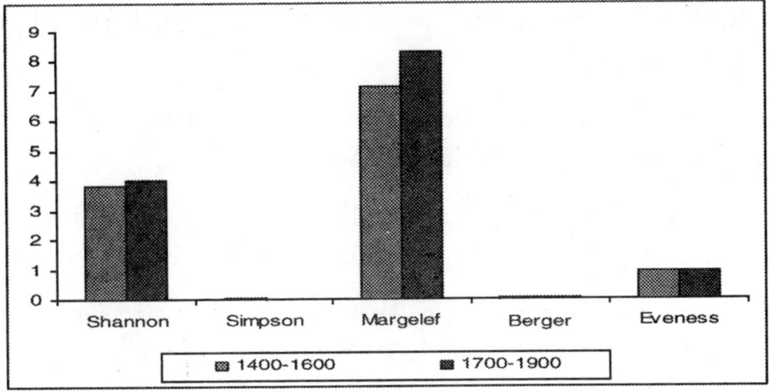

Fig.1: Different diversity indices of herbs in Phakot watershed

Discussion

Herbs were the highest contributors of plant richness. Herb richness increased with increase in altitude. One popular explanation for the decrease in species in relation to latitude is the decrease in productivity from equator towards the poles (Rohde, 1992). Jeetram and Bhatt (2004) also reported herb richness (2.0) for six forest types of Uttarakhand. Herb species diversity was found to be highest at altitudinal zone of 1700-1900 m asl. It was increasing towards increasing altitudinal range. Forest diversity is mostly influenced by topography, soil characteristics, climate, altitude, soil moisture, geographical location and intensity of biotic factors of the area (Joshi and Johari,1985; Kharakwal *et al.*, 2004). Whittaker (1972) has suggested that dominance of one stratum may affect the diversity of another.

References

Dhar, U., Vir, J. and Kachroo, P. 1994. Ladhak: A update on Natural Resources. In: *High altitudes of the Himalaya biogeography* (Y.P.S. Pangtey and R.S. Rawal, eds). *Ecology and Conservation*, 102-114.

Maikhuri, R.K., Rao, K.S. and Semwal, R.L. 2001. Changing scenario of Himalayan agro-ecosystems: loss of agrobiodiversity, an indicator of environmental change in Central Himalaya, India. *The Environmentalist*, 21: 23-39.

Mishra, R. 1968. *Ecology Work book.* Oxford and IBM publishing Co. Calcutta, 244.

Muller-Dombios, D.J. and Ellenberg, H. 1974. *Aims and Methods in Vegetation Ecology.* John Wiley and Sons. New York.

Jeetram, Kumar, A. and Bhatt, J. 2004. Plant diversity in six forest types of Uttaranchal, Central Himalaya, India. *Current Science,* 86: 975-978.

Joshi, H.C. and. Johari, S.C. 1985. *Working Plan of Western Bahraich Forest Division.* Eastern Circle of U.P., 1985-86 to 1994-95.

Kharakwal, G., Mehrotra, P., Rawat, Y.S, and Pangtey, Y.P.S. 2004. Comparative study of herb layer diversity in *Pine* forest stand at different altitudes of Central Himalaya. *Applied Ecology and Environmental Research,* 2(2): 15-24.

Samant, S.S., Rawal, R.S. and Dhar, U. 1993. Botanical Hot Spots of Kumaun; Conservation Perspectives for the Himalaya. In: *Himalayan Biodiversity and Conservation Strategies* (U. Dhar, eds.). Gyandaya Prakashan, Nainital. 377-400.

Whittaker, R.H. 1972. Evolution and measurement of species diversity. *Taxon.* 21: 213-251.

Biodiversity Conservation and Envir. Management (2012)
Editors: D.R. Khanna et al.
Pub. by Biotech Books. *ISBN: 978-81-7622-262-4*

Pages: 81-97

8

ENVIRONMENTAL MONITORING AND ASSESSMENT DURING THE CONSTRUCTION OF INDIAN SCIENTIFIC BASE (BHARTI STATION) IN ANTARCTICA

Pawan Kumar Bharti
30th Antarctica Expedition Member
Environment Protection Division
Shriram Institute for Industrial Research, Delhi (INDIA)
E:mail: gurupawanbharti@rediffmail.com

The Larsemann Hills (Latitude 69°20'–69°30'S and Longitude 75°55'–76°30'E) is an ice-free coastal oasis with exposed rock and low rolling hills. The Larsemann Hills contain hundreds of freshwater lakes of varying sizes, depth and biology. An environmental study was being conducted at Larsemann Hills in East Antarctica to evaluate the Ambient air quality, Lake and sea water quality, soil and sediment, Noise level monitoring, solid waste generation, handling and disposal practices. Geographically, the study area (Bharti Promontory) is situated on Latitude 69° 24' 00.0" S and 76° 10' 00.0" E on southern part of globe. The water, soil and sediment samples were collected from various locations of different Islands/Peninsulas like Bharti Island, Fisher Island, McLeod Island, Broknes Peninsula and Stornes Peninsula.

The aim of this study was to assess the general characteristics, metal content, pesticide, radiation contamination and bacteriological analysis of water, soil and sediment. The air quality of different islands was also conducted to assess the level of particulate matter, oxides of nitrogen, oxides of sulphur, carbon monoxide and volatile compounds in air. The present work was aimed towards developing base line data for the local environmental settings and to evaluate the impact of various activities on the environmental components during the construction work of third Indian Scientific Station (Bharti) in Antarctica.

Introduction

Antarctica is the coldest, windiest, driest, whitest, highest (averagely) and least accessible continent on the earth. Ice is covered here continuously for the last 25 million years. Antarctica is the 5th largest continent on the earth. Antarctica has no government and is considered no mans-land. Antarctica is the most fragile, vulnerable and pristine environment on the earth.

Globally, Antarctica is important for environmental point of view. It is Premier location to study the effects of global warming, climate change, ozone depletion and many environment global issues.

The Larsemann Hills is an ice-free area of approximately 50 km^2, located halfway between the Vestfold Hills and the Amery Ice Shelf on the south-eastern coast of Prydz Bay, Princess Elizabeth Land, East Antarctica (69°30'S, 76°19'58"E). The ice-free area consists of two major peninsulas

(Stornes and Broknes), four minor peninsulas and approximately 130 near shore islands. Nella Fjord further divides the eastern-most peninsula, broknes, into western and eastern components. The closest significant ice-free areas are the Bølingen Islands (69°31'58"S, 75°42'E) 25 km to the south-west and the Rauer Islands (68° 50'59"S, 77°49'58"E) 60 km to the north-east (http://www.esri.com/news/ arcnews). Broknes is one of very few coastal areas of Antarctica that remained partially ice-free through the glaciating period and sediments deposited there contain continuous biological and palaeoclimate records dating back 1,30,000 years.

The Larsemann hills area was first discovered by a Norwegian expedition led by Christensen in 1935. Subsequently, visits were made by several nations during the last 50 years, but human activity of a significant or sustained nature did not occur until the mid-1980s (Antarctic fact sheet, 2000).

However, from 1980 to 1990 rapid infrastructure development in the area: an Australian summer research base, a Chinese year-round research station (Zhongshan) and two Russian research stations (Progress I and Progress II) were established within approximately 3 km of each other on eastern Broknes (http://www.eia.doc.gov/emeu/cabs antarctica.html).

Antarctica is the most precious asset on the earth and is the last heritage of human kind. Antarctica is the only area on earth planet which is strictly devoted to scientific research and the continents of extremes come to be known as the "Continent of Science". It is the biggest laboratory on earth where no outside anthropogenic (human activities) interference has taken place over the centuries till recent times. Being at a unique geographic location, it offers unique opportunities for Scientists to conduct number of scientific research experiments. Antarctica is attracting world attention because of the tremendous biological species in surrounding seas and likelihood of vast hydrocarbons. Even though it is difficult to survive at Antarctica, still Scientists all around the worlds have been engaged in pursuing the exciting scientific research investigations. The investigations are essential not for the exploitation of natural resources buried under the region but for the preservation of environment and ecology on earth; especially in the light of climate change.

Study area

The Indian Research Station at Larsemann hills, east Antarctica is proposed to be located on an ice-free rocky area, situated between Quilty Bay on the east and Thala Fjord on the west. The Larsemann hills provide a unique opportunity for studying the environmental impacts of research facilities run by three nations (Australia, China and Russia) as well as for the upcoming Indian Scientific Base (Bharti Promontory).

Fig. 1: Map showing the lakes (L) (water sampling sites) at the Larsemann Hills, Ingrid Christensen coast, east Antarctica.

A major feature of the climate of the Larsemann Hills is the existence of persistent, strong katabatic winds that blow from the northeast in summer days. Daytime air temperatures from December to February frequently may exceed 4°C, with the mean monthly temperature a little above 0°C (Draft Comprehensive Environmental Evaluation, 2004). Mean monthly winter temperatures are between –15°C to –40°C. Pack ice is extensive inshore throughout summer and the fjords and bays are rarely ice-free. Precipitation occurs as snow and is unlikely to exceed 250 mm water equivalent annually (Report of the Norwegian Antarctic Inspection, 2001). Snow cover is generally deeper and more persistent on Stornes than Broknes, due to northeasterly prevailing winds and the perennial sea ice held in by the islands offshore from Stornes.

There are many regulatory authorities and regulations to protect the environment of Antarctica (Erich *et al.* 2000). Few of them are given here:

- Antarctic Treaty, 1959
- Protocol on Environmental Protection to the Antarctica: Madrid Protocol,.1991
- Antarctic Conservation Act, 1978
- Scientific Committee of Antarctic Research (SCAR)
- The Convention on the Conservation of Antarctic Seals and Whales, 1972
- The Convention on Conservation of Antarctic Marine Living Resources, 1980

Environmental research

A scientific study is being conducted in various island and peninsulas of Larsemann Hills. Ambient air quality monitoring was carried out to measure the pollution load in local environment. A respirable dust sampler (Envirotech 460NL) was installed to measure the load of respirable particulate matter, suspended particulate matter, SOx, NOx, CO *etc.* Water, soil, sediment samples were also collected from different locations. All the samples were collected, preserved, transported and analysed by using standard methods (APHA, 2005). Different Indian standards were followed for different study aspects. IS: 5182 for ambient air quality assessment IS: 3025 and IS: 10500 specifications for water research and IS: 2720 for edaphic quality.

These scientific tasks were performed during the expedition:

- Ambient Air Quality
- Indoor Air Quality
- Lake and Sea water quality
- Soil quality
- Study of bed sediments
- Generator Stack monitoring
- Indoor Noise level
- Ambient Noise level
- Machinery Noise level
- Drilling and Blasting Noise level
- Lichen patches and Moss community assessment

- Collection of Planktonic diversity and Benthos from lake water
- Fuel consumption and pollution load.
- Solid waste generation, handling, separation and disposal practices and % composition.
- Ash analysis
- Waste oil analysis

Fig. 2: Ambient air quality monitoring using RDS in Antarctica

Air environment

The study of air environment of Antarctica is very essential to measure the pollution status. The following main parameters were selected for study in Antarctica:

Table 1: List of parameters analysed for the indoor and ambient air.

S. No.	*Ambient parameters*	*Indoor air parameters*
1	SPM	Temperature
2	RSPM	Humidity
3	SOx	SOx
4	NOx	NOx
5	CO	CO
6	VOCs	VOCs
7	Ozone	CO_2

Fig. 3: Air quality monitoring by Palmtop based sensors

Table 2: National Ambient Air Quality Standards

MINISTRY OF ENVIRONMENT AND FORESTS
NOTIFICATION
New Delhi, the 16th November, 2009

S. No.	*Pollutant*	*Time Weighted Average*	*Concentration in Ambient Air*		
			Industrial, Residential, Rural & Other Areas	*Ecological Sensitive Area*	*Method of measurement*
(1)	**(2)**	**(3)**	**(4)**	**(5)**	**(6)**
1.	Sulphur Dioxide (SO_2)	Annual Average 24 hours	50 µg/m^3 80 µg/m^3	20 µg/m^3 80 µg/m^3	Improved West & Gaeke method Ultraviolet fluorescence
2.	Nitrogen Dioxides (NO_2)	Annual Average 24 hours	40 µg/m^3 80 µg/m^3	30 µg/m^3 80 µg/m^3	Modified Jacob & Hochheiser Na- Arsenite) method Chemiluminescence
3.	Particulate Matter less than 10 µm or PM_{10}	Annual Average 24 hours	60 µg/m^3 100 µg/m^3	60 µg/m^3 100 µg/m^3	Gravimetric TOEM Beta attenuation

contd...

S. No.	*Pollutant*	*Time Weighted Average*	*Concentration in Ambient Air*		
			Industrial, Residential, Rural & Other Areas	*Ecological Sensitive Area*	*Method of measurement*
4.	Particulate Matter less than 2.5 µm or $PM_{2.5}$	Annual Average 24 hours	40 µg/m³ 60 µg/m³	40 µg/m³ 60 µg/m³	Gravimetric TOEM Beta attenuation
5.	Ozone (O_3)	8 hours 1 hour	100 µg/m³ 180 µg/m³	100 µg/m³ 180 µg/m³	UV photometric Chemiluminescence Chemical method
6.	Lead (Pb)	Annual Average 24 hours	0.5 µg/m³ 1.0 µg/m³	0.5 µg/m³ 1.0 µg/m³	AAS/ICP Method after sampling using EPM 2000 or equivalent filter paper ED-XRF using Teflon filter
7.	Carbon Monoxide (CO)	8 hours 1 hour	2.0 mg/m³ 4.0 mg/m³	2.0 mg/m³ 4.0 mg/m³	Non dispersive infrared (NDIR) spectroscopy
8.	Ammonia (NH_3)	Annual Average 24 hours	100 µg/m³ 400 µg/m³	100 µg/m³ 400 µg/m³	Chemiluminescence Indophenol blue method
9.	Benzene (C_6H_6)	Annual Average	5 µg/m³	5 µg/m³	Gas chromatography based continuous analyser Adsorption & desorption followed by GC analysis
10.	Benzo (O)Pyrene (BaP)-particulate phase only	Annual Average	1 ng/m³	1 ng/m³	Gas chromatography based continuous analyser Adsorption & desorption followed by GC analysis
11.	Arsenic (As)	Annual Average	6 ng/m³	6 ng/m³	AAS/ICP Method after sampling using EPM 2000 or equivalent filter paper
12.	Nickel (Ni)	Annual Average	20 ng/m³	20 ng/m³	AAS/ICP Method after sampling using EPM 2000 or equivalent filter paper

Water environment

Water quality was assessed by analysing all parameters described in specification (IS: 10500). Besides, physico-chemical parameters, metals, pesticides & radiation contamination and biological parameters were also analysed.

A B

Fig. 4(A-B): Water sample collection from sea and frozen lake

Pesticide

Table 3: List of pesticides analysed in water.

S. No.	*Pesticide*	*Detection limits (mg/l)*	*Instrument*
1	Dichlorvos	0.0002	GC-MS
2	Propoxur	0.0002	GC-MS
3	Dimethoate	0.0002	GC-MS
4	Carbafuran	0.0002	GC-MS
5	BHC	0.0002	GC-MS
6	Atrazine	0.0002	GC-MS

contd...

S. No.	*Pesticide*	*Detection limits (mg/l)*	*Instrument*
7	Diazinon	0.0002	GC-MS
8	Lindane	0.0002	GC-MS
9	Phosphamidon	0.0002	GC-MS
10	Methyl parathion	0.0002	GC-MS
11	Femitrothion	0.0002	GC-MS
12	Aldrin	0.0002	GC-MS
13	Melathion	0.0002	GC-MS
14	Fention	0.0002	GC-MS
15	Parathion	0.0002	GC-MS
16	Endosulphan	0.0002	GC-MS
17	Dieldrin	0.0002	GC-MS
18	o,p, DDT	0.0002	GC-MS
19	Ethion	0.0002	GC-MS
20	p,p, DDT	0.0002	GC-MS
21	Captafol	0.0002	GC-MS
22	Phosalone	0.0002	GC-MS
23	Permethrin	0.0002	GC-MS
24	Cypermethrin	0.0002	GC-MS
25	Fenvalerate	0.0002	GC-MS
26	Detamethrin	0.0002	GC-MS
27	2,4, D dichlorophenylacitic acid	0.0001	LC-MS
28	Isoproturon	0.0001	LC-MS
29	Monocrotophos	0.0001	LC-MS

Table 4: Drinking water quality standard (Malik and Bharti, 2010)

Parameters	*BIS*	*CPCB*	*UPPCB*	*WHO*
Temperature (°C)	40	±5°C than receiving water	40	-
SS (mg/l)	100	100	100	5-25
pH	5.5-9.0	5.5-9.0	5.5-9.0	7.0-8.5
BOD_5 at 20°C (mg/l)	30	30	30	-
COD (mg/l)	250	250	250	-
Cl (mg/l)	-	-	-	75-200
Ca (mg/l)	-	-	-	75-200
Mg (mg/l)	-	-	-	30-150
Cd (mg/l)	2.0	2.0	2.0	0.1
Cu (mg/l)	3.0	3.0	3.0	0.05-1.5
Fe (mg/l)	-	3.0	-	0.1-1.0
Pb (mg/l)	0.1	0.1	0.1	0.1
Mn (mg/l)	-	2.0	-	0.05-0.5
Ni (mg/l)	2.0	3.0	2.0	-
Zn (mg/l)	5.0	5.0	5.0	5-15.0

Radiation

Radiation contamination was also monitored in water and soil samples. The radiation parameters are given below:

Table 5: List of radiation parameters analysed in water and soil samples

S. No.	*Radiation parameter*	*Requirement*
1	Gross Alpha (including Ra 226)	<0.5 bq/l
2	Gross Beta particle activity	< 1.9 bq/l
3	Cesium 137 content	< 1.9 bq/l

Edaphic environment

Soil and sediment samples were collected from different locations covering important aspects of physical and chemical properties of soil quality therein and by adopting established sampling procedures given in the IS-2720 specifications. Undisturbed soils and sediments samples of the area was collected by means of auger. Besides physico-chemical parameters many heavy metals and radiation contamination was also monitored.

Fig. 5: Soil and sediment samples inside the ship

Table 6: Noise level standards

Area Code	*Category of Area*	*Limits in dB(A) Leq*	
		Day time	*Night time*
A	Industrial Area	75	70
B	Commercial Area	65	55
C	Residential Area	55	45
D	Silence Zone	50	40

Acoustic environment

Noise level was also measured for 24 hours indoor and ambient conditions. Noise level monitoring was also done during the construction activity, especially for various vehicles, earthmovers, excavator, piston bully, dozer, stone crusher, drilling and blasting.

Noise pollution emanating from operation of generator, snow vehicles, incinerator, helicopter *etc.* shall have adverse impacts on the human beings, marine, life and also on Antarctic birds *i.e,* Skua and Penguin in long run.

A

B

Fig. 6 (A-B): Scene of drilling and blasting

Biodiversity

Ecosystem of Antarctica is not so complex. Few species of flora and fauna are present there permanently or in migratory period. Moss, Lichens, and few microbes are the dominant species, while Penguins, Seals Skua, Snow Petrel and few avian species may be studied for their behaviour, feeding habit and breeding.

Solid waste

To support the continuing scientific research and, stay of scientist and logistic staff, generators run throughout the study period to supply power to the station site. ATF is used as fuel to run generators, cranes, piston bullies, snow scooters and boilers. The supporting activities like storage of food material, cooking of food, maintenance of generators, cranes, snow scooter, piston bullies, supply of drinking water, temperature regulation inside temporary shelters, waste disposal *etc.* are regular phenomenon. These activities may pose a little extent of impact on surrounding environment, but environment regulations are to be strictly adhered in Antarctica. All the solid waste is stored in a separate container and brought back to the real land.

A

B

Fig. 7(A-B): Seggregation and storing of solid waste

There are few separate bins for degradable and non-degradable types of solid waste. After separating, waste was dumped into a container, which was inside the ship.

Fuel and Oils

Fuel and other oil quantity were also measured for various earthmovers, machines, helicopters, ships, *etc.* Pollution load can be calculated by fuel combustion. A prediction for SOx and NOx & other gases can be made using a formula.

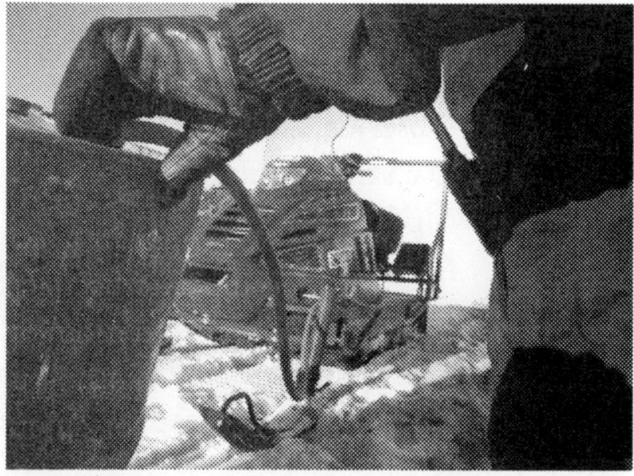

A

B

Fig. 8(A-B): Handling and spillage of fuel

The internal combustion engine converts the chemical energy contained in the fuel into mechanical power. During the operation of the Maitri station, air emission produced from generator, incineration, snow vehicles *etc* were estimated. The exhaust gas discharged from the operation of snow vehicles, generator, incinerator *etc* contain several gases and suspended particulate matter. These emissions comprise of Carbon dioxide (CO_2), Carbon mono oxide (CO), Oxides of Sulphur (SOx), Oxides of Nitrogen (NOx) and other gases. The fuel consumption along with estimated air exhaust gases emissions of all activities of the generator, incinerator and snow vehicles was measured. The estimation of air emission was done based on fuel consumption as per the guidelines: Emergency of Manual of Antarctica.

Environmental Impact Assessment

The environmental study is necessary for the construction of new scientific base in Antarctica. So, after completing these tasks, final report will be given to national institute. The main steps of EIA are:

- Site selection / Identification
- Sampling and monitoring
- Analysis
- Observations
- Data interpretation
- Modeling and prediction
- Mitigation and recommendations
- EIS

These are the many methodologies adopted for environmental study in Antarctica. We have adopted few of them and generally make a focus on these things:

- Action and impacts
- Environmental components
- Basic parameters
- Simple checklists
- Leopold matrix *etc.*

Conclusion

The environmental assessment studies carried out at Indian Scientific Base, located in Antarctica throws light on the status of air, water and soil quality and waste generated. After completion of analysis, the report should be submitted to Antarctica centre, which will be beneficial among the SCAR or other regulatory authorities of Antarctica. Furthermore, the environment assessment and monitoring work will be beneficial while predicting the long term impacts to Antarctica. The data will be the baseline and reference for the further study in future.

Acknowledgement

I want to thank the Secretary, MoES, Director, Shriram Institute for Industrial Research and Director, NCAOR for providing me opportunity to execute the environmental monitoring in Antarctica. Thanks to Voyage

Leader, members of XXX ISEA and all the crew members *M/V IVAN PAPANIN* who braved the tough challenges of the weather deserve special thanks.

References

Antarctica: Fact sheet, September 2000, page-14, http://www.antarctica.ac.uk

APHA, 2005: Standard methods for the examination of water and waste water. Washington.

Article 14 of the Protocol on Environmental Protection to the Antarctica Treaty, January 2001.

Draft Comprehensive Environmental Evaluation for the concept of upgrading the Norwegian. (2004).

Erich R, Gundlach; John J, Gallagher; John Hatcher and Tom Vinsor 2000. Planning and Hazards Oil http://www.eia.doc.gov/emeu/cabs antarrctica.html., http://www.esri.com/news/arcnews

Malik, D.S. and Bharti, P.K. (2010): Textile Pollution. *Daya Publishing House, Delhi*, pp: 383.

Report of the Norwegian Antarctic Inspection under Article VII of the Antarctica Treaty. 2001.

Biodiversity Conservation and Envir. Management (2012)
Editors: D.R. Khanna et al.
Pub. by Biotech Books. *ISBN: 978-81-7622-262-4*

Pages: 99-109

9

PHYSICO-CHEMICAL CHARACTERISTICS OF RIVER GANGA AT SAPTSAROVAR IN HARIDWAR

D.R. Khanna[1], Anuraag Mohan[2] and Sneh Lata[2] ✉
[1]Department of Zoology & Environmental Science,
Gurukul Kangri University, Haridwar, Uttarakhand (INDIA)
[2]Department of Chemistry, Bareilly College, Bareilly.
Uttar Pradesh (INDIA)

In the present investigation water quality monitoring of River Ganga at Saptsarovar, Haridwar was studied during the period January, 2009 to December, 2009. Parameters studied were Temperature, pH, free CO_2, D.O., B.O.D, T.D.S., Acidity, Alkalinity, Hardness, Calcium, Chloride and Conductivity. All the parameters studied were found within the prescribed limits but data reveals the alarming

✉ Corresponding author

position and it is recommended that proper measures should be taken to prevent River Ganga from pollution.

Keywords: Physicochemical parameter, River Ganga, Water quality, Hardness, Alkalinity.

Introduction

Water is one of the most vital natural resources present on earth. Compared with other resources, water is generally very utilizable resource. Hence we can say that water is one of the most important components for survival of any kind of living organism. It covers nearly three fourth of the surface of the earth. Fresh water is the most precious resource on earth. Today, the easy availability of fresh water is a major problem as 80% of the rivers are getting polluted.

Different rivers are present in India out of which Ganga is one of the holiest River of India. The River Ganges is the largest river of the Indian sub-continent with a total length of about 2525 km, originates from ice-cave '*Gaumukh*' (30°55' N / 70° 7' E) in the Garhwal Himalaya at an altitude of 4100 m and discharges into Bay of Bengal. At Haridwar Ganga cuts across the Shiwalik hills. Due to its religious importance million of people in India take holy dips in River Ganga especially on some auspicious occasions because it is believed that a holy dip in the same purges away all the sins. Day by day human's activities have polluted the River Ganga water. Due to unproportional growth of population and industries, water quality of River Ganga is degrading at a faster rate. Keeping this aspect in view, regular monitoring of River Ganga is essential. With this aim, various studies have been conducted in the past on fresh water related to various aspects. In the present investigation physicochemical parameter of River Ganga at Saptsarovar was done.

Materials and Method

Haridwar is one of the most holiest cities of India. It lies in the foothills of Shivalik range. The position of city on the globe is on latitude 29°58'N and longitude 78° 13' E.

Water samples were taken from Saptsarover, Haridwar on monthly basis during January, 2009 to December, 2009. Samples were collected between 7.00 a.m. to 9.00 a.m. in borosil glass bottles of 300 ml capacity and plastic containers. The analysis of all the parameters, was done following the standard methods of APHA (1998) and Khanna and Bhutiani (2004).

Results and Discussion

The monitoring of the river water is an essential step to mark the trend pattern of pollutants and their effect on living systems in today's developing life. On the basis of analysis the results of various physico-chemical parameters studied are given in Table 1.

Temperature is an important parameter, which is directly related with the chemical reaction in water and biochemical reactions in the living organisms. In the present investigation, the maximum temperature (22.00±0.26) was recorded in May while minimum temperature (17.50± 0.28) in December. Similar results regarding temperature was observed by Yadav and Kumar (2011) in the River Kosi at Rampur district. pH is one of the important tool to measure acidity or alkalinity in water. In this study, pH was observed maximum (7.60) in the month of June and July; while minimum (7.10) in the month of February, March, October and December. Similar observation was also observed by Khanna *et al.* (2007) in river Ganga. Free CO_2 comes in water due to activity of aquatic organism. In the present investigation the value of free CO_2 varies from 1.50±0.04 to 3.00±0.02. Similar findings have been reported by Khanna and Bhutiani (2005) in river Ganga. Acidity of water is quantitative capacity to react with a strong base to a designated pH. In the present study the concentration of acidity was found to be fluctuating between (6.87±0.04 to 13.00±0.40). This maximum (13.00±0.40) and minimum (6.87±0.04) concentration of Acidity was observed in the month of August and January respectively. A more or less similar trend was also observed by Rai *et al.* (2010) in river Ganga at Varanasi. Alkalinity is due to presence of Carbonate and Bicarbonate ions. Alkalinity of water found to be maximum (163.00±0.39) in the month of August and minimum (110.00±0.32) in the month of December. Sarkar *et al.* (2007) observed similar trend for Alkalinity. Hardness indicates the concentration of calcium and magnesium ions. Hardness of water is due to the presence of chloride, nitrate, sulphate and bicarbonate of calcium and magnesium (Kumar *et al.* 2010). The maximum (125.00±0.46) hardness of water was observed in the month of May while minimum (83.00±0.57) in month of September. Mishra and Joshi (2003) observed hardness in River Ganga at Haridwar and found more or less similar trend in their study. The concentration of Calcium was found to be maximum (38.48±0.06) in the month of November while minimum (21.78±0.12) in the month of August. Similar findings were observed by Khajuria and Dutta (2009) in the River Tawi, Jammu. In natural water sometimes chloride may be due to leaching of rocks. In the present study the concentration of chloride was found to be fluctuating between (7.18±0.34 to 27.20±0.43). This maximum (27.20±0.43) and minimum (7.18±0.34) concentration of chloride was observed in the month of July and December respectively. Similar observation were also observed by Vishnoi *et al.* (2008) in River Ganga at Kangri village. In the present investigation maximum (10.25±0.02) value of DO was recorded in the month of December while minimum (8.16±0.36) in the month of May. Khanna *et al.* (2009) reported the similar trend in river Panv Dhoi. Conductivity is the measure of the ability of an aqueous solution to carry electric current. Conductivity of Ganga river water fluctuates from (165.00±0.44 to 397.20±0.25). The concentration of Conductivity was found to be maximum (397.20±0.25) in the month of January while minimum (165.00±0.44) in the month of July. Khanna *et al.* (2007) also reported the similar trend in Song river water at Dehradun. Total dissolved solids or filterable residue are those solids, which left after evaporation of the filterable sample. TDS indicate total amount of inorganic chemicals in the solution. In the present study, the concentration of Total dissolved solids was found to be fluctuating between (115.20±0.51 to 256.30±0.11). This maximum (256.30±0.11) and minimum (115.20±0.51) concentration of TDS was observed in the month of May and March respectively. Khanna (1993) also observed similar trend for total dissolved solid. The biochemical oxygen demand is the amount of oxygen required to degrade the organic compound biologically. The concentration of BOD increases with the increase in chemical pollution of the water body. In the present investigation maximum (3.50±0.04) value of BOD was recorded in the month of June while minimum (1.00±0.02) in the month of September. These findings were supported by Khanna and Bhutiani (2005).

Table 1: Average value of physiochemical parameter of River Ganga from January, 2009 to December, 2009.

	Temp (°C)	*pH*	*CO_2 (mg/l)*	*Acidity (mg/l)*	*Alkalinity (mg/l)*	*Hardness (mg/l)*	*Calcium (mg/l)*	*Chloride (mg/l)*	*DO (mg/l)*	*Conductivity (μmhos/cm)*	*Total Dissolved Solids (mg/l)*	*BOD (mg/l)*
January	18.00 ±0.28	7.20 ±0.17	1.50 ±0.04	6.87 ±0.04	116.00 ±0.51	98.00 ±0.49	36.08 ±0.97	8.20 ±0.40	10.06 ±0.47	397.20 ±0.25	172.00 ±0.46	1.80 ±0.05
February	19.00 ±0.11	7.10 ±0.28	1.89 ±0.05	7.50 ±0.21	120.00 ±0.46	92.00 ±057	32.06 ±0.45	8.90 ±0.05	9.86 ±0.11	287.60 ±0.92	156.00 ±0.28	1.10 ±0.38
March	19.00 ±0.29	7.10 ±0.23	2.36 ±0.03	8.00 ±.40	119.00 ±0.49	84.00 ±0.56	24.04 ±0.50	7.85 ±0.02	9.68 ±0.21	187.20 ±0.40	115.20 ±0.51	1.90 ±0.05
April	21.00 ±0.32	7.30 ±0.04	2.45 ±0.02	8.00 ±0.40	133.00 ±0.34	120.00 ±0.46	28.05 ±0.51	11.50 ±0.19	8.98 ±0.15	387.00 ±0.34	192.00 ±0.11	2.60 ±0.04
May	22.00 ±0.26	7.20 ±0.01	2.40 ±0.04	9.00 ±0.46	148.00 ±0.56	125.00 ±0.46	30.06 ±0.51	17.00 ±0.44	8.16 ±0.36	248.90 ±0.51	256.30 ±0.11	1.40 ±0.05
June	21.50 ±0.28	7.60 ±0.04	2.96 ±0.03	10.00 ±0.40	150.00 ±0.51	96.00 ±0.49	34.48 ±0.27	19.00 ±0.48	8.68 ±0.39	178.00 ±0.55	162.00 ±0.40	3.50 ±0.04
July	21.10 ±0.50	7.60 ±0.03	3.00 ±0.02	10.50 ±0.17	150.50 ±0.28	98.00 ±0.53	36.08 ±0.45	27.20 ±0.43	8.96 ±0.10	165.00 ±0.44	178.00 ±0.48	2.20 ±0.04
August	21.00 ±0.40	7.20 ±0.01	2.90 ±0.03	13.00 ±0.40	163.00 ±0.39	87.00 ±0.01	21.78 ±0.12	23.00 ±0.55	8.90 ±0.11	190.80 ±0.48	147.00 ±0.29	1.20 ±0.43
September	20.50 ±0.26	7.20 ±0.02	2.76 ±0.06	11.00 ±0.51	116.00 ±0.54	83.00 ±0.57	22.09 ±0.49	17.60 ±0.18	9.45 ±0.16	247.60 ±0.32	152.90 ±0.46	1.00 ±0.02
October	20.00 ±0.40	7.10 ±0.00	2.45 ±0.13	8.00 ±0.51	133.00 ±0.49	90.00 ±0.08	33.05 ±0.81	9.80 ±0.05	9.65 ±0.11	186.00 ±0.28	159.50 ±0.57	1.70 ±0.04
November	19.00 ±0.25	7.20 ±0.01	2.10 ±0.05	7.80 ±0.04	128.00 ±0.49	110.00 ±0.51	38.48 ±0.06	8.40 ±0.11	9.99 ±0.06	392.00 ±0.28	186.00 ±0.48	1.44 ±0.02
December	17.50 ±0.28	7.10 ±0.06	1.98 ±0.04	6.90 ±0.04	110.00 ±0.32	89.00 ±0.11	33.38 ±0.11	7.18 ±0.34	10.25 ±0.02	348.00 ±0.55	168.00 ±0.38	1.33 ±0.04
Average	19.96 ±1.44	7.24 ±0.17	2.35 ±0.46	8.88 ±1.87	132 ±17.08	97.66 ±13.75	30.80 ±5.66	13.80 ±6.74	9.38 ±0.64	267.94 ±91.31	170.40 ±33.64	1.76 ±0.71

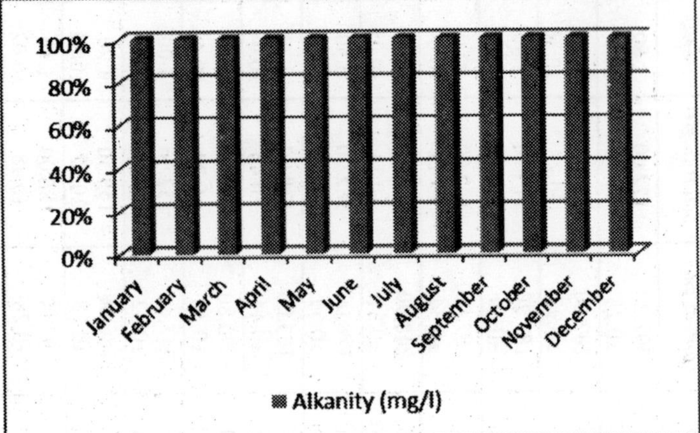

Fig.-1

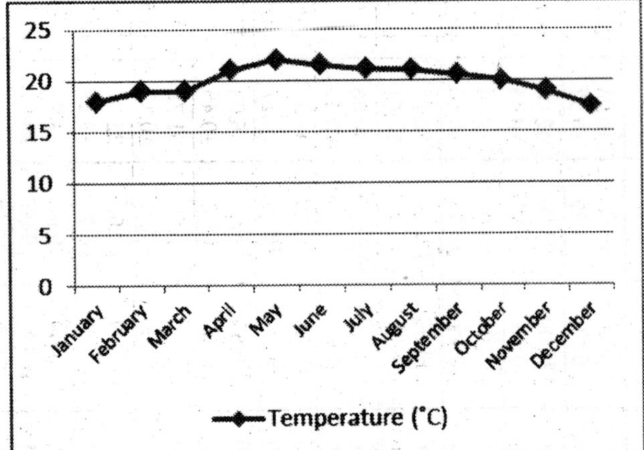

Fig -2

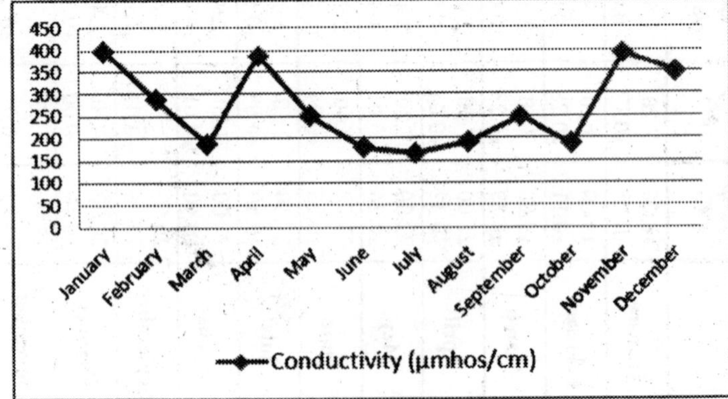

Fig.-3

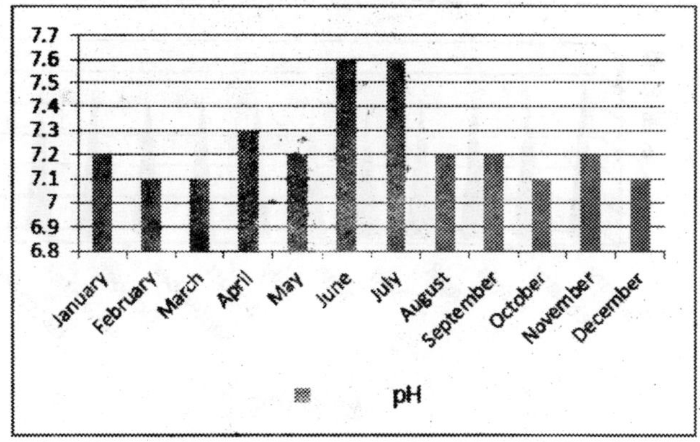

Fig.-4

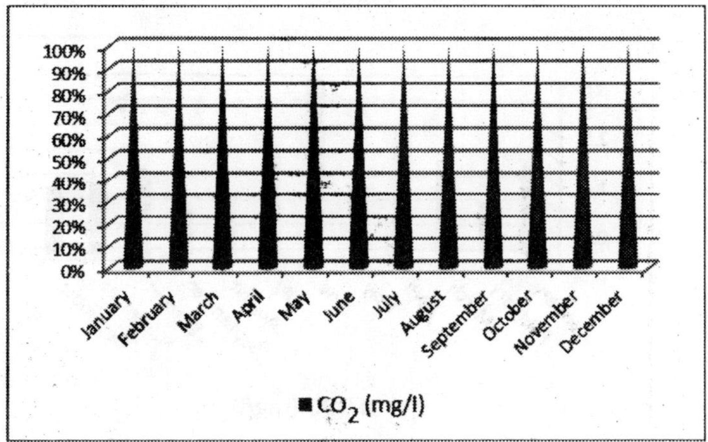

Fig.-5

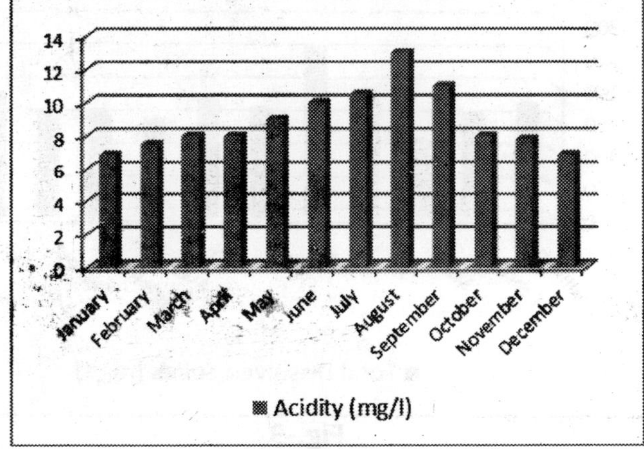

Fig.-6

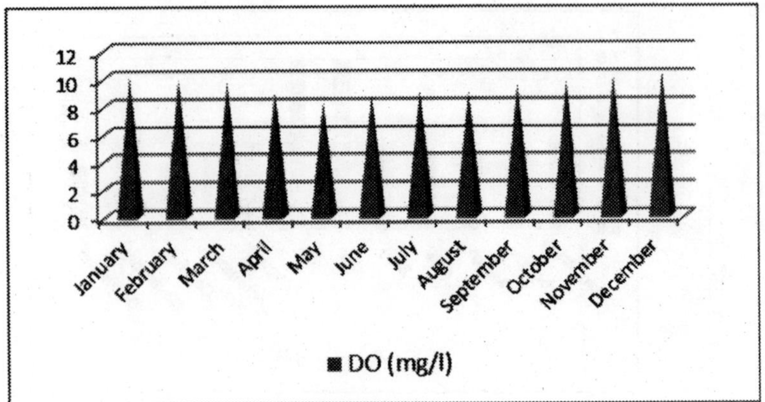

Fig.-7

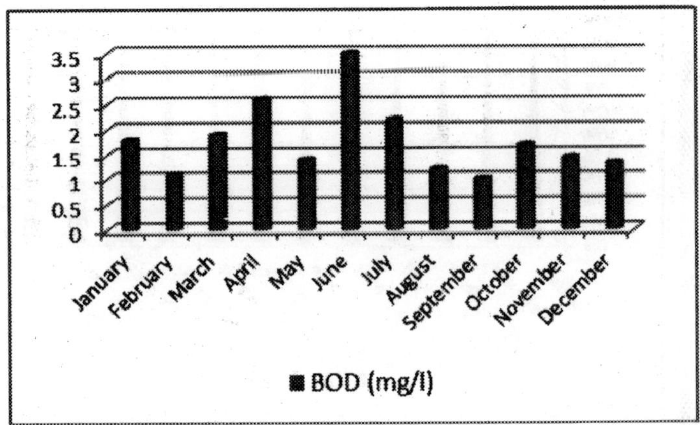

Fig.-8

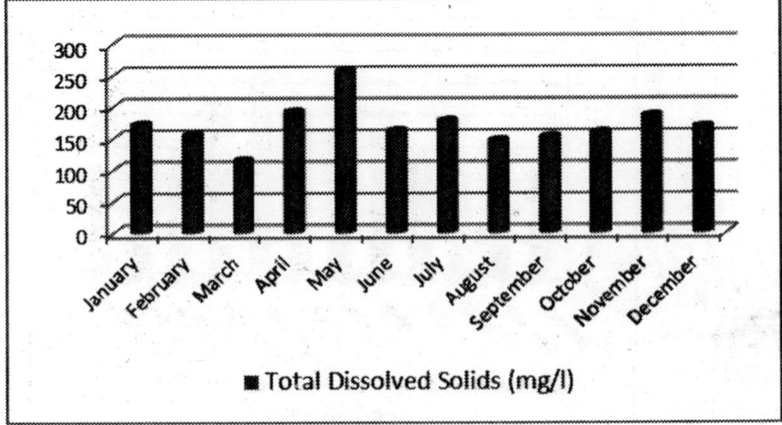

Fig.-9

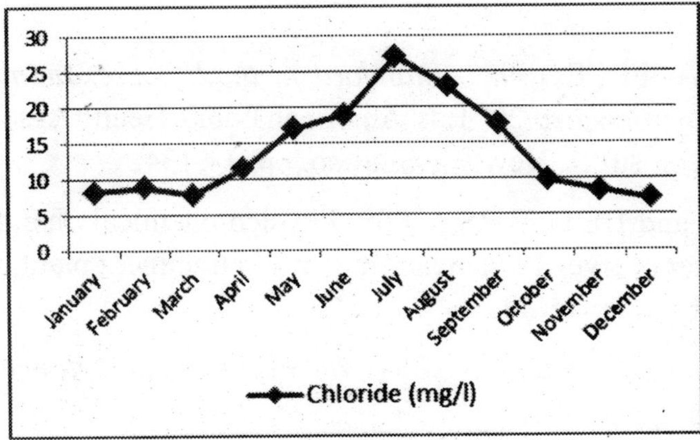

Fig.-10

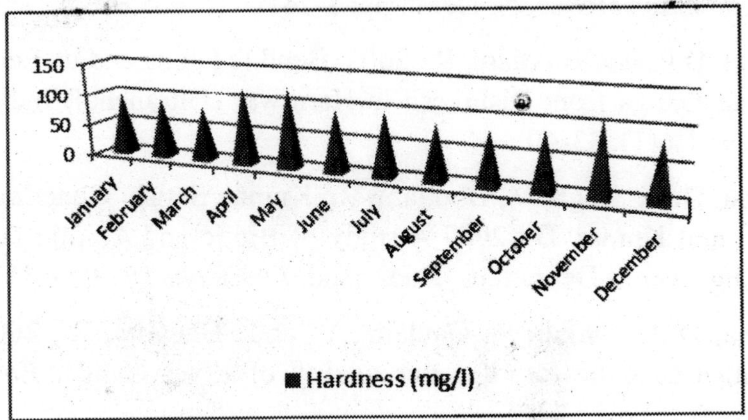

Fig.-11

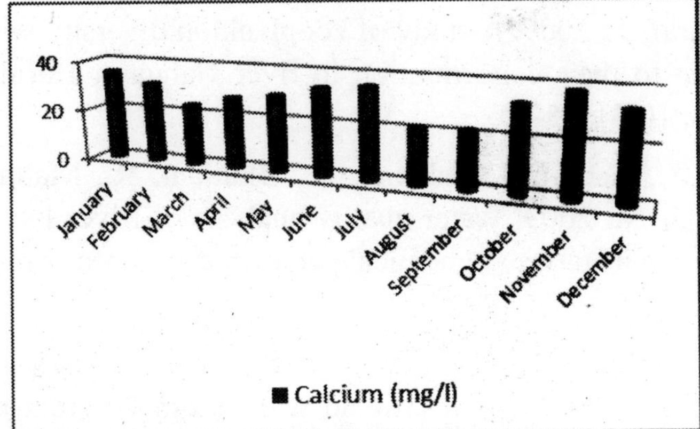

Fig.-12

Fig.1-12: Showing monthly fluctuation in physiochemical parameters of River Ganga at Saptsarovar.

References

APHA-AWWA-WPCF 1998. *Standard methods for examination of water and wastewater.* 20th Eds. American Public Health Association; 1015, Fifteen street, New Washington. pp: 1-1134.

Khajuria, M., and Dutta, S.P.S., 2009. Physico-chemical charateristics of raw water of river Tawi, near Sitlee water treatment plant, Jammu. *Env.Con.J.* 10(3): 45-52.

Khanna, D.R. and Bhutiani, R., 2004. *Water analysis at a glance.* ASEA Publication, pp: 1-115.

Khanna, D.R., 1993. *Ecology and pollution of Ganga river*. Ashish Publication House, N. Delhi. pp: 1-241.

Khanna, D.R. and Bhutiani, R., 2005. Benthic Fauna and its Ecology of river Ganga from Rishikesh to Haridwar (Uttranchal) India. *Env. Con. J.* 6(1): 33-40.

Khanna, D.R., Singh, V., Bhutiani, R., Kumar, Satish Chandra, Matta, G., and Kumar, D., 2007 A study of Biotic and Abiotic factors of Song river at Dehradun, Uttrakhand. *Env. Cons. J.*, 8(3): 117-126.

Khanna, D.R., Sarkar, P., Gautam, A., and Bhutaini, R., 2007. Fish scales as indicator of water quality of River Ganga. *Env.Monit. Assess.* 134: 153-160.

Khanna, D.R., Bhutiani, R., Matta, Gagan., Kumar, Dheeraj., Singh, V., and Ashraf, J., 2009. A study of zooplankton diversity with special reference to their concentration in river Ganga at Haridwar. *Env. Cons. J.*, 10(3): 15-20.

Khanna, D.R., Ashraf, J.,Chauhan, B., Bhutiani, R., Matta, Gagan., and Singh, V., 2009. Water quality analysis of River Panv Dhoi in reference to its physico-chemical parameters and heavy metals. *Env. Cons. J.*, 10(1&2): 159-169.

Kumar,A., Bisht, B.S., Talwar,A., and Chandel, D.,2010. Physicochemical and microbial analysis of Ground water from different regions of Doon valley. *Int. J. Appl. Env. Sci.* 5(3): 433-440.

Mishra, S., and Joshi, B.D., 2003. Assessment of water quality with few selected parameters of river Ganga at Haridwar. *Him. J. Env. Zool.* 17(2):113-122.

Rai, P.K., Mishra, A., and Tripathi, B.D., 2010. Heavy metal and microbial pollution of the river Ganga: A case study of water quality at Varanasi. *Aquatic Eco. Health and Management,* 13(4): 352-361.

Sarkar, S.K., Saha, M., Takada, H., Bhattacharya, A.K., Mishra, P. and Bhattacharya, B., 2007. Water quality management in the lower stretch of the river Ganges, east coast of India; an approach through environmental education. *J. of Cleaner. Production.* 15: 1559-1561.

Vishnoi, U., Kaur, J. and Pathak, S.K., 2008. Water quality of river Ganga in respect of physicochemical characteristics at Kangri village, District Haridwar. *Env. Cons. J.* 9 (1&2): 145-147.

Yadav, S.S., and Kumar, R., 2011. Monitoring water quality of Kosi river in Rampur District, U.P., India. *Applied Sci. Res.* 2(2): 197-201.

Biodiversity Conservation and Envir. Management (2012)
Editors: D.R. Khanna et al.
Pub. by Biotech Books. *ISBN: 978-81-7622-262-4*

Pages: 111-122

IN VITRO ANTIOXIDANT ACTIVITY OF SOME COMMON CITRUS FRUITS

Ajay Singh[1]✉, Harish Chandra[2] and Jatin Srivastava[3]
[1]Department of Chemistry, Uttaranchal Institute of Technology, Premnagar, Dehradun, Uttarakhand (INDIA)
[2] Department of Biotechnology, G.B.Pant Engineering College, Ghurdauri, Pauri, Garhwal, Uttarakhand (INDIA)
[3] Genentech Laboratory Pvt. Ltd., Biotech Park, Lucknow, Uttar Pradesh (INDIA)
✉E-mail: ajay21singh@yahoo.com

Antioxidant compounds in food play an important role as a health-protecting factor. Scientific evidence suggests that antioxidants reduce the risk for chronic diseases including cancer and heart disease. Primary sources of naturally occurring antioxidants are whole grains, fruits and vegetables. Plant sourced food antioxidants like

✉ Corresponding author

vitamin C, vitamin E, carotenes, phenolic acids, phytate and phytoestrogens have been recognized as having the potential to reduce disease risk. Most of the antioxidant compounds in a typical diet are derived from plant sources and belong to various classes of compounds with a wide variety of physical and chemical properties. In the present study, antioxidant activity of Citrus fruits were investigated by three methods *i.e.* (Superoxide dismutase activity, β-Carotene bleaching method and Thiocyanate Assay) and it was found that out of five citrus plants tested, lemon has significant antioxidant activity than rest of the plants tested, however protein content was lower as compared to other plants.

Keywords: Antioxidant, citrus fruit, protein content, lemon, vitamin C.

Introduction

Citrus is a common term and genus of flowering plants in the family Rutaceae, originating in tropical and subtropical Southeast Asia. The word citrus is derived from the Ancient Greek kedros or its Latin derivate cedrus, names applied by the ancient Greeks and Romans to several different trees with fragrant foliage or wood (compare the completely unrelated cedars). The taxonomy and systematic of the genus are complex and the precise number of natural species is unclear, as many of the named species are clonally-propagated hybrids, and there is genetic evidence that even some wild, true-breeding species are of hybrid origin (German, 1999). Cultivated Citrus may be derived from as few as four ancestral species. Natural and cultivated origin hybrids include commercially important fruit such as the Oranges, Grapefruit, Lemons, Mousambi. Some limes, and some tangerines. Citrus fruits are notable for their fragrance, partly due to flavonoids and limonoids contained in the rind and most are juice-laden. The juice contains a high quantity of citric acid giving them their characteristic sharp flavor. They are also good sources of vitamin C and flavonoids (Knight, 1998).

Citrus fruits and juices are an important source of bioactive compounds including antioxidants such as ascorbic acid, flavonoids, phenolic compounds and pectins that are important to human nutrition (Fernandez- Lopez *et al.,* 2005; Jayaprakasha and Patil, 2007 and Ebrahimzadeh *et al.,* 2004). Flavanones, flavones and flavonols are three types of flavonoids which occur in Citrus fruit (Calabro *et al.,* 2004). The main flavonoids found in citrus species are hesperidine, narirutin, naringin and eriocitrin (Mouly *et al.,* 1994; Schieber *et al.,* 2001). Epidemiological studies on dietary Citrus flavonoids improved a reduction in risk of coronary heart disease (Di Majo *et al.,* 2005; Hertog *et al.,* 1993) and is attracting more and more attention not only due to their antioxidant properties, but as anti-carcinogenic and anti-inflammatory agents because of their lipid anti-peroxidation effects (Stavric, 1993; Elangovan *et al.,* 1994; Martın *et al.,* 2002). The interest in these classes of compounds is due to their pharmacological activity as radical scavengers (Cotelle *et al.,* 1996).

Lemon (citrus × limon) is a hybrid in cultivated wild plants. It is the common name for the reproductive tissue surrounding the seed of the angiosperm lemon tree. Lemon juice is about 5% (approximately, 030 moles/liter) citric acid, which gives lemons a tart taste, and a pH of 2 to 3. This makes lemon juice an inexpensive, readily available acid for use in educational science experiments. An orange—specifically, the sweet orange—is the citrus fruit *Citrus sinensis* (syn. *Citrus aurantium* L. var. dulcis L. or *Citrus aurantium* Risso) and its fruit. The orange is a hybrid of ancient cultivated origin, possibly between pomelo (*Citrus maxima*) and tangerine (*Citrus reticulata*). Orange juice has strong antioxidant property. *Citrus limetta* is a species of citrus. Common names for varieties of this species include sweet limetta, sweet lemon and sweet lime. In India, it is commonly called mousambi (Jacob, 1996).

Antioxidants are substances that reduce, neutralize and prevent the damage done to the body by free radicals. Free radicals are the species containing simply electrons that are no longer attached to atoms. Instead of circling the nucleus of an atom (much like the earth circles the sun), free radicals are both free and radical enough to go careening through our cells (Meister, 1994). A process called oxidation creates free radicals and this process happens in the context of normal metabolic processes and our everyday exposure to environment. In other words, eating, breathing and going out in the sun all contribute to the the process of oxidation, free radical formation and resulting in damage that is caused to the cells of our bodies. It causes the deterioration of bone, joints and connective tissue; the wearing out of organs; the decline of the immune system; the irritating advance of the visible effects of aging; and even, possibly, to some extent, the aging process itself (Padayatty *et al.*, 2003). Early research on the role of antioxidants in biology focused on their use in preventing the oxidation of unsaturated fats, which is the cause of rancidity. Antioxidant activity could be measured simply by placing the fat in a closed container with oxygen and measuring the rate of oxygen consumption. However, it was the identification of vitamins A, C, and E as antioxidants that revolutionized the field and led to the realization of the importance of antioxidants in biochemistry of living organisms (Jeffrey, 1986).

Much like the immune system itself, which operates at a cellular level, the hardworking Vitamin C reaches every cell of the body. The concentration of vitamin C in both blood serum and tissues is quite high. In fact, this nutrient plays a major role in the manufacture and defence of our connective tissue, the elaborate matrix that holds the body together. It serves as a primary ingredient of collagen, a glue-like substance that binds cells together to form tissues (Padayatty *et al.*, 2003). As a water-soluble antioxidant, Vitamin C is in a unique position to "scavenge" aqueous proxy radicals before these destructive substances have a chance to damage the lipids. It works along with Vitamin E, a fat-soluble antioxidant, and the enzyme glutathione peroxidase to stop free radical chain reactions (Valko *et al.*, 2007).

Materials and Method

Materials for experiment (Orange, Mousambi, Lemon, Pineapple and Chakotra) were collected from local market of Dehradun. All the fruits were peeled off. After peeling juice was extracted with the help of Juicer & mixer in laboratory. Equal volume of the juices was taken and were mixed with 50% ethanol. They were centrifuged at 5000 rpm for 10 minutes. Supernatant was taken & pellet was discarded. Then protein content was estimated by using Lowry's Method & Bradford Method taken from Singh

& Sawhney's introductory practical Biochemistry book (Sawhney and Singh, 2007).

Superoxide dismutase (SOD) activity assay

In this method four solutions A, B, C & D were prepared.
Solution A: 0.1mM EDTA containing 50mM Na_2CO_3.
Solution B: 90 μM Nitroblue tetrazolium in solution A.
Solution C: 0.6% triton X in solution A.
Solution D: 20 mM hydroxylamine HCl (pH-6)

In the test tubes, 1.3 ml of solution A and 0.5ml solution B then 0.1 ml of solution C and 0.1 ml of solution D were added to the tubes and mixed well. 5μl juice extract was added to that reagent mixture and immediately absorbance at 560 nm at interval of 20 second was taken. SOD was calculated. Estimation of ascorbic acid (Vitamin C) of samples done by using method given in Sawhney and Singh (2007).

β-Carotene bleaching method

In this procedure the plant extracts, Vitamine E and BHA (butylated hydroxy anisole) were applied on TLC plates and after developing with a suitable solvent system, plates were sprayed with a betacarotene solution and exposed to daylight until discoloring of the background (6h) The active compounds were seen as orange color on the plate. Methanolic extracts of Citrus plants, Vitamin E and BHA were used as positive controls. Extracts which showed strong antioxidant activity were subjected to further tests. The same experiment was performed for the isolated fractions and compounds.

Thiocyanate assay

The peroxy radical scavenging activity was determined by thiocyanate method using Vitamine C as standard. Increasing concentration of the fractions in 0.5 ml of distilled water was mixed with 2.5 ml of 0.02 M linoleic acid emulsion (in 0.04 M phosphate buffer pH 7.0) and 2 ml phosphate buffer (0.04 M, pH 7) in a test tube and incubated in darkness at 37 °C. At intervals during incubation, the amount of peroxide formed was determined by reading the absorbance of the red colour developed at

500 nm by the addition of 0.1 ml of 30% ammonium thiocyanate solution and 0.1 ml of 20 mM ferrous chloride in 3.5% hydrochloric acid to the reaction mixture. The percentages scavenging inhibitions was calculated and were compared with the standard Ascorbic acid. A control was also prepared replacing water with plant extract.

Results and Discussion

Experimental results of protein content in fruit samples are given in Table 1. Ascorbic acid and SOD activities are given in Table 2. antioxidant activities by β-carotene bleaching method are given in Table 3 while Table 4 contain the results of Antioxidant activity by Thiocyanate assay method.

Table 1: Protein content in samples

S. No.	*Samples*	*Wt. of sample for 1ml of juice (gms)*	*Protein in Juice Extract μg/ml*	*Protein in sample μg/gm*
1.	Lemon	50/10 = 5.00	100	100/5=20
2.	Orange	50/12.2 = 4.09	125	125/4.09=30.5
3.	Mousambi	50/12.6 = 3.97	118	118/3.97=29.5
4.	Chakotra	50/12=4.17	115	115/4.17=27.6
5.	Pineapple	50/12.2=4.09	120	120/4.09=29.3

From above table, it can be estimated that for 1ml juice, 5.00 gm of lemon, 4.09 gm of Orange 3.97 gm of Mousambi, 4.17 gm of Chakotra and 4.09 gm of Pineapple was needed. Further Orange was found to have maximum protein content *i.e.* 125 μg/ml, followed by Pineapple Mousambi and if calculated on weight of sample (*i.e.* protein / gm of sample). It was observed as 30.5 μg/gm in Orange, 29.5 μg/gm in Mousambi, 29.3 μg/gm in Pineapple and 27.6 μg/gm in Chakotra.

Ascorbic Acid & SOD in Citrus fruits

Superoxide dismutase *i.e.* enzymatic antioxidant property was found to be maximum in Lemon (18.5 unit/mg of protein) followed by chakotra (17.6 unit/mg). Ascorbic acid in the citrus fruits was observed maximum in lemon juice (0.65 mg/ml) followed by Chakotra (0.62 mg/ml). Ascorbic acid on the basis of weight of fruit was observed maximum in Chakotra then in lemon and close result by Pineapple followed by Mousambi and Orange. Lemon and Pineapple were having nearly same enzymatic antioxidant property (Table 2).

Table 2: Ascorbic Acid & SOD in Citrus fruits

S.No.	*Fruit Sample*	*SOD Activity (unit/mg)*	*Ascorbic acid mg/ml of juice*	*Ascorbic acid in per gram of sample*
1.	Lemon	18.5	0.65	0.65/5=0.13
2.	Orange	14.6	0.49	0.49/4.09=0.11
3.	Mousambi	15.2	0.47	0.47/3.9=0.12
4.	Chakotra	17.6	0.62	0.62/4.17=0.14
5.	Pineapple	16.9	0.53	0.53/4.09=0.129

Lemons and other citrus fruits contain different chemical compounds like terpenes *i.e.* limonene is present in lemon which gives their characteristic taste and smell. Lemons contain significant amount of citric acid which provides low pH and sour taste to juice. They also contain Vitamin C (ascorbic acid), which is essential to human health. Various researches have shown that 100 ml of lemon juice contains approximately 55-65 mg of Vitamin C. In our findings this is about 65 mg per 100ml of juice, which is close with the findings of different other research scholars. Antioxidant property of citrus fruits is mainly due to Vitamin C (ascorbic acid), terpenes like limonene and flavonoids compounds. The results reported by Gavy and Singh also support our findings (Padayatty *et al.*, 2003).

β-Carotene bleaching method

Antioxidant activity by β-carotene bleaching method was determined by taking 10 & 20 µl of essential oils. Results are given in Table 3, in this method, Lemon has shown maximum activity followed by Chakotra, Mousambi, Orange and Pineapple. Similar trend was obtained when concentration of essential oil was increased upto 20 µl; but on very high oil concentration, antioxidant property might vary.

Table 3: Antioxidant activity by β-carotene bleaching method

S.No.	*Fruit Sample*	*Antioxidant activity when Quantity of essential oil -10 μL*	*Antioxidant activity when Quantity of essential oil - 20 μL*
1	Lemon	78.6	82.2
2	Orange	66.8	68.5
3	Mousambi	69.2	72.2
4	Chakotra	70.4	72.9
5	Pineapple	62.3	65.9

Antioxidant activity by Thiocyanate assay

Thiocyanate method also showed that lemon has max antioxidant property, followed by mousambi and nearby equal result has been shown by chakotra and orange.

Table 4: Antioxidant activity by Thiocyanate assay

S.N	*Fruit Sample*	*Antioxidant activity when Quantity of essential oil -25 μL*	*Antioxidant activity when Quantity of essential oil - 50 μL*
1	Lemon	46.22	49.04
2	Orange	32.22	40.20
3	Mousambi	35.45	44.32
4	Chakotra	32..40	55.22
5	Pineapple	28.12	36.44

Several studies have demonstrated the antibacterial and/or antioxidant properties of these plants, mainly using *in vitro* assays. Moreover, some researchers reported that there is a relationship between the chemical structures of the most abundant compounds in the plants and their above mentioned functional properties (Dean and Svoboda, 1989; Farag *et al.*,

1989). In addition, Citrus byproducts also represent a rich source of naturally occurring flavonoids (Horowitz, 1961). The peel which represents almost one half of the fruit mass, contains the highest concentrations of flavonoids in the Citrus fruit (Anagnostopoulou *et al.*, 2006; Manthley and Grohmann, 1996; 2001). Many papers have reported antioxidants in juice and edible parts of oranges of different origin and from different varieties (Miller and Rice-Evans, 1997; Rapisarda *et al.*, 1999; Roberts and Gordon, 2003; Vinson *et al.*, 2001). As far as the peel is concerned, extracts from this part of the fruit were found to have a good total radical antioxidative potential (TRAP) (Gorinstein *et al.*, 2001).

The present study suggests that citrus fruit juice has good antioxidant activity and regular consumption of juice can prevent the chronic disease.

References

Anagnostopoulou MA, Kefalas P, Papageorgiou VP, Assimopoulou AN and Boskou D (2006). Radical scavenging activity of various extracts and fractions of sweet orange peel (Citrus sinensis). *Food Chem.*, 94: 19-25.

Calabro ML, Galtieri V, Cutroneo P, Tommasini S, Ficarra P and Ficarra R (2004). Study of the extraction procedure by experimental design and validation of a LC method for determination of flavonoids in *Citrus bergamia* juice. *J Pharm and Biomed Anal.*, 35: 349-363.

Cotelle N, Bernier JL, Catteau JP, Pommery J, Wallet JC and Gaydou EM (1996). Antioxidant properties of hydroxylflavones. *Free Radic Biol Med.*, 20(1): 35-43.

Dean SG and Svoboda KP (1989). Antimicrobial activity of summer savory (Satureja hortensis L.) essential oil and its constituents. *J. Hortic. Sci.* 64: 205-210.

Di Majo D, Giammanco M, La Guardia M, Tripoli E, Giammanco S and Finotti E (2005). Flavanones in Citrus fruit: Structure-antioxidant activity relationships. *Food Res. Intern.*, 38: 1161-1166.

Ebrahimzadeh MA, Hosseinimehr SJ and Gayekhloo MR (2004). Measuring and comparison of vitamin C content in citrus fruits: introduction of native variety, *Chemistry: An Indian Journal,* 1(9): 650-652.

Elangovan V, Sekar N and Govindasamy S (1994). Chemoprotective potential of dietary bipoflavonoids against 20-methylcholanthrene-induced tumorigenesis. *Cancer Lett.,* 87: 107-113.

Farag RS, Daw ZY, Hewedi FM and El-Baroty GSA (1989). Antimicrobial activity of some Egyptian spice essential oils. *J. Food Prot.,* 52: 665-667.

Fernandez-Lopez J, Zhi N, Aleson-Carbonell L, Perez-Alvarez JA and Kuri V (2005). Antioxidant and antibacterial activities of natural extracts: application in beef meatballs. *Meat Sci.,* 69: 371-380.

German J (1999). Food processing and lipid oxidation. *Adv Exp Med Biol* 459: 23–50.

Gorinstein S, Martin-Belloso O, Park Y, Haruenkit R, Lojek A, Ciz M, *et al.* (2001). Comparison of some biochemical characteristics of different citrus fruits. *Food Chem.,* 74: 309-315.

Hertog MGL, Feskeens EJM, Holmann CH, Katan MB and Kromhout D (1993). Dietary antioxidant flavonoids and risk of coronary heart disease: The Zutphen elderly study. *Lancet,* 342: 1007-1011.

Horowitz RM (1961). The citrus flavonoids. *In*: Sinclair WB editor. The orange. Its biochemistry and physiology, CA, University of California, Division of Agricultural Science, Los Angeles, pp.334-372.

Jacob R (1996). Three eras of vitamin C discovery. *Subcell Biochem* 25: 1–16.

Jayaprakasha GK and Patil BS (2007). In vitro evaluation of the antioxidant activities in fruit extracts from citron and blood orange. *Food Chem.,* 101: 410-418.

Jeffrey Bland, Ph.D., The Nutritional Effects of Free Radical Pathology: 1966/A Year in Nutritional Medicine, Keats Publishing Inc., New Canaan, CT; 1986; p. 16.

Knight J (1998). Free radicals: their history and current status in aging and disease. *Ann Clin Lab Sci* 28 (6): 331-46.

Manthley JA and Grohmann K (1996). Concentrations of hesperidin and other orange peel flavonoids in citrus processing byproducts. *J. Agric. Food Chem.,* 44: 811-814.

Manthley JA and Grohmann K (2001). Phenols in citrus peel byproducts. Concentrations of hydroxycinnamates and polymethoxylated flavones in citrus peel molasses. *J. Agric. Food Chem.*, 49: 3268-3273.

Martın FR, Frutos MJ, Perez-Alvarez JA, Martınez-Sanchez F and Del Rıo JA (2002). Flavonoids as nutraceuticals: structural related antioxidant properties and their role on ascorbic acid preservation. *In*: Atta-Ur-Rahman (editor), Studies in natural products chemistry Elsevier Science. Amsterdam, pp.324-389.

Meister A (1994). Glutathione-ascorbic acid antioxidant system in animals. *J Biol Chem* 269 (13): 9397 – 400.

Miller NJ and Rice-Evans CA (1997). The relative contributions of ascorbic acid and phenolic antioxidants to the total antioxidant activity of orange and apple fruit juices and blackcurrant drink. *Food Chem.*, 60(3): 331-337.

Mouly PP, Arzouyan CR, Gaydou EM and Estienne JM (1994). Differentiation of citrus juices by factorial discriminant analysis using liquid chromatography of flavonone glycosides. *J. Agric. Food Chem.*, 42: 70-79.

Padayatty S, Katz A, Wang Y, Eck P, Kwon O, Lee J, Chen S, Corpe C, Dutta A, Dutta S, Levine M (2003). "Vitamin C as an antioxidant: evaluation of its role in disease prevention". *J Am Coll Nutr* 22 (1): 18-33.

Rapisarda P, Tomaino A, Lo Cascio R, Bonina F, De Pasquale A and Saija A (1999). Antioxidant effectiveness as influenced by phenolic content of fresh orange juices. *J. Agric. Food Chem.*, 47: 4718-4723.

Roberts WG and Gordon MH (2003). Determination of the total antioxidant activity of fruits and vegetables by a liposome assay. *J. Agric. Food Chem.*, 51(5): 1486-1493.

Sawhney, S.K and Singh, R. 2007. Introductory practical biochemistry, Fifth Reprint, 2007.

Schieber A, Stintzing FC and Carle R (2001). Byproducts of plant food rocessing as a source of functional compounds - recent developments. *Trends in Food Science and Technology,* 12: 401-413.

Stavric B (1993). Antimutagens and anticarcinogens in foods. *Food Chem. Toxicol.,* 32: 79-90.

Valko M, Leibfritz D, Moncol J, Cronin M, Mazur M, Telser J (2007). "Free radicals and antioxidants in normal physiological functions and human disease". *Int J Biochem Cell Biol* 39 (1): 44–84.

Vinson JA, Su X, Zubik L and Bose P (2001). Phenol antioxidant quantity and quality in foods: fruits. *J.Agric. Food Chem.,* 49: 5315-5321.

Biodiversity Conservation and Envir. Management (2012)
Editors: D.R. Khanna et al.
Pub. by Biotech Books. *ISBN: 978-81-7622-262-4*

Pages: 123-134

PATHOLOGICAL LESIONS IN THE GILLS OF CATFISH *HETEROPNEUSTES FOSSILIS* (BLOCH.) EXPOSED TO FLUORIDE

Sandeep Bajpai[✉] **and Madhu Tripathi**
Aquatic Toxicology Research Laboratory, Department of Zoology
University of Lucknow, Lucknow, Uttar Pradesh (INDIA)
✉E-mail: sandbajpai@gmail.com

Aquatic environment is under great threat by indiscriminate use of chemicals by humans, influencing the biology of inhabiting organisms. Fluoride is one of the major pollutants that is finding its way into water bodies through natural as well as anthropogenic sources causing high risk to non-target organisms. One of the most important non-target inhabitants of aquatic ecosystem affected by pollution are fish and their gills. These are well known target organ due to their wide surface area and permanent

✉ Corresponding author

contact with the surrounding medium. Present study was carried out to investigate the effect of fluoride on the gill histology of *Heteropneustes fossilis* after long term exposure. Fish were exposed to 38.60 mg/l and 77.20 mg/l ($1/10^{th}$ and $1/5^{th}$ of 96th hour LC_{50} value) of fluoride for a period of 45 and 90 days. The gills of the exposed fish showed various histopathological alterations such as necrosis, inflammation, cuticular eruption, swelling in primary and secondary epithelium, clubbing at gill tips, fusion of gill lamellae, epithelial uplifting, hyperplasia, nuclear pyknosis, hypertrophy in primary lamellar epithelial cells, reduction in inter-lamellar spaces and broken tissue debris near gill base. The severity of alteration was found concentration and duration dependent. It was concluded that fluoride affects the normal histoarchitecture of gills that play crucial and multifunctional role in homeostasis, respiration, osmoregulation and excretion in fishes. In aquatic toxicity studies fish are used as a monitor or as an animal model for the aquatic ecosystem. They are good bioindicator against slight change in their surrounding environment therefore can be helpful in environmental management programmes, necrosis.

Keywords: Fluoride toxicity, fish gills, structural alterations, catfish, necrosis.

Introduction

Environmental factors of both natural and anthropogenic origins are known to induce alteration of different magnitudes in the physiological status of animals (Vosyliene and Kazlauskiene, 1999). Variations in the chemical composition of a natural aquatic environment due to hazardous substances like chemicals, pesticides, trace elements, heavy metals and effluents from industries affect the behaviour, biochemistry and physiology of inhabiting fauna including fish (Bajpai *et al.*, 2009; Bajpai and Tripathi 2010; Pathan *et al.*, 2010). Therefore, biomarker parameter assessment is a means of environmental monitoring, with the advantage of providing quantitative response as valuable information on ecological relevance as well as on the adverse effects caused by water pollution (De la Torre *et al.*, 2005).

Water is the most precious natural resource on earth with wide range of benefits to humans including fisheries, wildlife, agriculture, industrial, urban and social development (Allan and Flecker, 1993). However, the uncontrolled release of industrial effluents and agricultural chemicals into water bodies have caused tremendous environmental problems to all classes of organisms in the aquatic habitat causing great threat to rich aquatic biodiversity (Rahman *et al.*, 2002). Among other contaminants, fluoride is highly reactive ion. It is highly soluble in water and therefore can be easily injested by aquatic organisms including fish (Camargo, 2003).

In aquatic ecosystem, fish represents the largest and most diverse group of vertebrates. A number of characteristics make them excellent experimental models for toxicological study. Their gills are important target organ for toxicants due to large surface area in contact with the external environment. Thus, gills come immediately in contact with pollutants dissolved in water provoking morphological and functional disturbances. Alteration in gill structure affects the normal functioning of vital physiological processes such as ion and gas exchanges, osmoregulation, excretion of nitrogenous wastes and acid-base equilibrium (Wendelaar Bonga and Lock, 2008). On this basis, gills arise as an appropriate model organ for studying the effects of environmental stressors on fish (Vigliano *et al.*, 2006).

Histopathological alterations are used as biomarkers in the evaluation of the fish health exposed to contaminants. It provides the real picture of the toxic effects of xenobiotics in vital functions of a living organism. One of the major advantages of using histopathological biomarkers in environmental monitoring is that this category of biomarkers allows examining of specific target organ responsible for particular vital function in fish (Gernhofer *et al.*, 2001) and serve as warning signal of damage to animal health. Hence, the present study was aimed to analyze the histopathological alterations in gill of catfish, *Heteropneustes fossilis* chronically exposed to sublethal concentrations of fluoride.

Materials and Method

Healthy specimens of *H. fossilis* measuring 13.09±0.20 cm length and 12.04±0.61 gm weight were collected from nearby areas of Lucknow and brought to the laboratory in large plastic containers to minimize stress. The fish were reared in the laboratory using aquaria containing 40 *l* dechlorinated tap water with continuous aeration. Fish were fed with dried prawn pieces once a day and were acclimated to laboratory conditions for about 30 days prior to experimentation. LC_{50} concentration and sublethal concentrations were derived for fluoride based on APHA (2005). For experimentation, fish were divided into three groups having ten fish in each. Group one served as control, without any treatment and maintained only in dechlorinated tap water. Group two and three were exposed to 38.60 mg/*l* and 77.20 mg/*l* ($1/10^{th}$ and $1/5^{th}$ of 96th hour LC_{50} value) as lower and higher concentration of sodium fluoride (NaF) for a period of 45 and 90 days. At the end of exposure periods the fish from both control and exposed groups were sacrificed, dissected carefully to isolate gill and fixed in alcoholic Bouin's fluid. After 24 hours they were processed following standards technique (Culling, 1974). Tissues were embedded in paraffin wax and serial section of 4-6 µm thickness were cut, deparaffinised and stained in hematoxylin and counterstained with eosin. The sections were examined under light microscopy and photographed using a digital camera.

Results and Discussion

The gill section of control fish showed primary and secondary lamellae. The primary lamellar epithelium was multilayered. Various cells such as mucus cells, acidophilic cells, basophilic cells and chloride cells were visible along the primary lamellar epithelium, especially at the base of the secondary lamellae (Plate-1, 2; Fig.-1). Secondary gill lamellae were composed of a single layer of epithelial cells supported by pillar cells.

Lower concentration exposed group- After 45 days of exposure to fluoride, severe inflammatory changes occurred. Swellings in primary and secondary lamellar epithelium along with clubbing on the gill tips of few secondary lamellae were observed. Primary lamellar epithelium was thickened and ruptured at some places. Fusion of gill lamellae was also marked at some places. Pillar, chloride and mucus cells showed pre-necrotic and degenerative changes. Epithelial necrosis, nuclear pyknosis, lifting and swelling of epithelium were also noticed at this exposure period (Plate-1; Fig-2).

After 90 days of fluoride exposure, most of the gill lamellae were stouted, clubbed and fused together. Degenerative and necrotic changes were more severe. Primary lamellar epithelial cells showed hypertrophy, vacuolization and necrosis. Secondary lamellae were fused with primary lamellar epithelium and broken tissue debris was found in interlamellar spaces. Migration of fibroblast like cells in blood channels was also noticed at this stage. Heavy degenerative, necrotic and lytic changes produced disorganization of gill lamellae (Plate-2; Fig-2).

Higher concentration exposed group- After 45 days of exposure, lifting in primary lamellar epithelium, vacuolization in mucus cells and epithelial cells were seen. Fusion and degeneration of secondary lamella at various places was also marked along with hypertrophy in chloride and pillar cells. Secondary lamellae also lost its epithelium and showed necrosis at many places. Almost complete structural loss of gill tissue was found at this stage (Plate-1; Fig.-3).

After 90 days of exposure, gill showed cellular damage, atrophy, bulging of gill tip, hypertrophy and hyperplasia of interlamellar epithelium, necrosis and separation of epithelial layer from pillar cells was increased. Reduction of respiratory surface area of secondary lamellae, sticking and vacuolization was severe. Primary lamellae showed thinning, disruption and detachment from the axis. Complete loss of lamellar arrangement along with necrosis, hyperplasia and vacuolization in epithelial cells, pillar cells and chloride cells was also noticed (Plate-2; Fig.-3).

The gills are very sensitive to adverse environmental conditions and changes in gill epithelia have been considered as good indicator of the effects of xenobiotics on fish.

In the present study, gills of *H. fossilis* exposed to lower and higher concentration of fluoride showed thickening of the primary lamellar epithelium, odema and fusion of secondary lamella. These observations are in accordance with the findings of Gupta (2003) in *Channa punctatus*, Kumar (2005) in *Clarias batrachus* and Regar and Bhatnagar (2007) in *Labeo rohita,* who have reported shortening, swelling and fusion of secondary gill lamellae, hypertrophy and hyperplasia in mucus cells, swelling and vacuolization in chloride cells after exposure to fluoride for

PLATE 1

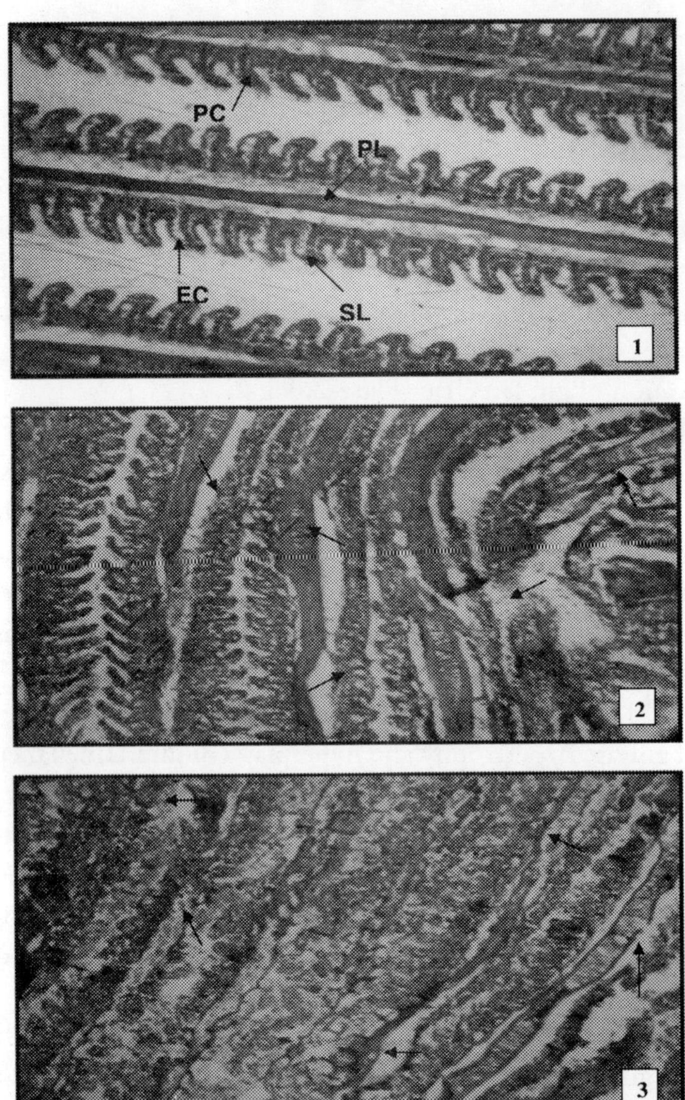

Photomicrograph of T.S. of Gills of *H. fossilis* after 45 days exposure (HEx100)

Fig. 1: Gills of control group showing well organized primary and secondary gill lamella with continuous epithelial lining (arrow).

Fig. 2: Gill of low concentration exposed group showing ruptured epithelium, necrosis swelling and fusion of the gill lamellae (arrow).

Fig. 3: Gill of high concentration exposed group showing necrosis, hypertrophy, hyperplasia, atrophy and loss of cellular organization (arrow).

PL=Primary lamellae; SL= Secondary lamellae; PC= Pillar cells; EC= Epithelial cells

PLATE 2

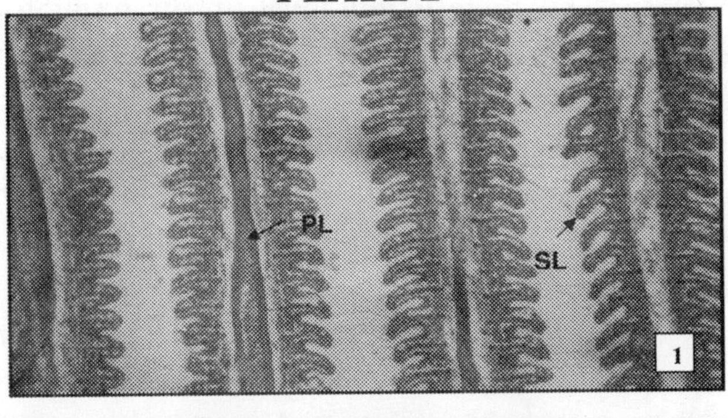

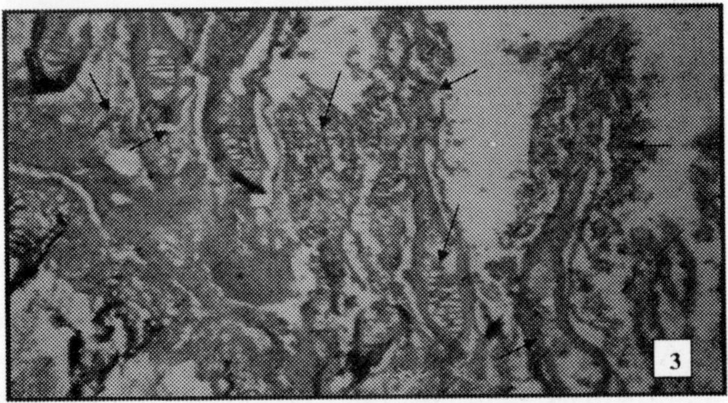

Photomicrograph of T.S. of Gills of *H. fossilis* after 90 days exposure (HEx100)

Fig. 1: Gills of control group showing normal histoarchitecture of gill filament with distinct epithelial lining and clear and prominent inter lamellar regions (arrow).

Fig. 2: Gill of low concentration exposed group showing bulging, curling and clubbing of the gill tips, necrosis, atrophy, hyperplasia and vacuolization (arrow).

Fig. 3: Gill of high concentration exposed group showing complete cellular degeneration, damaged primary and secondary lamellae, swollen tips, hypertrophy, vacuolization and broken tissue debris in intra lamellar spaces (arrow).

PL= Primary lamellae; SL= Secondary lamellae.

different durations. All these lesions may impair respiratory functions. Lifting of epithelium or hyperplasia of epithelium results in increase of diffusion distance, thus affecting the exchange of gases and the fusion of lamellae caused a decrease in the total respiratory areas of gills, resulting in a decreased oxygen uptake capacity. Hypertrophy and hyperplasia of chloride cells following fluoride exposure are considered compensatory responses to maintain ion balance.

The gills of higher concentration exposed fishes showed epithelial detachment, hyperplasia, clubbing of gill filaments and swollen distal tips. It is well known that hyperplasia decreases the surface area of gill and if it is severe, causes gill fusion and consequently the ability of oxygen absorption is hampered (Gardener and Yevich, 1970). The detachment of epithelial lining is supposed to be an inflammatory reaction of pollutants. Hyperplasia and hypertrophy of mucus and chloride cells and abnormalities in blood vessels (lamellar telangiectasia and mucoid metaplasia) observed in the present investigation are similar to the observation of Kumar (2005) in *C. batrachus* and Bhatnagar *et al.* (2007) in *Labeo rohita* after exposure to fluoride.

The secondary gill lamellae of exposed fish showed large number of mucus cells in comparison to control fish, indicating the response of fish to reduce the toxicity of fluoride. This is one of the defensive mechanism adapted by fishes and has been supported by the observations of other workers who have reported similar defensive and compensatory responses in fish exposed to different pollutants including fluoride (Nowak and Barbara, 1992; Cerqueira and Fernandes, 2002; Olojo *et al.*, 2005; Athikesavan *et al.*, 2006; Al-Attar, 2007; Saenphet *et al.,* 2009; Parikh *et al.*, 2010). Hyperplasia of primary filaments and secondary lamellae were reported in the gills of *Puntius ticto* exposed to industrial sewage (Chauhan and Pandey, 1987) and *Rasbora daniconius* exposed to paper mill effluents (Pathan *et al.*, 2010).

Loss of gill epithelium lining at many places and fusion of gill lamella observed in this study is also supported by findings of Tripathi *et al.* (2008) in *C. punctatus* after sublethal exposure to fluoride. Cellular changes in gill tissues have shown two types of responses *i.e.* defensive (inflammatory) and compensatory (cell proliferation, mucus secretion) responses. According to Mallat (1985) and Takashima and Hibiya (1995), both these responses help to block the entry of toxicants and prevent them from reaching the blood stream and to some extent prevent damages caused by the direct effects of toxicants (necrosis, peeling of gill filaments).

Results obtained during this study revealed that histology of gill of *H. fossilis* appears to be a sensitive monitoring tool for aquatic environment as it reflects the fish health. Histopathological studies on fish therefore can make valuable contribution in monitoring of aquatic ecosystems and may be an important part of environmental impact assessment criteria and biodiversity conservation in the environmental management programmes.

Acknowledgement

The authors are thankful to Prof. A.K. Sharma, Head, Department of Zoology, University of Lucknow for providing necessary laboratory facilities to carry out this study.

References

Al-Attar, A.M. (2007). The influences of nickel exposure on selected physiological parameters and gill structures in teleost fish, *Oreochromis niloticus*. J. Biol. Sci., 7(1): 77-85.

Allan, J.D. and Flecker, A.S. (1993). "Biodiversity conservation in running waters." BioScience. 43(1): 32-43.

APHA, AWWA and WEF (2005). Standard methods for the examination of water and waste water, 21st Edition. American Public Health Association. Washington D.C.

Athikesavan, S., Vincent, S., Ambrose, T. and Velmurugan, B. (2006). Nickel induced histopathological changes in the different tissues of freshwater fish, *Hypophthalmichthys molitrix* (Valenciennes). J Environ. Biol., 27(2): 391-395.

Bajpai, S. and Tripathi, M. (2010). Effect of fluoride on growth bioindicators in stinging catfish, *Heteropneustes fossilis* (Bloch). Fluoride. 43(4): 232-236.

Bajpai, S., Tewari, S. and Tripathi, M. (2009). Evaluation of acute toxicity levels and behavioural responses of *Heteropneustes fossilis* (Bloch) to sodium fluoride. Aquacult., 10(1): 37-43.

Bhatnagar, C., Bhatnagar, M. and Regar, B.C. (2007). Fluoride-induced histopathological changes in gill, kidney and intestine of fresh water teleost, *Labeo rohita*. Fluoride., 40(1): 55-61.

Camargo, J.A. (2003). Fluoride toxicity to aquatic organisms: A review. Chemosphere., 50(3): 251-261.

Cerqueira, C.C.C. and Fernandes, M.S. (2002). Gill tissue recovery after copper exposure and blood parameters responses in the tropical fish, *Prochilodus scrofa*. Ecotoxicol. Environ. Saf., 52: 83-91.

Chauhan, M.S. and Pandey, A.K. (1987). Histopathological changes in the gills of *Puntius ticto* (Ham) following Khan river water exposure under ambient laboratory conditions. J. Environ, Biol., 8(2): 121-127.

Culling, C.F.A. (1974). Handbook of histopathological and histochemical techniques, 3rd Eds. London, Butterworth.

De la Torre, F.R., Saliba, A. and Ferrair, L. (2005). "Biomarkers of native fish species (*Cnesterodon decemaculatus*) application to the water toxicity assessment of a per-urban polluted river of Argentina." Chemosphere. 59(4): 577-583.

Gardener, G.R. and Yevich, P.P. (1970). Histological and hematological response of estuarine teleosts to cadmium. J. Fish. Res. Board. Can., 27: 2185-2196.

Gernhofer, M., Pawet, M., Schramm, M., Muller, E. and Triebskorn, R. (2001). Ultrastructural biomarkers as tools to characterize the health status of fish in contaminated streams. J. Aquatic Ecosystem. Stress and Recovery. 8: 241-260.

Gupta, R. (2003). Pathophysiological consequences to freshwater fish, *Channa punctatus* induced by fluoride. Ph.D. Thesis, University of Lucknow, Lucknow.

Kumar, A. (2005). Evaluation of fluoride toxicity on reproductive system of freshwater fish. Ph.D. Thesis, University of Lucknow, Lucknow.

Mallat, J. (1985). Fish gill structural changes induced by toxicants and other irritants. A statistical review. Can. J. Fish Aquat. Sci., 42: 630-648.

Nowak and Barbara (1992). Histological changes in gills induced by residues of endosulphan. Aquatic toxicol. (AMST). 23(1): 65-83.

Olojo, E.A.A., Olurin, K.B., Mbaka, G. and Oluwemimo, A.D. (2005). Histopathology of the gill and liver tissues of the African catfish *Clarias gariepinus* exposed to lead. African J. Biotechnol., 4(1): 117-122.

Parikh, P.H., Rangrez, A., Adhikari-Bagchi, R. and Desai, B.N. (2010). Effects of dimethoate on some histoarchitecture of freshwater fish *Oreochromis mossambicus* (Peters, 1952). The Bioscan. 5(1): 55-58.

Pathan, T.S., Thete, P.B., Shinde, S.E., Sonawane, D.L. and Khillare, Y.K. (2010). Histopathological changes in gill of freshwater fish, *Rasbora daniconius* exposed to paper mill effluent. Iranica J. Energy & Environ., 1(3): 170-175.

Rahman, M.Z., Hossain, Z., Mollah, M.F.A. and Ahmed, G.U. (2002). "Effect of diazinum 60 EC on *Anabas testudineus, Channa punctatus* and *Barbodes gonionotus* (Naga)" The ICLARM Quaterly. 25: 8-12.

Regar, B.C. and Bhatnagar, C. (2007). Histopathological changes in the gill architecture of freshwater teleosts, *Labeo rohita* exposed to sodium fluoride. J. Herbal Med. Toxicol., 1(1): 35-41.

Saenphet, S., Thaworn, W. and Saenphet, K. (2009). Histopathological alterations of the gills, liver and kidneys in *Anabas testudineus* (Bloch) fish living in an unused lignite mine, Li District, Lamphun Province, Thailand. Southeast Asian J. Trop. Med. Pub. Hlth., 40(5): 1121-1126.

Takashima, F. and Hibiya, T. (1995). An atlas of fish histology, normal and pathological features. 2nd (Eds) Kodansha. Tokyo.

Tripathi, M., Gupta, R. and Sharma, U.D. (2008). Recovery of adverse effects induced by fluoride after ascorbic acid treatment in *Channa punctatus* (Bloch). J. Ecophysiol. Occup. Hlth., 8: 147-152.

Vigliano, F.A., Aleman, N., Quiroga, M.I. and Nieto, J.M. (2006). "Ultrastructural characterization of gills in juveniles Argentinian Silverside, *Odontesthes bonariensis* (Valenciennes, 1835) (Teleostei: Atheriniformes), Anatomy histology and Embryology. 35: 76-83.

Vosyliene, M.Z. and Kazlauskiene, N. (1999). "Alterations in fish health status parameters after exposure to different stressors." Acta Zoologica Lituanica Hydrobiol., 9(2): 82-95.

Wendelaar Bonga, S.E. and Lock, R.A.C. (2008). The osmoregulatory system in Di Giulio, R.T. and Hinton, D.E. (Eds): The toxicology of fishes, CRC Press- Taylor and Francis Group, Boca Raton, FL, pp.401-405.

Biodiversity Conservation and Envir. Management (2012)
Editors: D.R. Khanna et al.
Pub. by Biotech Books. *ISBN: 978-81-7622-262-4*

Pages: 135-145

POPULATION ECOLOGY OF THE INDIAN TORRENT CATFISH *AMBLYCEPS MANGOIS* (HAMILTON- BUCHANAN) FROM GARHWAL, UTTARAKHAND, INDIA

A.K. Dobriyal[1✉], Ram Krishan[2], K.L. Bisht[3], R. Kumar[4], P. Bahuguna[4] and H.K. Joshi[5]

[1]Department of Zoology, H.N.B. Garhwal, Central University Campus, Pauri Garhwal, Uttarakhand (INDIA)

[2]Department of Zoology, Government S P College, Srinagar, J & K (INDIA)

[3]Department of Zoology, P.D.B.H. Government PG College, Kotdwara Garhwal, Uttarakhand (INDIA)

[4]Department of Zoology, L.M.S. Government PG College, Pithoragarh, Uttarakhand (INDIA)

[5]Department of Zoology, Government Degree College, Chaubattakhal, Uttarakhand (INDIA)

✉E-mail: anoopkdobriyal@rediffmail.com

✉ Corresponding author

This chapter deals with population study of Indian torrent catfish *Amblyceps mangois* (Ham-Buch) from River Mandal in between Rathuadhab and Banjadevi (longitude 78°17'15"E-78°55'20" E and latitude 29°45°N-29°55'40"N) during January, 2008 to December, 2010 in district Pauri Garhwal, Uttarakhand. Sex ratio from a sample size of 114 specimens were analysed month wise and also season wise to see whether there is any disturbance in population or not for being a factor of its low population. The significance was tested by Chi-square test. It was observed that during breeding season (Monsoon months), the sex ratio differed significantly being 1 male: 3.17 female ($\chi^2_{0.01} = 6.67$). It clearly indicates that the low size of male population may lead into low fertility of the species and hence care should be taken during conservation efforts of the species. Though the fish is thinly populated yet it is well flourishing in river Mandal. In other streams of Garhwal Himalaya, it has rare occurrence. The sex composition is discussed month wise and season wise. The ecological parameters (temperature, dissolved oxygen and pH) of the habitat were also discussed.

Keywords: *Amblyceps mangois*, Habitat ecology, Population structure, River Mandal, Sex ratio

Introduction

A large number of rivers, rivulets and streams form a vast network in the central himalayan mountains (Garhwal and Kumaun region; latitudes 29°5'-31° 25' N and longitudes 77°45'-81° E) and have a large number of indigenous fish species. About 65 species of fish have been reported from Garhwal region (Singh *et al.*, 1987). Major rivers of Garhwal are the Alaknanda, Mandakini, Bhagirathi, Asiganga, Bhilangana, Ganga, Nayar (Eastern and Western Nayar), Song, Suswa, Khoh, Mandal and

Table 2: Seasonal Population structure (Sex Ratio) of *Amblyceps mangois* in the river Mandal during January, 2008 to December, 2010.

Season	*Total No. of fish*	*No. of Male*	*No. of Female*	*% of Male*	*% of Female*	*Ratio*		χ^2	*Remark*
						Male	*Female*		
Winter (Dec-Feb)	19	8	11	42.11	57.89	1.00	1.37	0.472	NS
Spring (Mar-Apr)	16	6	10	37.50	62.5	1.00	1.67	1.0	NS
Summer (May-Jun)	21	11	10	52.38	47.62	1.1	1.00	0.048	NS
Monsoon (Jul-Aug)	25	6	19	24.00	76.00	1.00	3.17	6.76	*
Autumn (Sep-Nov)	33	12	21	36.36	63.64	1.00	1.75	2.46	**
Pooled Data	114	43	71	37.72	62.28	1.00	1.65	6.876	*

*** Significant at 1 % level ($\chi^2_{0.01}$ = 6.63), ** Low significance, NS= Non significant**

Conservation of the species

It was observed that apart from biological facts there are some anthropological, social and agricultural issues which are needed to be properly taken care of. Rural folk usually practice overfishing by the use of illegal fishing techniques which damage the ichthyofauna specially these minor species. Stream bed which is important for feeding and breeding of fish is considerably disturbed by anthropological activities. The river is embanked by agricultural fields which are fertilized with urea and other chemicals. These chemicals generally leak into stream and damage biota.

To conserve the species like *Amblyceps mangois*, it is essential that these problems are solved through ecological awareness. Plantation should be encouraged on the banks of stream so that the riparian vegetation can be improved which indirectly contributes towards productivity of rivers. Attempts should be made of induced breeding which is easy in case of minor species in aquariums. By introducing fish into its natural habitat may solve the problem of differential sex ratio also. By these conservation efforts the entire aquatic resource can be conserved.

References

Badola, S.P. and Singh H.R. 1980. Food and feeding habits of fishes of the genera *Tor, Puntius* and *Barilius*. Proc Indian. natn. Sci. Acad. B 46(1): 58-62.

Bahuguna, P. 2008. Fish biology of *Puntius conchonius* (Ham. Buch) From Garhwal, Central Himalaya. D. Phil. Thesis, HNB Garhwal University Srinagar Garhwal.

Bahuguna, P.K., Joshi, H.K. and Dobriyal, A.K. (2007). Fecundity and sex ratio in *Puntius conchonius* (Pisces: Cyprinidae) from Garhwal Himalaya. *Environment Conservation J.* (1-2): 37-43.

Bahuguna, P., Kumar, R. and Joshi, H.K. (2010). Studies on the reproduction capacity and Sex ratio in a hill – stream loach fish *Noemacheilus denisoni* Day from river Mandal of Garhwal Himalaya, Uttarakhand. *Uttar Pradesh J. Zool.* vol (30): (Accepted in Press)

Day, F. (1889). The fauna of British India including Ceylone and Burma. Dawson and Sons Ltd., London.

Dobriyal, A. K. and Singh, H. R. 1987: The reproductive biology of a hillstream minor carp *Barilius bendelisis* (Ham.) from Garhwal Himalaya, India.Vest cs. Spolec. Zool.51: 1-10.

Dobriyal, A.K. and Singh, H.R. 1990. Ecological studies on the age and growth of *Barilius bendelisis* (Ham.) from India. Arch. Hydrobiol. 118: 93 –103.

Dobriyal, A. K. and Singh, H. R. 1993: Reproductive biology of a Hillstream catfish *Glyptothorax madraspatanum* (Day) from Garhwal Himalaya, India. Aquaculture and Fisheries Management: 24: 699-706.

Dobriyal, A.K., Kumar, N., Bahuguna, A.K. and Singh, H.R (2000): Breeding ecology of some coldwater minor carps from Garhwal Himalayas. *Cold water aquaculture and fisheries.* (Eds H. R. Singh and W.S. Lakra), Narendra Publishing House, Delhi. 177-186.

Dobriyal, A.K., Negi, K.S., Joshi, H.K. and Bisht, K.L. (2004). Breeding capacity of *Crossocheilus latius latius* (Pisces: Cyprinidae) in the river Mandakini of Garhwal, Uttaranchal. *Flora and Fauna* Vol 10: 151-153.

Jameela Beevi, K.S. and Ramachandran, A. (2005). Sex ratio in *Puntius vittatus* Day in the fresh water bodies of Ernakulam District, Kerala. *Zoos Print Journal* 20(9): 1989-90.

Kumar, K., Bisht, K.L., Dobriyal, A.K., Joshi, H.K., Bahuguna, P.K., Goswami, S., Balodi, V.P. and Nautiyal, P. (1982). Some aspects of bioecology of *Tor putitora* in relation to hydrobiology of some Garhwal hillstreams. D.Phil.Thesis.HNB Garhwal University, Srinagar Garhwal.

Nikolsky, G.V. (1980).Theory of fish population dynamics. Bishen singh and Mahendar Pal Singh, India and Ottokoeltz Science Publishers (West Germany), pp. 317.

Panwar, B.A. and Mani, U.H. (2006): Sex ratio of *Macrones bleekeri* (Blecker) from Sadatpur Lake, Ahmednager, District Maharashtra. *J.Aqua. Biol.* **21**(2): 182-185 (2006).

Rautela, K.K. 1999. Ecological studies on the spawning biology of some coldwater fishes from the Khoh stream. D. Phil. Thesis, HNB Garhwal University, Srinagar Garhwal.

Singh, H.R., Badola, S.P. and Dobriyal, A.K. (1987): Geographical distributional list of ichthyofauna of the Garhwal Himalay a with some new records. J. Bombay nat. Hist Soc. 84 : 126-132.

Sobhana, B. and Nair, N.B. (1976). Observation on the maturation and spawning of *Puntius sarana subnasutus* (Valenciennes). *Indian J. Fish.* 21(2): 357-359.

Talwar, P.K. and Jhingran, A. G. (1991): Inland fishes of India and adjacent countries. Oxford & IBH Publ. Co. Pvt. Ltd., New Delhi. pp. 250-295.

Welch, P.S. (1948). *Limnological Methods.* Mc Graw- Hill Book Co. NY, Toronto, London.

Biodiversity Conservation and Envir. Management (2012)
Editors: D.R. Khanna et al.
Pub. by Biotech Books. *ISBN: 978-81-7622-262-4*

Pages: 147-155

IN-VITRO ANTIBACTERIAL ACTIVITY OF EXTRACTS OF *JATROPHA CURCAS* AND *HIBISCUS ROSASINENSIS*

Shozeb Jawed[1], Fouzia Ishaq[2] and Amir Khan[3]✉

[1] Department of Biotechnology, Beehive College of Advance Studies, Dehradun, Uttarakhand (INDIA)

[2] Department of Zoology and Env. Science, Gurukul Kangri University, Haridwar, Uttarakhand (INDIA)

[3] Department of Biomedical Science and Biotechnology Dolphin P.G. Institute of Biomedical and Natural Sciences, Dehradun, Uttarakhand (INDIA)

✉E-mail: amiramu@gmail.com

Use of plant based drugs and chemicals for curing various ailments and personal adornment is as old as human civilization. Plants and plant-based medications are the basis of many of the modern pharmaceuticals that we use today

✉ Corresponding author

for our various ailments. The extracts of *Jatropha curcas* and *Hibiscus rosasinensis* by using petroleum ether, chloroform and ethanol as solvents were screened for antibacterial activity. The chloroform extract of *Jatropha curcas* leaves was found to be the most active, showing activity against *B. cereus*, the most susceptible gram-positive bacteria followed by *B. subtilis*, *E. coli*, *S. aureus* and *M. luteus* and thus displayed highest inhibitory zone as compared with the extracts obtained from *Hibiscus rosasinensis*. *In vitro* antimicrobial activity was performed by agar disc diffusion. The medicinal value of this plant could be attributed to the presence of one or more of the metabolites. The results obtained in the present study suggests that *Jatropha curcas* leaves can be used in treating diseases caused by the test organisms.

Keywords: Plant extract, Inhibition zone and Antimicrobial activity, *Jatropha curcas*, medicinal value.

Introduction

There are many approaches to search new biologically active principles in higher plants (Farnsworth and Loub, 1983). Many efforts have been made to discover new antimicrobial compounds from various kinds of sources such as soil, microorganisms, animals and plants. One of such resources is folk medicine and systematic screening of the same may result in the discovery of novel effective compounds (Janovska *et al.*, 2003). The substances from plants & animal sources are being used as food since antiquity. Later on, these substances were differentiated as food stuffs & therapeutic agents, which man tried to explore and utilize to treat disorders. Undoubtedly, while the plant kingdom still holds many plant species containing substances of medicinal value which are yet to be discovered; large number of plants are constantly being screened for their possible medicinal values, particularly for their anti-inflammatory, hypotensive, hypoglycemic, amoebicidal, antifertility, cytotoxic, antibiotic and anti-parkinsonism properties. The wealth of uninvestigated material available is illustrated by the fact that in 1985, it was reported that natural product research elicited some 3500 new chemical structures of which more than 2600 were from higher plants. The increasing failure of chemotherapeutics and antibiotic resistance exhibited by pathogenic microbial infectious agents have led to the screening of several medicinal plants for their potential antimicrobial activity (Colombo and Bosisio, 1996; Iwu *et al.*, 1999). It is anticipated that phytochemicals with adequate antibacterial efficacy will be used for the treatment of bacterial infections (Balandrin *et al.*, 1985). Since time immemorial, man has used various parts of plants in the treatment and prevention of various ailments (Tanaka *et al.*, 2002). The development of drug resistance as well as appearance of undesirable side effects of certain antibiotics (WHO, 2002) has led to the search for new antibacterial agents; in particular from medicinal plants. A number of reports concerning the antibacterial screening of plant extracts of medicinal plants have appeared in the literatures (Salvet *et al.*, 2001). Scrutiny of the drugs obtained from plants reveals that the majority of drugs are derived from seed bearing plants (spermatophytes). Among the spermatophytes the angiosperms (flowering plants) have yielded a good number of useful medicinal plants than the gymnosperms (non-flowering plants). The gymnosperms are useful source of oils, resins and the alkaloid ephedrine. Within the angiosperms both monocots and dicots provide many useful drugs, among the dicots Cinchona, Ipecac, Rauwolfia *etc.* are some of the important drugs from higher plants.

Materials and Method

Collection of plant material

Fresh *Jatropha curcas* leaves and seeds were collected randomly from Forest Research Institute, Dehradun, India and the leaves and flowers of *Hibiscus rosasinensis* from our prestigious college Beehive Group of Colleges, Selaqui, Dehradun, India for investigation of antimicrobial property. Fresh plant materials were washed under running tap water, then with distilled water dried in an oven at 37 °C for a day and then homogenized to fine powder and stored in airtight bottles. Dried and powdered parts of the plants under investigation were extracted with different solvents (ethanol, petroleum ether and chloroform) by Soxhlet apparatus.

Preparation of Extracts

The material taken in a suitable comminute form is usually packed in thimble, made of absorbing cotton and placed in the extractor. Solvent is placed in a round bottom flask and boiled. The vapors are allowed to pass through the side tube to the condenser, where they get condensed and fall back to the packed material, through which it percolates and extract out the active constituents as the volume of the solvent in the extractor increases. The level of the liquid in the siphon also increases till it reaches the maximum point from where it is siphoned out in the flask. On further heating, the solvent gets vaporized while the dissolved active parts are finally extracted from the material and get collected in the flask.

Bacterial Strains

In vitro antimicrobial activity was examined for chloroform, petroleum ether and methanol extracts from 2 medicinal plants used by traditional healers. Microorganisms were obtained from the IMTECH, Chandigarh and Gurukul Kangri University, Haridwar, India. Amongst five microorganisms investigated, four Gram-positive bacteria were *Bacillus cereus* (MTCC No. 443); *Bacillus subtilis* (MTCC No. 441). *Micrococcus luteus* (Gurukul Kangri University) and *Staphylococcus aureus* (Gurukul Kangri University) while one Gram-negative bacteria *Escherichia coli* (MTCC No. 443). All the microorganisms were maintained at 4 °C on nutrient agar slants.

Antibacterial Assay

The antimicrobial assay was performed by agar disc diffusion method (Parekh *et al.,* 2005) for aqueous extract of solvent. The molten Mueller Hinton agar was inoculated with 100 μl of the inoculum (1×10^8 CFU/ml) and poured into the Petri plate. Discs of 6 mm diameter were prepared from Whatman No.1 filter paper. They were sterilized by autoclaving and subsequently dried at 80 °C for an hour. After drying the discs were placed on the MHA agar plates with flamed forceps and gently pressed down to ensure contact with the agar surface. The discs were spaced far enough to avoid both reflection waves from the edges of the petriplates and overlapping rings of inhibition. The plates were incubated overnight at 37 °C in an inverted position. Microbial growth was determined by measuring the diameter of zone of inhibition. For each bacterial strain, controls were maintained where Dimethyl Sulphoxide (DMSO) was used instead of the extract. The result was obtained by measuring the zone diameter.

Results and Discussion

The data reported in Table 1 and Table 2 presents the antibacterial activity of the extracts of *Jatropha curcas* and *Hibiscus rosasinensis.* The results indicate that the extracts from the medicinal plants studied showed inhibition of growth of some of the tested microorganisms. The chloroform extract of *Jatropha curcas* leaves was found to be the most effective antimicrobial agent as compared to the other extracts in different solvents. *B. cereus,* was the most susceptible gram-positive bacteria followed by *B. subtilis. E. coli.*, *S. aureus* and *M. luteus.* There were no antibacterial report on the extracts of *Jatropha curcas* seeds and *Hibiscus rosasinensis* leaves in the tested strains in different solvents.

Successful prediction of botanical compounds from plant material is largely dependent on the type of solvent used in the extraction procedure. The traditional healers or practitioners make use of water primarily as a solvent, but our study showed that chloroform extracts of these plants were certainly much better and powerful. This may be due to the better solubility of the active components in organic solvent (De Boer *et al.*, 2005). These observations can be rationalized in terms of the polarity of the compounds being extracted by each solvent and in addition to their intrinsic bioactivity, by their ability to dissolve or diffuse in the different media used in the assay. The growth media also seem to play an important role in the determination of the antibacterial activity. Lin *et al.*, (1999) reported that Muller-Hinton agar appears to be the best medium to explicate the antibacterial activity and the same was used in the present study. Amongst the gram-positive and gram-negative bacteria, gram-positive bacterial strains were more susceptible to the extracts as compared to gram negative bacteria. This is in agreement with previous reports that plant extracts are more active against gram-positive bacteria than gram-negative bacteria (Vlietinck *et al.*, 1995; Rabe and Van Staden, 1997).

The results of present study supports the traditional usage of the studied plants and suggests that *Jatropa curcas* extract possesses compounds with antimicrobial properties that can be used as antimicrobial agents in new drugs for the therapy of infectious diseases caused by pathogens. The most active extracts can be subjected to isolation of the therapeutic antimicrobials and carry out further pharmacological evaluation. The higher resistance of Gram-negative bacteria to plant extracts has previously been documented and related to thick murein layer in their outer membrane, which prevents the entry of inhibitor substances into the cell (Martin, 1995; Brantner *et al.*, 1996; Palombo and Semple, 2001; Tortora *et al.*, 2001; Matu and Van Staden, 2003). Similarly, our results indicated that the antibacterial activities of the extracts were more pronounced on Gram positive than on Gramnegative bacteria.

Table 1: Zone of inhibition produced by *Hibiscus rosasinensis* flower extract against test bacterial strains at different concentrations.

S.No	*Microbes*	*Petroleum Ether*	*Ethanol*	*Chloroform*
1	*B. subtilis*	Ctrl – 100% -6mm 50% -6mm 25% -	Ctrl – 100%- 6mm 50% - 25% -	Ctrl – 100%- 50% - 25% -
2	*S. aureus*	Ctrl – 100% - 50% - 25% -	Ctrl – 100%- 6mm 50% - 25% -	Ctrl – 100% - 50% - 25% -
3	*B. cereus*	Ctrl – 100% -6mm 50% -6mm 25% -	Ctrl – 100%- 6mm 50% -6mm 25% -	Ctrl – 100% - 50% - 25% -
4	*E. coli*	Ctrl – 100% - 50% - 25% -	Ctrl – 100%- 6mm 50% - 25% -	Ctrl – 100% - 50% - 25% -
5	*M. luteus*	Ctrl – 100% - 50% - 25% -	Ctrl – 100%- 50% - 25% -	Ctrl – 100% - 50% - 25% -

Table 2: Zone of inhibition produced by *Jatropha curcas* leaf extract against test bacterial strains at different concentrations.

S.No	*Microbes*	*Petroleum Ether*	*Ethanol*	*Chloroform*
1	*B. subtilis*	Ctrl – 100% - 50% - 25% -	Ctrl – 100%- 6mm 50% - 7mm 25% -	Ctrl – 100%-9mm 50% -8mm 25% -6mm
2	*S. aureus*	Ctrl – 100% - 50% - 25% -6mm	Ctrl – 100%- 6mm 50% - 25% -	Ctrl – 100%-8mm 50% -8mm 25% -6mm
3	*B. cereus*	Ctrl – 100% - 50% -6mm 25% -8mm	Ctrl – 100%- 6mm 50% -6mm 25% -	Ctrl – 100%-9mm 50% -8mm 25% -7mm
4	*E. coli*	Ctrl – 100% - 50% - 25% -	Ctrl – 100%- 50% -8mm 25% -	Ctrl – 100%-9mm 50% -8mm 25% -6mm
5	*M. luteus*	Ctrl – 100% - 50% -7mm 25% -8mm	Ctrl – 100%-6mm 50% - 25% -	Ctrl – 100%-9mm 50% -7mm 25% -6mm

References

Balandrin, M.F., Kjocke, A.J. and Wurtele, E. (1985). Natural plant chemicals as sources of industrial and mechanical materials. *J. Science* 228: 154-160

Brantner, A., Males, Z., Pepeljnjak, S. and Antolic, A. (1996). Antimicrobial activity of *Paliurus spina-christi* Mill (Christ's thorn). *J. Ethnopharmacol.* 52: 119-122.

Colombo, M.L. and Bosisio, E. (1996). Pharmacological activities of *Chelidonium majus L* (Papaveraceae). *Pharmacol Res* 33: 127-134.

De Boer, H.J., Kool, A., Broberg, A., Mziray, W.R., Hedberg, I. and Levenfors, J.J. (2005). Antifungal and antibacterial activity of some herbal remedies from Tanzania. *J Ethnopharmacol*; 96:461-469.

Farnsworth, N.R. and Loub, W.D. (1983): *Information gathering and data bases that are pertinent to the development of plant-derived drugs in Plants: The Potentials for Extracting Protein, Medicines, and Other Useful Chemicals.* Workshop Proceedings. OTA-BP-F-23. U.S. Congress, Office of Technology Assessment, Washington, D.C., pp. 178-195.

Iwu, M.W., Duncan, A.R. and Okunji, C.O. (1999). *New antimicrobials of plant origin. In:Janick J. ed. Perspectives on New Crops and New Uses. Alexandria, VA*: ASHS Press: pp. 457-462.

Janovska, D., Kubikova, K.and Kokoska, L. (2003): Screening fo antimicrobial activity of some medicinal plant species of traditiona Chinese medicine. Czech. *J. Food Sci.* 21: 107-111.

Lin, J., Opoku, A.R., Geheeb-Keller, M., Hutchings, A.D., Terblanche S.E., Jager, A.K. and Van Staden, J. (1999). Preliminary screenin of some traditional zulu medicinal plants for anti-inflammatory an antimicrobial activities. *J. Ethnopharmacol.* 68: 267-274.

Martin, G.J. (1995). *Ethnobotany: A Methods Manual. London*: Chapma and Hall.

Matu, E.N. and Van Staden, J. (2003). Antibacterial and anti-inflammator activities of some plants used for medicinal purposes in Kenya. *Ethnopharmacol.* 87: 35-41.

Palombo, E.A. and Semple, S.J. (2001). Antibacterial activity of tradition Australian medicinal plants. *J. Ethnopharmacol.* 77: 151-157.

Parekh, J., Nair, R. and Chanda, S. (2005). Preliminary screening of son folklore medicinal plants from western India for potential antimicrobi activity. *Indian J Pharmacol* 37: 408-409.

Rabe, T. and Van Staden, J. (1997). Antibacterial activity of South Afric plants used for medicinal purposes. *J Ethnopharmacol*; 56: 81-7.

Salvet, A., Antonnacci, L., Fortunado, R.H., Suarez, E.Y. and Godoy, 2001). Screening of some plants from northern Argentina for th anti-microbial activity. *Journal of Ethnopharmacology*; 32: 293-297.

Tortora, G.J., Funke, B.R. and Case, C.L. (2001). *Microbiology: An Introduction.* San Francisco: Benjamin Cummings.

Tanaka, H., Sato, M. and Fujiwara, S. (2002). Antibacterial activity of isoflavonoidsisolatedfromErythrinavariegataagainstmethicillinresistant *Staphylococcus aureus. Lett Appl Microbiol* 35: 494-498.

Vlietinck, A.J., Van Hoof, L. and Tott, J. (1995). Screening of hundred Rwandese medicinal plants for antimicrobial and antiviral properties. *J. Ethnopaharmacol* 46: 31-47.

WHO (2002). *Traditional Medicine Strategy* 2002-2005. WHO Publications. 1-6.

Biodiversity Conservation and Envir. Management (2012)
Editors: D.R. Khanna et al.
Pub. by Biotech Books. *ISBN: 978-81-7622-262-4*

Pages: 157-172

MITIGATION OF UV-B INDUCED DELETERIOUS EFFECTS ON CHLOROPHYLL CONTENTS OF *BRASSICA CAMPESTRIS* PT-303 (BROWN SARSON) BY SOME PLANT GROWTH REGULATORS

Sanjeev Lal[1], G.K. Dhingra[2✉], Praveen Kumar[1], Ram Das[2], Amrita Gupta[2], Swati Kuriyal[2] and Rajeev Gautam[1]

[1]Uttaranchal College of Science & Technology, Dehradun, Uttarakhand (INDIA)

[2]Department of Botany, R.C.U. Govt. P.G. College, Uttarkashi,Uttarakhand (INDIA)

During the past decade, reduction in the stratospheric ozone layer due to accumulation of green-house gases *viz.* anthropogenic chlorofluorocarbons (CFCs), carbon dioxide (CO_2), methane (CH_4) & nitrous oxide (N_2O)

✉ Corresponding author

has resulted in an increase of UV-radiation, reaching on the earth's surface and may lead to global warming. Ozone (O_3) depletion by anthropogenic gases has increased the atmospheric transmission of solar UV-B (280-315 nm). These undergo sunlight and induced photochemical changes and could alter natural balance of creative and destructive process and in stratosphere. Every CFC molecule atom can destroy ozone (O_3) molecules. Taking this in view the present study was conducted and it was observed during course of study that IAA was found most effective in (10^{-7} M), Kn in (10^{-5} M) and $GA_{3\ in}$ (10^{-6} M) in *Brassica campestris* PT-303 respectively. Therefore, for the field studies only these concentrations were taken to assess for the chlorophyll estimation.

Keywords: *Brassica campestris* PT-303, Chlorophyll a, b and Protochlorophyll, IAA, Kn, GA_3 and UV-B exposure.

Introduction

UV-radiation may cause adverse effect on biological system & penetrate to a depth of 10% of euphotic zone, but such penetration may reach to a depth of 23 m or greater in clear ocean water. UV-B radiation is considered to be a lethal factor in aquatic system even for submerged organisms. It is known to affect a wide range of functional aspects including genetic variation, cytological, biochemical and physiological, behavioral (motility) and ecological system in photosynthetic organisms. The photosynthesis is more sensitive to UV-B in phytoplankton than in terrestrial plants, probably owing to less effective screening in phytoplankton. Growth of terrestrial plants reduction has been observed and may increase in magnitude over successive years. Aquatic productivity is often compromised by short-term exposures to enhanced UV-B radiation and long-term assessments are complicated by dynamic-nature of aquatic systems and by non-linear responses. The effects of UV-B radiation on multiple tropics levels suggest that the outcomes will be diverse and difficult to predict (Thomas and Patrick, 2002).

Dynamic changes in plant populations and communities are a product of intra- and inter-specific competition for resources needed for growth and productivity of competing species, as influenced by abiotic and biotic environmental factors (Treshow, 1968). Atmospheric concentrations of chemicals, which cause ozone depletion have peaked in the late 1990s and others are expected to peak in the early years of the 21st century. Ozone depletion is predicted to reach its peak about 15 years later than the peak halogen loadings because of coupling between stratospheric climate change and ozone chemistry. The maximum amount of depletion and its timing is uncertain due to complexity of this interaction. However, the future abundance of ozone will be influenced by changes in other atmospheric gases and by interactions with the climate system. Climatic change producing warmer weather and reductions in the capacity of ozone layer to reduce UV-penetration to the earth's surface.

Disturbance in the thermal structure of atmosphere probably causes changes in atmospheric circulation. The life time of ozone (O_3) destroying substance is very long and they may continue to deplete the ozone layer long. Although the use of most CFCs molecules has been passed out, ozone depletion is currently near or at its minimum (McKenzie *et al.*, 2003). The prospects of increased solar UV-B radiation as a result of stratospheric ozone depletion have attracted a great deal of attention. One of the reasons is that clouds have a substantial effect on the amount of UV-B reaching on the earth surface. Summer cloud cover decreases and an increase in UV-B can be expected, with a decrease in cloud cover of 4% likely to be associated with ~2% increase in ambient UV-B levels. Patterns of environmental changes in biosphere include concurrent and sequential combinations of increasing CO_2 levels; long term changes are resulting mainly from stratospheric O_3 depletion, greater troposphere O_3 photochemical synthesis and increasing CO_2 emissions. Effects of selected combinations were evaluated in tomato (*Lycopersicum esculentum*) seedling, using sequential exposures to enhanced UV-B radiation and O_3 in differential CO_2 concentrations. UV-B exposure, increased leaf chlorophyll and UV-absorbing compounds, but decreased leaf area and root/shoot ratio (Hao *et al.*, 2000).

Plant species are frequently contradictory, when comparing the effects of UV-B exposures in growth chambers or green-houses to those under field conditions (Dumpert and Knacker, 1985). Such differences may well arise not only from differences in microclimatic radiant and heat energy budgets extent in different exposure systems at the time of exposure, but also because of differences induced by environmental conditions. In these plants the cuticle may act as a barrier to UV-B (Steinmüller and Tevini, 1985) and green-house grown plants are known to have a much thinner and less well-developed cuticle than field-grown plants (Martin and Juniper, 1970) and thus might exhibit greater sensitivity.

Enhanced UV-B radiation can deleteriously effect overall growth and biomass accumulation in the plant species (Tevini, 2000). The plants contain large number of UV-B sensitive targets as nucleic acids, lipids, proteins and quinines (Jordan, 1996), which must be protected to ensure the normal growth and development of plants. Failure to do so may lead to alternations in all over morphology and physiology of many plants exposed to UV-B. Therefore, UV-B exposure is one of the major factors, which is responsible for the low productivity of crop plants and natural vegetation, so has become an increasing threat for agriculture.

Effects of UV-B radiation on plants observed by Demchik & Day (1996) and Smith *et al.*, (2000), green-house and growth chamber studied number of crop plants exposed to an enhanced UV-B radiation to determine their susceptibility to photosynthetic impairment. As these species were included as "sensitive" *viz.* pea (*Pisum sativum*), mustard (*Brassica*), soyabean (*Glycine max*) and oat (*Avena sativa*) & "moderately sensitive" tomato (*Lycopersicum esculentum*), sorghum (*Sorghum bicolar*), rye *(Secale cereale)*, rice (*Oryza sativa*), tolerant corn (*Zea mays*), pearl millet (*Pennisuetum americium*), pea nut (*Arachis hypogaea*) in respect to their sensitivity to UV-B radiation (Van *et al.*, 1976).

The general principal in study to determine, deleterious effects of UV-B on crop plants that involve the use of UV-source (lamp) coupled with different types of filters to exclude bands of UV-wavelength. The effects of UV-B radiation on crop plants, includes reduction in yield and quality, alteration in species competition, photosynthetic activity, susceptibility to disease and changes in plant structure and pigmentation (Tevini and Teramura, 1989). The physiological properties of plants such as damage to Photosystem-II (Heinrich *et al.*, (1999) and reduction in the photosynthetic rate (Basseman *et al.*, 2003 and Feng *et al.*, 2003) increased, activity of antioxidant enzymes such as catalase and ascorbate peroxides (APX) observed by Kim *et al.*, (1996) and Mazza *et al.*, (1999). UV-B radiation damage to genetic material *viz.* DNA, RNA has also been reported by Hidema *et al.*, (2000).

The UV-B radiation resulted in a decrease of adaxial stomatal conductance by approximately 65% increasing stomatal limitation of CO_2 uptake by 10 to 15%. The growth in UV-B radiation resulted in large reduction of leaf area and plant biomass, which were associated with a decline in leaf cell numbers, cell divisions and also inhibited epidermal cell expansion of exposed surface of leaves. Photo-repair activity of DNA is enhanced and synthesis of UV-absorbing compounds increased. Most of these responses are thought to play some role in mitigating the hazardous or deleterious impacts of UV-B radiation (Bilger *et al.*, 2001 and Cerovic *et al.*, 2002).

It has been well established that, plant growth regulators (PGRs) such as IAA, Kn and GA_3, influence the growth and development of plants. These chemical substances are able to coordinate growth among different plant parts or different physiological and biochemical processes. Plant growth regulators have been tried to improve growth and ultimately yield (Ram *et al.*, 1973; Patil *et al.*, 1987 and Kumar *et al.*, 1996), by the foliar application of some plant growth regulators (Chhonkar and Jha, 1963). IAA, Kn and GA_3, which are most important growth regulators, has a profound effect on crop production, through increase in the stem length, leaf area, flower induction, shelling percentage, yield and weight of crops.

Materials and Method

Laboratory and field experiments were conducted in R.C.U Govt. P.G. College Uttarkashi. Geographically, the District Uttarkashi is located between the central Himalayan region at 30° 28' to 31° 28'N latitude and 77° 49' to 79° 25'E longitude at an altitude of 1140 m above mean sea level. The seeds of *B. campestris* PT-303 were procured from Seed centre of G.B. Pant University of Agriculture and Technology Pantnagar (Uttarakhand) for the research study.

General Experimental Design: During laboratory studies, following sets were taken into consideration:

(A) Control: Seeds of mustard crop (*Brassica campestris* PT-303) were soaked for 24 hrs. in distilled water and placed on moistened filter paper in Petridishes.

(B) UV-B: UV-B radiation was supplied for 3-hrs daily by sunlamps (300W) filtered with quartz interference filters (320 NM, ORIEL, USA).

(C) Growth Regulators: Test solution of IAA, Kn and GA_3 were prepared in three concentrations *viz.* 10^{-7}, 10^{-5} & 10^{-6} M (molarities) in *B. campestris* PT-303. Seeds of *B. campestris* PT-303 were soaked for 24 hrs in different concentrations of growth regulators, soaked seeds were placed in paired petridishes lined with moistened filter paper. One set of petridish containing soaked seeds was allowed to grow without any UV-B exposure.

(D) Growth Regulators + UV-B: In second set, one from each concentration of different growth regulators was sprayed with UV-B radiation, for 3-hrs daily.

Treatments	Control	UV-B	IAA			Kn			GA_3			IAA+ UV-B			Kn + UV-B			GA_3+ UV-B		
Concentration		(3-hrs)	10^{-7}	10^{-6}	10^{-5}	10^{-7}	10^{-6}	10^{-5}	10^{-7}	10^{-6}	10^{-5}	10^{-7}	10^{-6}	10^{-5}	10^{-7}	10^{-6}	10^{-5}	10^{-7}	10^{-6}	10^{-5}

Field study: During field study, mustard crop of *B. campestris* PT-303 were grown in field and the plots were divided by black paper sheets into five blocks. Each field block was given treatments as follows:

Treatments of field plots:

1. In plot-A, mustard plant (*B. campestris PT-303*) species was taken as control. No treatments was given to crop of this plot.
2. Plot-B was exposed to 3-hrs daily UV-B radiation (24.23 Jm^{-2} Z^{-1}) by Sunlamps (300W) filtered with quartz interference filters (320 nm, ORIEL, USA).
3. Plot-C was sprayed with IAA (10^{-7} M) concentration daily, along with 3-hrs supplemental UV-B radiation using the same source.
4. Plot-D was sprayed with Kn (10^{-5} M) concentration daily, along with 3-hrs. supplemental UV-B radiation by using the same source.
5. Plot-E was sprayed along with GA_3 (10^{-6} M) along with 3-hrs. supplemental UV-B radiation, using the same source as above.

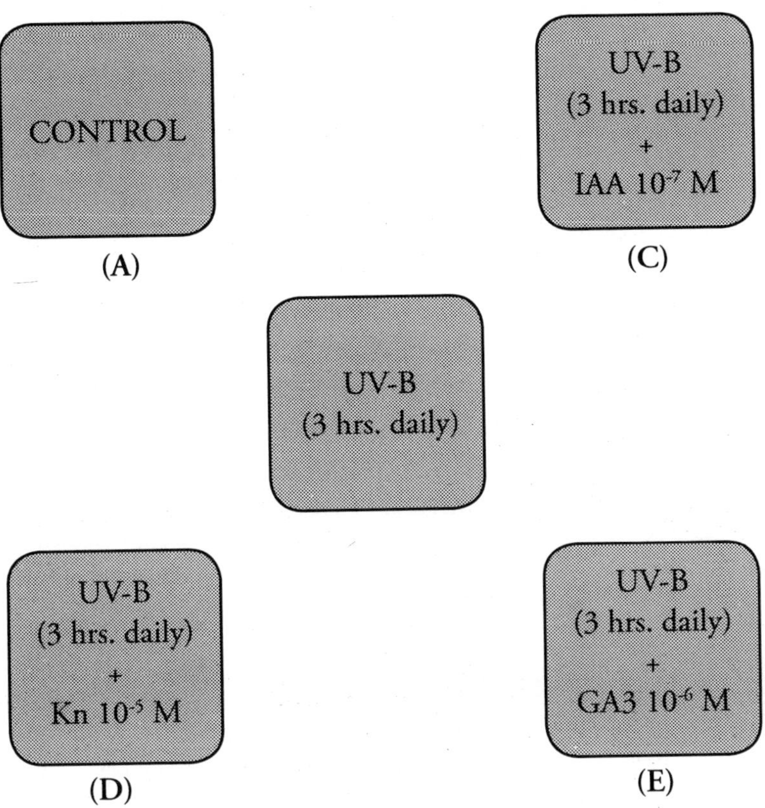

Fig.1.1: Experimental design in the field plot.

Results

For Chlorophyll analysis during seedling growth in laboratory, the surface sterilized seeds of mustard crop *B. campestris* PT-303 were imbibed in water for 6-hrs and then seeds were washed by distilled water and transferred to petridishes, for seed germination and seedling growth studies and were exposed with UV-B radiation (3-hrs. daily), alone and along with various concentration of plant growth regulators. Chlorophyll a, b and protochlorophyll were measured after 7 days of growth in different treatments.

Data showed in Table 1.1 revealed that, chlorophyll a (mg/pl), chlorophyll b (mg/pl), protochlorophyll (mg/pl) and ratio of chlorophyll a and b in control set were 0.45±0.05, 0.39±0.03, 0.50±0.07 and 1.15 respectively. When, germinating seedling growth was studied with UV-B radiation only, marked decline in contents of various chlorophyll pigments.

The inhibition was observed as ca. 23%, 17%, 6% & 8% of chlorophyll a, chlorophyll b, protochlorophyll and a/b ratio respectively as compared to control. When sets C, D and E were observed with (PGRs+UV-B), a general promotion was recorded in these pigments as compared to individual treatment of UV-B radiation (Set-B) only. The IAA was observed to be most effective PGR to counteract the UV-B induced deleterious impacts for all studied chlorophyll pigments. The result observed as ca. 14.7%, 12.1%, 10% and 3.3% increase over the UV-B treatment (Set-B) in chlorophyll a, b, protochlorophyll and a/b ratio respectively.

Table 1.1: Chlorophyll content at the seedling stage after 7-days of germination as affected by UV-B radiation (3 hrs daily), individually and in combination of IAA, Kn and GA_3 in *Brassica campestris* PT-303.

Treatments	*Chlorophyll a*	*Chlorophyll b*	*Protochlorophyll*	*a/b ratio*
A	0.45±0.05	0.39±0.03	0.50±0.07	1.15
B	0.38±0.02	0.30±0.01	0.47±0.04	0.78
C	0.39±0.03	0.33±0.02	0.49±0.03	0.84
D	0.34±0.07	0.35±0.03	0.48±0.02	0.97
E	0.39±0.03	0.34±0.01	0.50±0.06	1.14

Effects of UV-B exposure alone and along with some plant growth regulators on chlorophyll development during crop growth were also observed in same variety of mustard crop, which were grown for growth patterns studies. For the chlorophyll estimation, plants were selected regularly at 15 days interval from the seedling emergence upto maturity. Data presented in Table 1.2 and Fig. 1.2 (a), (b) & (c) showed that in control plot (A), UV-B plot (B), alone and along with different plant growth regulators (PGRs) affected the different chlorophyll pigments in *B. campestris* PT-303.

In plot-A (control), the values of chlorophyll a, chlorophyll b, protochlorohyll and chlorophyll a/b ratio were observed at 15 days interval of crop growth recorded as **0.35**±0.03, **0.39**±0.05, **0.40**±0.03 mg/pl and noticed a consistent increase up to 135th day stage of crop growth and amounted **1.12**±0.10, **0.94**±0.07, **1.11**±0.11 mg/pl and 1.191 for chlorophyll a, chlorophyll b, protochlorophyll and chlorophyll a/b ratio respectively.

The plants of plot-B (UV-B alone) were examined and noticed a reduction in the chlorophyll pigment as compared to control plot (A). The maximum inhibition of chlorophyll a, chlorophyll b, protochlorophyll and chlorophyll a/b ratio were recorded at 30th day stage and 135th day stage and reduced by ca. 20%, 17%, 18%, 5%; ca. 6%, 7%, 9% and 10% respectively, as compared to control (plot-A). When the plot C, D and E was studied with UV-B exposure along with PGRs, a general promotion were recorded in content of chlorophyll a, chlorophyll b, protochlorophyll and chlorophyll a/b ratio. Plot-C, showed the maximum amount of chlorophyll content as chlorophyll a, chlorophyll b and protochlorophyll at the 15th day stage of crop growth and a/b ratio at maturity stage of crop growth and estimated as ca. 19%, 45%, 5% and 20% respectively, as compared to UV-B exposure alone. The plot-D showed maximum amount of chlorophyll a, chlorophyll b, protochlorophyll and a/b ratio observed at the 60th day and increased by ca. 6%, 9%, 6%, 4%; at 90th day ca. 7%, 5%, 6%, 7% and at maturity ca. 6%, 8%, 17%, 11% respectively, as compared to UV-B treatment. The plot-E also showed the maximum promotion of chlorophyll-a, chlorophyll-b, protochlorophyll contents and chlorophyll a/b ratio noticed at the 30th day stage of growth and promoted by ca. 15%, 10%, 7%, 7%; at the 90th day ca.9%, 5%, 6%, 4% and at maturity as ca. 4%, 10%, 16%, 10% respectively, as compared to the over UV-B exposure.

Table 1.2: Chlorophyll contents as affected by UV-B radiation (3 hrs. daily), individually and in combination of IAA, Kn and GA_3 in field grown crop of *Brassica campestris* PT-303.

Treatments	*Chlorophyll*	*15*	*30*	*45*	*60*	*75*	*90*	*105*	*120*	*135*
	Chlorophyll a	0.35 ±0.03	0.47 ±0.03	0.058 ±0.04	0.065 ±0.07	0.097 ±0.08	1.04 ±0.04	1.06 ±0.06	1.09 ±0.08	1.12 ±0.10
A	Chlorophyll b	0.39 ±0.05	0.53 ±0.04	0.063 ±0.03	0.072 ±0.08	0.088 ±0.05	0.97 ±0.04	0.94 ±0.04	0.84 ±0.05	0.94 ±0.07
	Protochlorophyll	0.40 ±0.03	0.50 ±0.02	0.068 ±0.04	0.68 ±0.08	0.83 ±0.05	0.086 ±0.07	0.91 ±0.01	0.99 ±0.01	1.11 ±0.11
	a/b ratio	0.89	8.87	0.92	0.902	1.102	1.072	1.127	1.297	1.191
	Chlorophyll a	0.27 ±0.05	0.38 ±0.05	0.48 ±0.03	0.060 ±0.01	0.83 ±0.02	0.90 ±0.05	1.01 ±0.05	1.09 ±0.06	1.06 ±0.10
B	Chlorophyll b	0.24 ±0.03	0.44 ±0.04	0.53 ±0.05	0.68 ±0.03	0.75 ±0.04	0.84 ±0.04	0.89 ±0.02	0.072 ±0.03	0.89 ±0.02
	Protochlorophyll	0.38 ±0.04	0.46 ±0.06	0.58 ±0.06	0.61 ±0.03	0.72 ±0.03	0.78 ±0.05	0.81 ±0.04	0.74 ±0.03	0.91 ±0.04
	a/b ratio	1.125	0.86	0.905	0.882	1.106	1.114	1.134	1.513	1.235
	Chlorophyll a	0.32 ±0.02	0.42 ±0.09	0.52 ±0.06	0.62 ±0.03	0.89 ±0.04	0.99 ±0.09	1.04 ±0.11	1.05 ±0.05	1.09 ±0.09
C	Chlorophyll b	0.35 ±0.03	0.46 ±0.06	0.60 ±0.05	0.70 ±0.07	0.85 ±0.08	0.91 ±0.01	0.90 ±0.08	0.89 ±0.09	0.99 ±0.07
	Protochlorophyll	0.39 ±0.05	0.50 ±0.03	0.65 ±0.04	0.63 ±0.04	0.78 ±0.06	0.82 ±0.08	0.88 ±0.07	1.02 ±0.01	1.09 ±0.09
	a/b ratio	0.914	0.913	0.866	0.885	1.04	1.08	1.15	1.17	1.10
	Chlorophyll a	0.34 ±0.03	0.44 ±0.04	0.51 ±0.05	0.63 ±0.04	0.87 ±0.3	0.97 ±0.07	1.03 ±0.13	1.04 ±0.11	1.08 ±0.01
D	Chlorophyll b	0.36 ±0.04	0.47 ±0.07	0.59 ±0.04	0.69 ±0.09	0.83 ±0.08	0.90 ±0.09	0.89 ±0.09	0.88 ±0.08	0.97 ±0.06
	Protochlorophyll	0.38 ±0.06	0.49 ±0.08	0.64 ±0.03	0.62 ±0.02	0.76 ±0.05	0.81 ±0.01	0.88 ±0.08	0.91 ±0.99	1.07 ±0.06
	a/b ratio	0.944	0.936	0.864	0.913	1.04	1.07	1.15	1.18	1.11
	Chlorophyll a	0.33 ±0.03	0.44 ±0.04	0.50 ±0.05	0.62 ±0.09	0.88 ±0.08	0.98 ±0.09	1.02 ±0.12	1.05 ±0.22	1.09 ±0.02
E	Chlorophyll b	0.36 ±0.06	0.48 ±0.07	0.61 ±0.01	0.68 ±0.07	0.84 ±0.04	0.91 ±0.01	0.90 ±0.09	0.89 ±0.21	0.98 ±0.07
	Protochlorophyll	0.37 ±0.07	0.49 ±0.08	0.63 ±0.03	0.61 ±0.06	0.75 ±0.05	0.82 ±0.02	0.87 ±0.07	0.92 ±0.03	1.06 ±0.06
	a/b ratio	0.91	0.92	0.81	0.91	1.04	1.07	1.13	1.17	

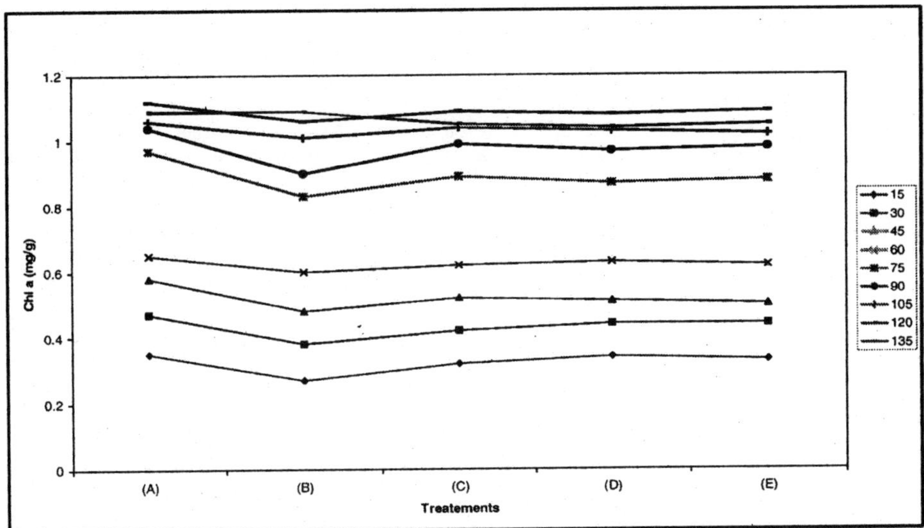

Fig.1.2(a): Chlorophyll a (mg/g) of *Brassica campestris* PT-303 as affected by different treatments.

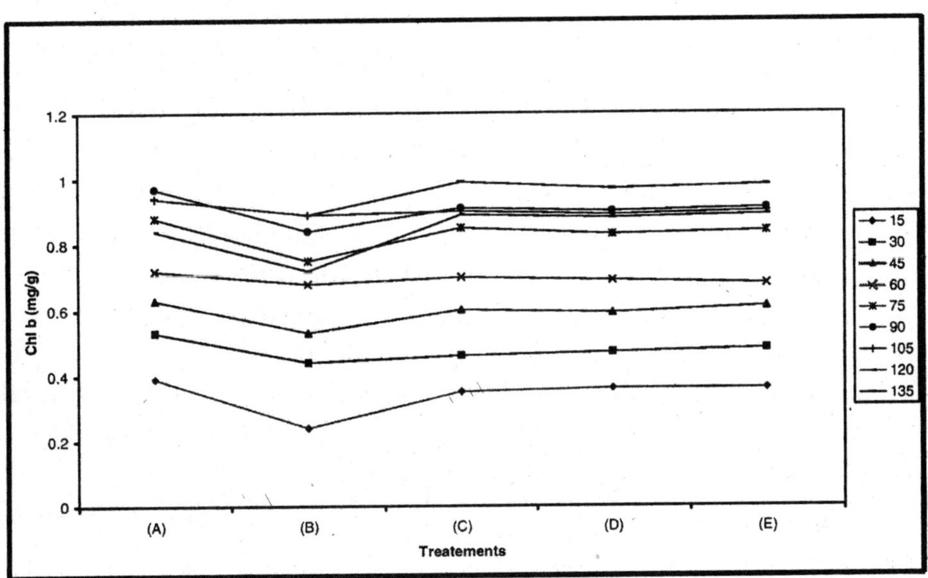

Fig.1.2(b): Chlorophyll b (mg/g) of *Brassica campestris* PT-303 as affected by different treatments.

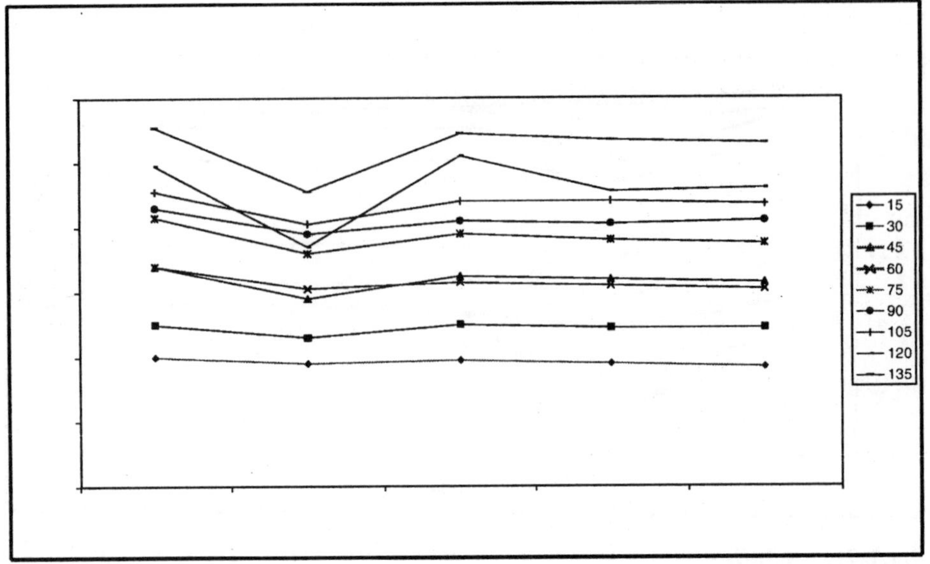

Fig.1.2(c): Protochlorophyll (mg/g) of *Brassica campestris* PT-303 as affected by different treatments.

Discussion

The present study was carried out in the laboratory and field. The destruction of chlorophyll a, b, protochlorophyll and chl a/b ratio were noticed, when mustard crop was treated with UV-B radiation. In this crop, chlorophyll a and chlorophyll b were found almost equally reduced due to 3-hrs daily treatment of UV-B exposure. When the crop was supplemented with PGRs in addition to UV-B radiation, a promotory effect was noted. IAA and GA_3 were found most promising growth regulators, when compared with Kn. Significant reductions in different chlorophyll pigment by UV-B exposure were also reviewed by Jain and Goyal (1990), Sharma *et al.*, (1988) and Ambrish (1992).

Chlorophyll content was also analyzed in field grown crops under the influence of various treatments. In general, it was noticed that UV-B inhibits the chlorophyll development throughout the crop age. However, more reduction was recorded in early stages of growth and at maturity. Kn, when applied with UV-B radiation, it was found to enhance the different chlorophyll pigments in *B. campestris* PT-303, however the other PGRs also mitigate the adverse effects of UV-B, marginally. These findings showed that lethal effects of UV-B towards chlorophyll development and repaired by Kn (10^{-5} M). This effect was found variable with the crop species. Vu *et al.* (1981) reported that chlorophyll a/b ratio decreased due to UV-B radiation in soybean but increased in pea. Tevini *et al.* (1981) concluded that UV-B radiation inhibited the biosynthesis of chlorophyll b more than chlorophyll a. Jain and Goyal (1990), while working with lentil crop under field conditions, reported similar results. They also emphasized that interconversion of protochlorophyll to chlorophyll was retarded. As Kn was found to improve the synthesis of chlorophyll even under increased radiation energy (Purohit, 1988), an improvement in different chlorophyll contents was reported in the present study under similar conditions.

Conclusion

As noted, individual treatment of UV-B radiation in *B. campestris* PT-303, inhibited the chlorophyll pigments *viz.* chlorophyll a chlorophyll b, protochlorophyll and chlorophyll a/b ratio. When treated with IAA, Kn and GA_3, showed promotion or enhancement in all chlorophyll pigments. Therefore, plant growth hormones (PGRs), would show the maximum mitigation against UV-B induced deleterious or hazardous effects on chlorophyll a chlorophyll b, protochlorophyll and chlorophyll a/b ratio.

Acknowledgement

We are thankful to Dr. Hema Prasad, Principal R.C.U. Govt. P.G.College, Uttarkashi for support and facilities provided for this research work. We are also highly thankful to UCOST and State Biotech Program for indirect support through Dr. Dhingra, District coordinator UCOST and Nodal Person Plant Tissue Culture lab for extending lab facilities and financial help.

References

Basseman, J.H, G.E. Edwards and R. Robberecht (2003). Photosynthesis and growth in seedlings of five forest tree species with contrasting leaf anatomy subjected to supplemental UV-B radiation. For sci 49: 176-187.

Chhonkar, V.S. and Jha, R.N. (1963). The use of plant growth regulators in transplanting of cabbage and their response on growths yield. Indian J. Hort. 20:123-128.

Demchik, S.M. and Day, T.A. (1996). Effects of enhanced UV-B radiation on pollen quantity, quality & seed yield in Brassica rapa (Brassicaceae). American journal of Botany 93: 573-579.

Dumpert, K. and Knacker, T (1985). A comparison of the effects of enhanced UV-B radiation on some crop plants exposed to green-house and field conditions. Biochem. Physiology. 180 (8): 599-612.

Feng, H.L and Chen. W. Qiang. S. X.V. M. Xiao, X. Wang and G. Cheng (2003). The effects of enhanced UV-B radiation on growth, photosynthesis and stable carbon isotope composition of two soybean cultivars (Glycine Max) under field conditions. Environ Exp. Bot. 49: 1-8.

Hao, X., B.A. Hale, D.P. Ormrod, A.P. Papadopoulos, (2000). Effects of pre-exposure to UV-B radiation on responses of tomato (Lycopersicum esculentum) to ozone in ambient and elevated CO2. Environ Pollution. 110: 217-224.

Heinrich, G.H.K., C. Schmude, H. Garden O.Y. Koroleva and K. Winter (1999). Effects of solar UV-B radiation on the potential efficiency of photosystem-II in leaves of tropical plants. Plant physiol. 121: 1349-1358.

Hidema, J.T. Kumagai and B.M. Suterland (2000). UV-radiation sensitive Norin-1 Rice contains defective cyclobutane pyrimidine dimmer photolyase plant cell 12: 1569-1578.

Jain, V. K. and A. K. Goyal, (1990). Chlorophyll development response to supplemented UV–B in Lentil crop under field conditions. Symposium on Photophysiology and Photomedicine, New Delhi.

Jordon B.R (1996). The effects of ultraviolet-B radiation on plants: a molecular perspective. Advances in Botanical Research 22: 97-161.

Kim, H.Y., K. Kobayashi and T. Yoneyama. (1996). Differential influences of UV-B radiation on antioxidant and related enzymes between rice

(*Oryza sativa*) and cucumber (*Cucumis sativus*) leaves. Environ. Sci. 9: 55-63.

Kumar, D.K.D, Paliwar, R and Kumar, D., (1996). Yield and yield attributes of cabbage as influenced by GA and NAA Crop. Res. Hisar, 12 (i):120-122.

Martin, J.T. and Juniper, B.E. (1970). The cuticles of Plants. St. Martins, New York. 347 p.

Mazza, C.A, D. Battista A.M. Zima, and C.L. Ballare (1999). The effect of UV-B radiation on the increased DNA damage and antioxidant responses plant cell Environ. 22: 61-70.

McKenzie, R.L., B. jorn. L.O. and Bais A, (2003). Change in biologically active UV-radiation reaching the earth surface. Ohotochem. Photobiol. Sci. 2: 5-15.

Patil, A.A., Manipur, S.M. and Nalwadi, U.G., (1987). Effects of GA and NAA on growth and yield of cabbage. South Indian Hort., 35: 393-394.

Purohit, S. S., (1988). Hormonal regulation of plant growth and development. Agro Botanical Publishers (India), Vol. IVth.

Ram, K., Varma, A.N. and Sharma, P.K., (1973). Effects of NAA on growth, protein and ascorbic acid content of cabbage. Plant science. 5: 150-153.

Smith, J.L., D.J. Burrit & P. Bannister, (2000). Shoot dry weight, chlorophyll and UV-B absorbing compounds as indicators of plant sensitivity to UV-B radiation. Annals of botany 86: 1057-1063.

Steinmüller, D. and Tevini, M. (1985). Action of ultraviolet radiation (UV-B) upon cuticular waxes in some crop plants. Planta (Berl) 164 (4): 557-564.

T.K. Van, L.A. Garrand and S.H. West., (1976). Affects of UV-B radiation on net Photosynthesis of some crop plants 16: 715-718.

Tevini M. and A.H. Teramura, (1989). UV-B effects on terrestrial plants Photochem. Photobiology. 50: 479-487.

Tevini, M., (2000). UV-B Effects on plants. PP 83-97, In: S.B. Agarwl & M. Agarwal. Environmental pollution and plant responses, Lewis Publishers, Boca Ratio. USA.

Tevini, M., W. Iwanzik and U. Thomas, (1981). Some effects of enhanced UV–B irradiation on the growth and composition of plants. Planta 153: 388–394.

Thomas A. Day and Patrick J. Beale., (2002). Effects of UV-B radiation on terrestrial and aquatic primary producers Vol. 33: 371-396.

Treshow, M. (1968). The impact of air pollutants on plant populations. Phytopathology 58: 1108-1113.

Vu. C.V., Allen. L.H. and Garrard, L.A., (1981). Effect of supplementary UV-B radiation on growth and photosynthesis of soybean. Physiol. Plant. 52: 353-362.

Biodiversity Conservation and Envir. Management (2012)
Editors: D.R. Khanna et al.
Pub. by Biotech Books. *ISBN: 978-81-7622-262-4*

Pages: 173-182

TRADITIONAL USE OF SOME FERN AND FERN-ALLIES IN TARAI AND BHAWAR OF KUMAUN OUTER HIMALAYA, UTTARAKHAND, INDIA

Bhasker Joshi[1⊠], Pramod Kumar[1] and S. C. Pant[2]
[1]Department of Botany, R. H. Govt. P. G. College, Kashipur (Kumaun University, Nainital), Uttarakhand, (INDIA)
[2]Govt. Degree College Gairsain (H. N. B. University, Garhwal) Chamoli, Uttarakhand (INDIA)
⊠E-mail: bhaskerjoshiphd@yahoo.com

Ayurveda *i.e.* Charak Samhita (1000 BC- 400 AD) is perhaps the world's oldest medical system with well documented medicinal uses of plants. It is often defined as the "science of life and longevity". The tribal societies of different regions with long history and traditions have developed their own indigenous medicinal systems. They used plant and their parts for curing diseases. Indian Himalayan region is as

⊠ Corresponding author

considered one of the richest sources of medicinal plants and its tribal societies have developed indigenous medicinal systems, which have built in ecosystems of conservation and utilization of natural resources. In present chapter, the traditional uses of 14 pteridophytes (ferns and fern allies) in the treatments of various diseases are described, with botanical name, family, plant parts and mode of ethnomedicinal uses. Local communities, especially, older age class, including women heavily use these traditionally available ferns and fern allies for health and they have no side effects.

Keywords: Ethnomedicinal, Fern, Fern allies, Kumaun Himalaya, Medicinal.

Introduction

Man from ancient time to the present day has used the plants as a source of medicine. Indian traditional medicine is based on different systems such as Ayurveda, Siddha and Unani used by various tribal communities (Gadgil, 1996). Indian Himalayan region is one of the best habitats for medicinal plants in the globe, extending from 27°50'-37°06' N latitude and 72°30'-97°25' E longitude is one of the major repositories of biodiversity including medicinal plants (Samant and Palni, 2010). The pteridophytic flora of world comprises about 9800 species (Mebberley, 1997) and the pteridophytic flora of India comprises about 125 genera with 800-1000 species (Dixit, 1984; Bir, 1992). Chowdhary (1973) published an account of pteridophytes from Upper Gangetic plains, which include parts of Uttarakhand, plains of Uttar Pradesh, Bihar and part of West Bengal. Khullar (1994, 2000) in his illustrated fern flora of Western Himalaya included 360 species of ferns. Pande and Pande (2002) reported 350 species of ferns and fern allies from Kumaun Himalayas. Dixit and Kumar (2002) listed 487 species and 32 intra specific taxa belonging to 108 genera under 50 families.

Medicinal value of pteridophytes is known to man for more than 2000 years. Theophrastus (327-287 B.C.) and Dioscorides (100 A.D.) have referred to medicinal attributes of certain ferns. Sushruta and Charaka (100 A.D.) mentioned medicinal uses of *Marsilea minuta* L. and *Adiantum capillus-veneris* L. in their Samhitas (Chandra, 2000; Parihar *et al.*, 2004). Singh *et al.* (2005) conducted broad study of ethnomedicinal uses of eight pteridophytes of Amarkantak. Parihar and Parihar (2006) studied medicinal importance of 16 pteridophytes at Rajasthan. Benjamin and Manickam (2007) studied 61 medicinal pteridophytes of Western Ghats. Rout *et al.* (2009) studied medicinal uses of 33 pteridophytes species belonging to 21 families in Similipal Biosphere Reserve of Orissa. Upreti *et al.* (2009) studied ethnomedicinal uses of 30 pteridophytes of Kumaun Himalaya. In present time due to excessive dependence of human beings on allopathic medicine, the traditional uses of herbal plants has become dormant. However, in rural and tribal areas of India local natives use these plants for medicinal and other economical purposes. This study documents the traditional use of fern and fern allies in Tarai and Bhawar of Kumaun, Uttarakhand.

Study Area

The local communities carried out the present study in some interior areas of districts of Nainital and Udham Singh Nagar of Uttarakhand to collect the information of traditionally used ferns and fern allies. The study site was situated in Tarai and Bhawar of Kumaun adjacent to Kashipur, at (29° 14-43.6)–(29° 19-50.5) E longitude and (79° 03-22.6)–(79° 04-23.2) N latitude at an elevation of 253.4–265.5 meter above sea level.

Materials and Method

In present study, the information about plants was obtained by frequent field visits, from experience of personals of forest department and the local natives (older household and women). Plant species were indentified with the help taxonomists and from literature given by Khullar (1994, 2000), Khullar *et al.* (1991) and Pande and Pande (2002).

Results

All known 14 species (Fig. 1 to 14) of Pteridophytes were encountered in Kashipur, of Kumaun Outer Himalaya. Botanical names, family name, mode of use and plant parts used are given below:

***Fig. 1: Adiantum capillus-veneris* L.**

***Fig. 2: Adiantum incisum* Forsk.**

***Fig. 3: Adiantum philippense* L.**

***Fig. 4: Ceratopteris thalictroides* Brong**

***Fig. 5: Cheilanthes albomarginata* C.B. Clarke**

***Fig. 6: Diplazium esculentum* (Retz.) Sac.**

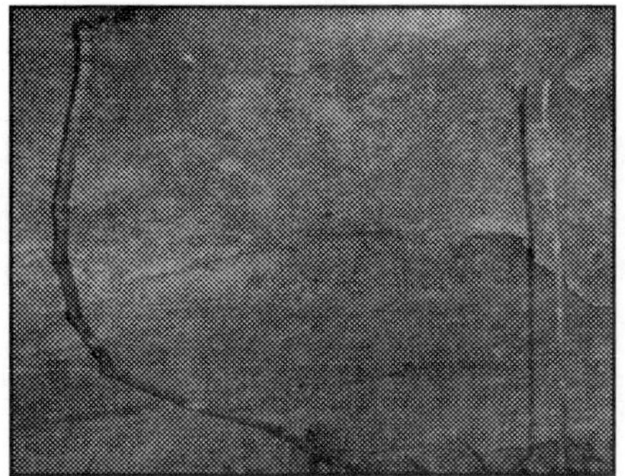

***Fig. 7: Equisetum diffusum* D.Don.**

***Fig. 8: Helminthostachys zeylanica* (L.) Hook.**

***Fig. 9: Lygodium flexuosum* (L.) Sw.**

***Fig. 10: Ophioglossum reticulatum* L.**

***Fig. 11: Pteris vittata* L.**

***Fig. 12: Selaginella subdiaphana* Spring**

***Fig. 13: Thelypteris dentata* (Forssk.) St. John**

***Fig. 14: Thelypteris prolifera* Retz.**

1. ***Adiantum capillus-veneris* L.**; Family: Adiantaceae; Vernacular Name: Maiden hair fern, Hansraj, Hanspadi and Dumtulli.

Use: *Adiantum capillus-veneris* L. is used as demulcent, expectorant, diuretic, tonic and febrifuge. Decoction of leaves is useful in curing of acute bronchitis and fever. Young fronds chewed for the treatment of mouth blisters.

2. ***Adiantum incisum* Forsk.**; Family: Adiantaceae; Vernacular Name: Hanspadi.

Use: The leaf powder of *Adiantum incisum* Forsk. is mixed with butter and used for controlling the internal burning of the body. Also, used in cough, diabetes, fever and skin diseases.

3. ***Adiantum philippense* L.**, Family: Adiantaceae; Vernacular Name: Kali-Jhant.

Use: *Adiantum philippense* L. is used as demulcent and astringent. It is used in treatment of cough, bronchitis, dysentery, epileptic fits, ulcer and erysipelas, asthma, fever, leprosy and hair falling.

4. ***Ceratopteris thalictroides* Brong.**; Family: Parkiaseae.

Use: Fronds of *Ceratopteris thalictroides* Brong. is used as poultice in skin complaints. Young fronds are used as green vegetables as salad and cooked vegetable.

5. ***Cheilanthes albomarginata* C.B.Clarke**; Family: Sinopteridaceae; Vernacular Name: Silver Fern.

Use: *Cheilanthes albomarginata* C.B.Clarke is used as a general tonic for children and weak people. The tribal people suffering from tuberculosis take the extract of the leaves mixed with honey after meal.

6. ***Diplazium esculentum* (Retz.) Sac.**; Family: Athyriaceae.

Use: Rhizomes of *Diplazium esculentum* (Retz.) Sac. used as an insect repellent. Young fronds are used as green vegetables as salad and cooked vegetable.

7. ***Equisetum diffusum* D. Don.**; Family: Equisetaceae; Vernacular Name: Horse Tail.

Use: Plant paste of *Equisetum diffusum* D.Don. is applied in bone fracture. The plant is diuretic, haemostatic and haemopoietic. Ashes of plants is useful in treatment of acidity of stomach and in dyspepsia.

8. ***Helminthostachys zeylanica* (L.) Hook.**; Family: Helminthostachyaceae; Vernacular Name: Kamraj.

Use: *Helminthostachys zeylanica* (L.) Hook. is used in curing of leucorrhea and worm infection in abdomen (anthelmintic). Also whole plant is considered as intoxicant, anodyne and used in sciatica. Fronds is used as aphrodisiac.

9. *Lygodium flexuosum* (L.) Sw.; Family: Lycopodiaceae; Vernacular Name: Kali jar.

Use: Extract of stems and rhizome is taken orally twice a day for a week for curing sexual diseases like gonorrhea and spermatorrhoea. Rhizome powder is used in skin diseases. Plants are used as expectorant, rheumatism, sprains, scabies, eczema and cut wounds. Fresh root is used in carbuncles and rheumatism diseases. The paste of fresh leaves is applied on the piles. Infusion of plant is used in treatment of menorrhagia.

10. *Ophioglossum reticulatum* L.; Family: Ophioglossaceae; Vernacular Name: Adder's tongue fern, Brahmi and Dhimraj.

Use: Paste of *Ophioglossum reticulatum* L. is applied on burns as a cooling agent. The extract of leaf is used as vulnerary and as remedy for wounds.

11. *Pteris vittata* L.; Family: Pteridaceae.

Use: The rhizome of *Pteris vittata* L. is used as an astringent, anthelmintic. Decoction of rhizomes and fronds given in chronic disorders of viscera and spleen, used to absorb *arsenic* from the soil. Plant extract is used as demulcent, hypotensive, tonic, antiviral and antibacterial.

12. *Selaginella subdiaphana* (Wall. ex Hook & Grev.) Spring; Family: Selaginellaceae; Vernacular Name: Sanjeevani and Spike Moss.

Use: Plant juice is used as an antibacterial agent in healing of wounds. Plant is used as diuretic and in gonorrhoea.

13. *Thelypteris dentata* (Forssk.) St. John; Family: Thelypteridaceae.

Use: Plant paste is used as an antiseptic agent in the treatment of wounds and cuts.

14. *Thelypteris prolifera* Retz.; Family: Thelypteridaceae.

Use: Plant paste is used as an antiseptic agent in the treatment of wounds and cuts.

Conclusions

India is also one of the leading countries in Asia in terms of the wealth of traditional knowledge systems related to the use of plants species. Of the total, 17643 species of angiosperm flora, nearly 8000 species are known for their medicinal value. So far, about 8000 species of angiosperms, 44 species of gymnosperms and 600 species of pteridophytes have been reported in the Indian Himalaya (Shiva, 1996; Singh and Hajara, 1996 and Singh *et al.* 2009). The Himalayas have a great wealth of medicinal plants and traditional medicinal knowledge. This region supports about 1,386 medicinal plant species, out of which 1,338 are used to treat human diseases and disorders and about 364 plant species are used for veterinary diseases by the people of Uttrakhand (Pande *et al.*, 2004). According to WHO approximately 80% of World population in developing countries depends on traditional medicines for primary healthcare (WHO, 2002) and in modern medicine too, nearly 25% are based on plant derived drugs (Tripathi, 2002). All plants are widely used by the local people of the Kumaun Himalaya for ethnomedicinal purpose. The study highlights that in absence of modern healthcare facility, people in the study area are dependent on plants for various medicinal purposes. This data may be useful for phytochemists and pharmacologists to determine their true therapeutic compounds. It may bring to light new sources of drugs of herbal origin because many medicinal plants are reported to be threatened to extinction. The present study was purposeful to understand traditional use of ferns and fern allies in Kashipur of Kumaun, Uttarakhand.

Acknowledgment

I gratefully thank Prof. Y. P. S. Pangtey, Kumaun University, Nainital for their help in identification in plants and Dr. Lalit Tewari, Kumaun University, Nainital for their valuable suggestions for this study. I am also thankful to forest department of Tarai West Forest Division, Ramnagar and local native of study area.

References

Benjamin and Manickam, V.S. (2007): Medicinal pteridophytes from the Western Ghats. *Indian Journal of Traditional Knowledge,* 6(4): 611-618.

Bir, S.S. (1992): Keynote address on ferns of India, their wealth exploitation, diversity, growth conditions and conservation. *Indian Fern J.,* 9: 4-6.

Chandra, S. (2000): Ferns of India. International Book Distributors, Dehradun, India.

Chowdhary, N.P. (1973): *The Pteridophytic flora of the Upper Gangetic Plain.* Navyug Traders, New Delhi.

Dixit, R.D. (1984): *A census of the Indian Pteridophytes.* Botanical Survey of India.

Dixit, R.D. and Kumar, R. (2002): Pteridophytes of Uttaranchal. *A check list.* Bishen Singh Mahendra Pal Singh, Dehradun.

Gadgil, M. (1996): Documenting diversity: An experiment. *Current Science,* 70:1- 36.

Khullar, S.P. (1994, 2000): *An Illustrated Fern Flora of the West Himalaya.* Vol. I & II. Dehradun.

Khullar, S.P., Pangtey, Y.P.S., Samant, S.S., Rawal, R.S. and Singh, P. (1991): *Ferns of Nainital.* Bisen Singh Mahendra Pal Singh, Dehradun.

Mebberley, D. (1997): *The Plant Book* (2nd Edition). U.K.: Cambridge University Press.

Pande, H.C. and Pande, P.C. (2002): An *Illustrated Fern Flora of Kumaun Himalaya* Vol. I & II. Bisen Singh Mahendra Pal Singh, Dehradun.

Pande, P.C., Tiwari, L. and Pande, H.C. (2004): Inventory of Folk Medicine and Related aspect in Uttaranchal. Bishen Singh Mahindra Paul Singh, Dehradun. 19-40.

Parihar, P, Parihar L. and Bohra, A. (2004): Antifungal activity of *Cheilanthes albomarginata* Clarke and *Marsilea minuta* L. against Aspergillus flavus. ***Indian Fern Journal,*** 21:140-143.

Parihar, P. and Parihar, P. (2006): Some pteridophytes of medicinal important from Rajasthan. *Natural Product Radiance,* 5(4):297-301.

Rout, S.D., Panda, T. and Mishra, N. (2009): Ethnomedicinal studies on some pteridophytes of Similipal Biosphere Reserve, Orissa, India. *Indian Journal of Medicine and Medical Sciences,* 1(5):192-197.

Samant, S.S. and Palni, L.M.S. (2010): Status and conservation issues of medicinal plants in the Indian Himalayan region. National Seminar, *Medicinal plants of Himalaya: Potential and* Prospect. 1-5.

Shiva, V. (1996): Protecting Our Biological and Intellectual Heritage in the Age of Biopiracy. The research foundation for science, Technology and Natural Resources Policy, New Delhi, India.

Singh, D.K. and Hajara, P.K. (1996): Floristic diversity. *In Biodiversity Status in the Himalaya New Delhi*: British Council. pp. 23-38.

Singh, P., Singh, B.K., Joshi, G.C. and Tewari, L.M. (2009): Veterinary Ethno-medicinal Plants in Uttarakhand Himalayan Region. *Nature and Science,* 7(8): 44-52.

Singh, S., Dixit, R.D. and Sahu, T.R. (2005): Ethnomedicinal uses of pteridophytes of Amarkantak, Madhya Pradesh. *Indian Journal of Traditional Knowledge,* 4(4):392-395.

Tripathi, G. (2002): Indegenous Knowledge and Traditional Practices of Some Himalayan Medicinal Plants. In: Samant SS, Dhar U, Palni LMS (eds) Himalayan Medicinal Plants Potential and Prospects, Gyanodaya Prakashan, Nainital. 151-156.

Upreti, K., Jalal, J.S., Tewari, L.M., Joshi, G.C., Pangtey, Y.P.S. and Tewari, G. (2009): Ethnomedicinal uses of pteridophytes of Kumaun Himalaya, Uttarakhand, India. *Journal of American Science,* 5(4):167-170.

WHO. (2002): WHO Traditional Medicine Strategy 2002-2005. World Health Organization, Geneva.

Biodiversity Conservation and Envir. Management (2012)
Editors: D.R. Khanna et al.
Pub. by Biotech Books. *ISBN: 978-81-7622-262-4*

Pages: 183-189

BACTERIAL DIVERSITY OF 'SAHASTRADHARA' – A SULPHUR SPRING IN DEHRADUN

Seema Rawat and Amir Khan✉
Department of Biomedical Science, Dolphin (PG) Institute of Biomedical & Natural Sciences, Dehradun, Uttarakhand (INDIA)
✉E-mail: amiramu@gmail.com

Sulfur springs have long been credited with unique healing powers for certain diseases both in the early eastern and western medical sciences. In these springs the concentration of elemental sulfur, sulfate, thiosulfate and dissolved sulfide is high and both oxic as well as anoxic conditions are maintained in the water and underlying sediments through the spring. Enrichment culture technique was employed to study the diversity of Sahastradhara. The population profile varied significantly with *Pseudomonas* and *Micrococcus* as the dominant microflora. Maximum structural diversity

✉ Corresponding author

was observed in the adjoining soil sample which harboured an assemblage of soil and surface water microflora.

Keywords: Sulfur springs, Thiosulfate, Enrichment culture technique, *Pseudomonas*, *Micrococcus.*

Introduction

Sulphur springs have long been credited with unique healing powers for certain diseases both in the early eastern and western medical sciences. These springs are spread world-wide. In the spring, concentration of sulphur, sulfate, thiosulfate and dissolved sulfide concentration in the emergent water is high and maintains oxic as well as anoxic conditions in the water and underlying sediments through the spring. These springs generally have acidic pH but few springs have neutral pH; can be a hot water spring or cold water spring (Yamamoto *et al.*, 1998; Reysenbach *et al.*, 1994; Skirnisdottir *et al.*, 2000; Nubel *et al.*, 2001).

The prevailing ecological conditions in these springs like pH, temperature, sulfide, sulphur or sulphate concentration, redox conditions, presence of other electron acceptors, light availability and organic content influence the community succession in these springs. The carbonate caves associated with these springs also harbours complex microbial communities (Kelly and Wood, 1998).

In sulphur and sulphide rich environments *viz.*, springs, hydrothermal vents, anaerobic zones of lakes, and shallow marine and intertidal systems, utilization and cycling of sulphur species play a major role in energy production and the maintenance of microbial community (Douglas and Douglas, 2001).

The wide range of ecological conditions existing in these springs make them an interesting ecosystem which has always drawn attention of microbial ecologists. The structural diversity of microbial form alongwith genetic, physiological and functional diversity is believed to provide complete understanding of these springs. The microbial community has been extensively studied in several springs by many workers (Skirnisdottir *et al.*, 2000; Martin, 2002; Tseke and Stahl, 2002). They harbour number of mesophilic, thermophilic and hyperthermophilic microflora. These springs comprise of a wide assemblage of bacteria belonging to alpha, beta, gamma and delta subdivisions of proteobacteria.

Springs comprise of sulfur oxidising bacteria, sulfur and sulfate reducing bacteria. The former comprises of aerobic bacteria *viz.*, species of genera *Alcaligens*, *Beggiatoa*, *Paracoccus*, *Pseudomonas*, *Thiothrix*, *Thioploca*, *Thiobacillus*, *Thiomicrospira*, *Thiosphaera* and *Xanthobacter* (Kelly and Wood, 1998) while latter comprises of anaerobic bacteria *viz.*, *Desulfovibrio*, *Desulfatomaculum*, *Desulfuromonas*, *Desulfobulbus*, *Desulfobacter*, *Desulfococcus*, *Desulfosarcina* and *Desulfonema* (Kelly and Wood, 1998; Stanier *et al.*, 2003).

Materials and Method

Collection of Samples: A total of 7 samples *viz.*, Normal Water (NW), Sulfur Water (SW), Cave Water (CW), Normal water mixed with sulfur water (NSW), Stalactite (STC), Soil sample (S), Soil sample mixed with sulfur water (SSW) were aseptically collected in sterile containers from Sahastradhara. These samples were transported on dry ice to the lab.

Recovery of Isolates: Microbial population dynamics was studied by enrichment culturing on Thiobacillus broth and Starkey's broth. Plating was done on the same medium after 24h, 72h, 120h, 168h and 264h.

Different morphotypes were purified by restreaking on Nutrient agar and *Starkey*'s medium and pure culture.

Biochemical Characterization: The various biochemical tests *viz.*, Indole, Methyl red, Voges Proskaeur, Citrate Utilization Test, Triple Sugar, Iron Agar Test and Nitrate Reduction Test, Oxidase Test, Catalase Test and Urease Test were carried out for characterization of isolates.

Results and Discussion

A significant difference in the population count was observed in water and soil samples of Sahastradhara (Table-1). The variation in the population count ($\log_{10}$ Cfu) of recovered mesophilic bacteria from NW to SW was 4.23±0.35 to 8.14±0.23.

During enrichment on Thiobacillus broth and Starkey's broth, changes in characteristics of broth were observed with time period. After 3 days in Thiobacillus broth, all samples changed colour of broth to dark red except sulfur water in which there was marginal change. Significant change in colour was observed for NSW, which changed the colour of broth to yellow. There was no change in colour of broth of all other samples. No further change in colour of broth was observed with further incubation to 11 days. Yellow colour depositions were observed in NW, SW, NSW and STC samples upto 11 days while black colour in CW and SSW upto 7 days; black and yellow colour in soil sample upto 7 days. Cave water depositions changed to yellow colour after 11 days while that of soil, to yellow and of SSW, to black and yellow. The changes in pH were non-significant. A significant decrease in pH was observed with sulfur water; pH of broth decreased to 3.5 after 11 days.

After 3 days, no change in colour of Starkey's broth was observed in all samples. White colour depositions were observed in SW upto 11 days while brown colour in S, STC and SSW upto 3 days. In other samples no depositions were present upto 3 days. In NW and CW, white colour depositions were observed after 7 days while in NSW, after 11 days.

Table1: Population Count of Bacteria Recovered From Sahastradhara.

Name of Samples	*Population Count (Log_{10} Cfu)*
SW	8.14±0.23
NW	4.23±0.35
NSW	5.68±0.14
CW	4.98±0.28
S	4.35±0.35
SSW	4.95±0.19
STC	5.56±0.22

± = standard deviation

Distinct bacterial morphotypes were selected on the basis of their colony morphology. Morphotypes were identified based on their morphological characteristics and biochemical characteristics according to Bergey's manual of Systematic Bacteriology (Kreig, 1984).

The overall distribution of recovered microflora species in Sahastradhara was *Pseudomonas* (26%), *Micrococcus* (28%), *Proteus* (11%), *Staphylococcus* (2%), *Klebsiella* (8%), *E. coli* (5%), *Alcaligens* (1%), *Bacillus* (3%), *Shigella* (1%), *Serratia* (1%), *Morexella* (6%) and unidentified (8%).

Maximum diversity was observed in SSW followed by Stalactite sample. *Pseudomonas* was documented to be the dominant microflora of CW (70%), NW (43%), NSW (40%) and SW (34%). It was also found in S (19%) and in SSW (14%). *Micrococcus* was recovered from all water and soil samples. *Morexella* was recovered only from stalactite sample and SSW. *Staphylococcus* was recovered only from sulfur water.

The distribution of other bacteria in different sample was: *Staphylococcus* (22%), *Proteus* (11%), *Micrococcus* (22%) and unidentified (11%) in SW; *E. coli* (20%), *Klebsiella* (20%) and *Micrococcus* (20%) in NSW; *Klebsiella* (10%), *Micrococcus* (10%), *Alcaligenes* (10%) in CW; *Klebsiella* (14%), *Micrococcus* (14%) and *E. coli* (29%) in NW; *Micrococcus* (30%), *Proteus* (25%), *Pseudomonas* (19%), *Bacillus* (13%) and unidentified (13%) in S; *Micrococcus* (37%), *Klebsiella* (10%), *E. coli* (5%), *Proteus* (10%), *Shigella* (5%), *Morexella* (5%) and unidentified (14%) in SSW and *Proteus* (14%), *Klebsiella* (14%), *Morexella* (29%), *Micrococcus* (29%), *Bacillus* (7%) and *Serratia* (7%) in STC.

Bacteria surviving in sulfur springs with neutral pH often form mats whose appearance seems to be governed by prevailing physical and chemical conditions (Tseke and Stahl, 2002). In Sahastradhara yellow, black, brown and white colour depositions were observed. The yellow colour depositions were probably sulfur granules as many sulfur oxidizers excrete internal or external sulfur in the form of hydrophilic spherical globules (Steudel *et al.*, 1999) that can serve as an energy reserve when environmental concentrations of hydrogen sulphide substantially decrease. Brown depositions were indicative of presence of ferrihydrite while black deposition observed was because of formation of iron sulfide. Yellow along with black colour depositions were observed in soil samples which could be due to concomitant accumulation of sulfate with iron sulfide. Similar observations were reported by Finster *et al.* (1998) in marine sediments. Brock (1978) reported that the thick bacterial mats could be spectacularly white or brown yellow from precipitated sulfur which are formed in neutral sulfur springs. The change in colour, depositions and pH in broth with time period were consequence of the microbial growth and activity.

The population profile varied significantly amongst different samples of Sahastradhara. *Pseudomonas* and *Micrococcus* were documented to be the dominant microflora. The dominance of *Pseudomonas* amongst heterotrophic population of springs had been reported by Kelly and Woods (1998).

The biasedness shown by cultivation approach due to intrinsic selectivity of each component of the cultivation technique nearly always resulted either in enhancement or decrease or even inhibition of growth of some specific members and therefore cultivation-dependent methods does not give complete diversity picture of a habitat.

References

Brock, T.D. (1978). Thermophilic microorganisms and life at high temperatures. Springer-Verlag, New York.

Douglas, S. and Douglas, D.D. (2001). Structural and geomicrobiological characteristics of a microbial community from a cold sulfide spring. *Geomicrobiol.* 18: 401-422.

Finster, K., Liesack, W. and Thamdrup, B. (1998). Elemental sulfur and thiosulfate disproportionation by *Desulfocapsa sulfoexigens* sp. Nov., a new anaerobic bacterium isolated from marine surface sediment. *Appl. Environ. Microbiol.* 64: 119-125.

Kelly, D.P. and Wood, A.P. (1998). Microbes of the sulfur spring. *In*: Techniques in Microbial Ecology (Eds: R.S. Burlage, R. Atlas, D. Stahl, G. Geesey and G. Seyler). Oxford Univ. Press, New York. pp. 31-56.

Kreig, N.R. (1984). Bergey's manual of Systematic bacteriology. Williams and Wilkins, Baltimore, vol. 1.

Martin, A.P. (2002). Phylogenetic approaches for describing and comparing the diversity of microbial communities. *Appl. Environ. Microbiol.*, **68**: 3673-3682.

Nubel, U.U., Bateson, M.M., Madigan, M.T., Kuhl, M. and Ward, D.M. (2001). Diversity and distribution in hypersaline microbial mats of bacteria related to *Chloroflexus* spp. *Appl. Environ. Microbiol.*, **67**: 4365-4371.

Reysenbach, A.L., Longnecker, K. and Kirshtein, J. (1994). Novel bacteria and archaeal lineages from an *in situ* growth chamber deployed at a mid-atlantic ridge hydrothermal vent. *Appl. Environ. Microbiol.*, **66**: 3798-3806.

Skirnisdottir, S., Hreggvidsson, O.G., Hjorleifsdottir, S., Marteinsson, V.T., Petursdottir, S.K., Hoist, O. and Kristjansson, J.K. (2000). Influence of sulfide and temperature on species composition and community structure of hot spring microbial mats. *Appl. Environ. Microbiol.*, **66**: 2835-2841.

Stanier, R., Ingraham, J.L., Wheelis, M.L. and Painter, P.R. (2003). Gram-negative anaerobic eubacteria. *In*: General Microbiology (Eds: R. Stanier, J.L. Ingraham, M.L. Wheelis and R.R. Painter). Mac Millan Press Ltd., London. pp. 404-407.

Steudel, R., Dahl, C. and Hormes, J. (1999). *In situ* analysis of sulfur in the sulfur globules of phototrophic sulfur bacteria by X-ray absorption near edge spectroscopy. *Biochim. Biophys. Acta.,* 1428: 446-454.

Tseke, A. and Stahl, D.A. (2002). Microbial mats and biofilms: evolution, structure and function of fixed microbial communities. *In*: Biodiversity of microbial life (Eds: J.T. Stanley and A.L. Reysenbach). Wiley-Liss Press, New York.

Yamamoto, H., Hirashi, A., Kato, K., Chiura, H.X., Maki, Y. and Shimizu, A. (1998). Phylogenetic evidence for existence of novel thermophilic bacteria in hot spring sulfur-turf microbial mats in Japan. *Appl. Environ. Microbiol.,* **64:** 1680-1687.

Biodiversity Conservation and Envir. Management (2012)
Editors: D.R. Khanna et al.
Pub. by Biotech Books. *ISBN: 978-81-7622-262-4*

Pages: 191-198

SOME VEGETATIVE METHODS AS A NATURE'S FREE TOOL FOR CONTROLLING SOIL EROSION LEADING TO FLOODS

Nitin Kamboj
Department of Zoology & Environmental Science
Gurukula Kangri Vishwavidyalaya, Haridwar, Uttarakhand (INDIA)
E-mail: kambojgurukul@gmail.com

The Uttarakhand state is particularly sensitive to land disturbing activities. Steep slopes, high rainfall and weak geology of the Uttarakhand state accentuate the land degradation and soil erosion process at much faster rate than in the plains. The rivers with flash flows and high sediment loads are known as torrents. Thus, the real problem of torrent formation lies in geomorphology of lesser Himalaya further aggravated due to misuse and mismanagement of upstream catchments. The problem of torrent menace has been increasing in Uttarakhand with the rise in population pressure and related mismanagement of upstream watersheds. The vegetative control measures were provided a base root technology by selecting the plant species *viz Arundo donax*, *Vitex negundo*, *Ipomoea carnea*, *Napier*, *Saccharum munja* and on reclaimed land along torrents, tree species like *Acacia catechu*, *Dalbergia sisoo*, *Bamboo*, *Acacia nilotica* and *Mulberry* for conservation and management of relative watershed in Shiwalik Himalayan region.

Keywords: Vegetative methods, Torrent, Watershed, Shiwalik region, Slope.

Introduction

The Himalayan eco-system is one of the most important life support system on the earth. The origin of torrents in Shiwalik Himalaya due to geological and existing climatic conductions created constant threat to the natural resources as well as human settlements. Torrents are the seasonal streams, which run along the steep mountain slopes down to the foothills and carry heavy bed load and flash flows during monsoon period. Flash floods with high sediment load fill up the channel beds and flow keeps on shifting its course and damaging foothills forest land area, fertile lands, agriculture pasture land and habitations. Shiwalik region comprises of lower hill ranges of lesser Himalayas spreading mainly over the states Uttarakhand, Punjab, Himanchal Pradesh, J&K, Chandigarh (U T), Haryana, Uttar Pradesh and extended up to Assam in the north east. These ranges are mostly affected by seasonally hilly streams popularly called 'Choes' or 'Raos' causing flash flood during the rainy season. Naturally equilibrium being disturbed in these hills by heavy denudation due to increasing demand for fodder and fuel, resulting in the loss of vegetation cover. According to National Agriculture Commission (1976), the devastation due to choes (torrent) in the states of Punjab alone was measured 250000 hectares of land. The problem of torrent formation in Shiwalik Himalayan in fact, due to misuse and mismanagement of upstream catchments. The problem of torrent menace has been rising in Shiwalik Himalayan with the rise, in population pressure and related mismanagement of upstream watersheds.

Materials and Method

Study area

The Sabhawala watershed is located in the Doon Shiwalik range of the Himalaya in tehsil Vikasnagar, block Sahaspur of district Dehradun. Sabhawala torrential watershed originates from Shiwalik Himalaya foothill regions at the longitude 77° 48' E and latitude 30° 20' N. The total length of torrent is 7.85 km (5.45 km under forest and 2.40 km is under agricultural and habitation area) in Sabhawala watershed.

Methodology

For the analytical work of the soil and hydrochemistry of the watershed, the sampling network was designed to cover a wide range of determinants characteristics at selected zones. Four sampling zones were selected in the torrent watershed covering a stretch of about 2.5 km (Figure 1).

Fig. 1: Satellite view of sampling zones at Sabhawala watershed.

- S1-Forest area
- S2-Forest area to agricultural land areas
- S3-Habitation and settlement of village areas
- S4-Agriculture flood plane areas

Morphological characteristics of the watersheds such as area, maximum length of flow and physico-chemical analysis was followed by APHA (1998). Soil analysis as organic carbon, moisture content, pH and available NPK were determined by Anderson and Ingram (1993), Walkley and Black (1934).

Results and Discussion

The area reported under torrent bed in Uttaranchal Shiwalik by RRSSC regional remote sensing service center recorded is 413.65 km^2, out of 53484 km^2 total area of state (Table 1). Vegetative reinforcement of structure and torrent bank protection was affected by planting different soil conservation plant species (Table 2). The overall performance of vegetative considering the factor of survival, site suitability, hydraulic properties *etc.* were in order of *Ipomoea* (Besharam), *mulberry* (Shahtoot), *Vitex negundo* (Shimalu), *Bamboo* (Baash), *Napier* (Hathi ghaas) *Arundo donax* (Narkul) did not survive well, may be because the site did not have the needed fine soil and moisture for its stabilization. Wani *et al.* (2006) reported that *Jatropha curcas* in erosion prone area is used as a soil conserving plant but does not survive well due to poor management and imperfect agronomic practices. Juyal *et al.* (2004) carried out extensive field cum laboratory studies on spurs in the hydraulic flume and based on that constructed prototype structures in the field, which are performing well *Arundo donax* (Narkul or Nada) has been found ideal for reinforcement of the structures. Juyal *et al.* (1995) reported that the roots of Munj, *Eulaliuopsis binata* and *Vetivaria zizanionides* reached the maximum depth followed by Khus and Bhabbar. Hettinger (1976) described the torrent control measures in the mountains with respect to the tropics and recommended various plant species to prevent the torrent problems. Hettinger (1976) recommended the list of plant species for the torrent control in the mountain with respect to the tropics like Nada (*Arundo donax*) Banah (*Vitex negundo*) Kana (*Saccharum munja*) Akra (*Ipomea carnea*) *Hybrid Napier* etc. Bio-engineering remedial measures works such as trees, shrubs and grasses are basically aimed to control meandering tendency of the torrent by restricting it in a defined and narrow course. If this is achieved, problems of overtopping and undermining of banks and debris deposition in the flow channel can be eliminated. Vegetative methods are nature's free tool to rehabilitate for controlling soil and torrential erosions and siltation leading to floods.

Table 1: Total area under Choes and rivers as Torrential structures in Shiwalik Himalayan region under different states

S. No.	*States*	*Area (km^2)*
1.	Uttarakhand	413.650
2.	Himanchal Pradesh	399.050
3.	Punjab	322.135
4.	Jammu & Kashmir	45.135
5.	Chandigarh	3.576
	Total	**1184.101**

Vegetative barrier is one of the important conservation techniques, which played a vital role in reducing soil and water losses and improving soil fertility, crop productivity and fodder availability. The usefulness of vegetative barrier were exceedingly realized as an alternative of earth bunds for soil and water conservation and erosion control, being relatively cheaper, eco-friendly and pro farmer. This technology mainly utilized local material, skill and indigenous knowledge and therefore has great potential for increasing production and conserving resources. Vegetative barriers and live hedges are used for erosion control purpose in agricultural lands as well as non-arable lands. Vegetative barriers established on farmer's field are expected to perform as erosion diminishing mechanism and also provide fodder, fruits and other minor products. Spurs are structures used for torrent training work. Singh *et al.*, (1971) worked on the torrent control measures and describe the vegetative and engineering measures for torrent training and stream bank protection in Doon valley in the state of Uttarakhand.

Table 2: Plant species suitable for degraded lands in watershed area at Shiwalik Himalaya.

Species & Common name	*Climatic zone*	*Sowing time*	*Planting time*	*Method of planting*	*Type of plant*	*Uses*
Acacia catechu (Kuthha) (Khair)	Dry to moist tropical	May/June	July	Seed sowing/ entire or Stump planting	Tree	Commercial fuel and Soil cons
Agave sislanal *Agave* sp., A.	Tropcial to Subtropical	Bulbils sown in	July (1-1/2 yr)	Transplanting of bulbils/	Shrub	Fibre, medicinal soil conservation
Arundo donax (Narkul)	Sub-tropical humid	April	July-Aug.	Cutting/ root stock	Shrub	Soil Cons., matting & bobbin pipes
Bauhinia purpurea (Khairwal)	Moist tropical to Sub-tropical	April-May	July-Aug.	Direct sowing, entire or stamp planting	tree	Gum, fodder, fuel, timber
B.variegeta (Kachnar)	-do-	-do-	-do-	-do-	tree	Fodder
Bombex ceiba (Semul)	Moist tropical to sub-tropical	June	July	Entire Stump planting Branch cutting	tree	Soil Cons., fodder wood, packing cases
Chrysopogon Fulvus (Gorda)	Sub-tropical	Feb-March	July	Direct sowing, - rooted Slips/clump	grass	Soil Cons., fodder
Dalbergia sissoo (Shisham)	Dry to moist tropical	March- April	July	Direct sowing entire / Stump	tree	Soil Cons., timber Fodder and fuel
Dendrocalamus Strictus (Bamboo)	Dry tropical, tropical moist	April-June	June-July	Direct sowing, entire / Stump	tree	Industrial timber, small timber poles, fodder, gum, soil Cons. and medicinal

Contd...

Species & Common name	*Climatic zone*	*Sowing time*	*Planting time*	*Method of planting*	*Type of plant*	*Uses*
Pennisetum	Wet tropical	June	July-August	By stem cuts	Grass	Soil *Eriophrus*
Cosmosum (Pseudobhabar)	Dry sub-tropical	April	July	sowing of seeds root cutting	Grass	Soil Cons. on shallow soils forage
Erythrina Suberosa (Dhok Dhak)	Moist tropical	June-July	July	root suckers, stump planting, branch cutting	Grass	Soil conservation and fuel
Greivia species *G.optiva, G.elastica*	Sub-tropical to temperate	June-July	July after 1 yr.	entire planting, direct sowing	Tree	Fodder, fuel, fibres & soil cons.
Ipomea camea (Besharam)	Tropical sub-tropical		July-August	Cutting	Shrub	Soil cons.
Jatropha curcas conservation	Tropical	June-July	July	direct sowing/stump	Shrub	fuel & soil
Saccharum munja	Arid areas	-do-	-do-	stump planting entire planting or rooted slip	Grass	making baskets ropes. & medicinal value
Hybrid Napier	-do-	June	July-August	Planting rooted slips/ clump, direct sowing	Grass	fodder & soil cons

References

Anderson, J.M., and Ingram, J.S.I. (1993). Tropical Soil Biology and fertility- a handbook of Method. 2nd edi., CAB International, Wallingford, U.K.

APHA (1998). Standard methods for the examination of water and waste water 20th ed. Washington D.C.

Hettinger, Hubert. (1976). "Torrent control in the Mountains with respect to the Tropics. FAO, Rome" Conservation Guide No.2: 119-134.

Juyal, G.P., Bankey Bihari, Ghosh B.N and Rathor A.C (2004). Cost effective technology for torrents treatment in Uttaranchal Shiwaliks. In Proceeding of *research conservation technology* for social uplift.

Juyal, G.P., Katiyar, V.S., Dadhwal K.S., Loshie P., and Arya R.K., (1995). Mined area rehabilitation for torrent control in Himalayas, In Torrent Menace – Challenges and Opportunities, Eds. G.Sastry, V.N. Sharda, G.P. Juyal and J.S. Sharma, *CSWCRTI,* Dehradun, pp. 119-145.

Singh, B., Mathur, H.N. Gupta, S.K. (1971). Vegetative and engineering measures for torrent bearing and stream bank protection in Doon valley. *Indian Forests,* 9: 47-54.

Walkley, A. and Black, C.A. (1934). An examination of the digestion method for determining soil organic matter and a proposed modification of the chromic acid titration method. Soil Sci. 37: 29-38.

Wani SP, Osman M, Emmanuel D'Silva and Sreedevi TK. (2006). Improved livelihoods and environmental protection through biodiesel plantations in Asia. *Asian Biotechnology and Development,* 8(2):11-29.

Biodiversity Conservation and Envir. Management (2012)
Editors: D.R. Khanna et al.
Pub. by Biotech Books. *ISBN: 978-81-7622-262-4*

Pages: 199-211

STUDIES ON PHYSICO-CHEMICAL, MICROBIOLOGICAL PROPERTIES OF RHIZOSPHERIC AND NON RHIZOSPHERIC SOILS IN COMPARTMENT NO. 792 OF GOREWADA FOREST

Kirti V. Dubey[1✉], Praveen N. Charde[1], Ashish Lambat[1] and Gagan Matta[2]
[1]Sevadal Mahila Mahavidyalaya, Sakkardara Square,
Umrer Road Nagpur, Maharashtra (INDIA)
[2] Gurukula Kangri Vishwavidyalaya, Haridwar, Uttarakhand (INDIA)
E-mail: kirtivijay_dubey@yahoomail.com

Study of Gorewada forest was carried out to find out physicochemical as well as microbiological properties of soil collected from the rhizosphere of different plant species. Studies have shown that the forest present in compartment No. 792 is of dry deciduous type having

✉ Corresponding author

trees, shrubs, herbs, climbers and grasses and some barren patches of land. Soil of the different plant species had clayey texture, porosity and water holding capacity was high. The electrical conductivity, organic carbon and cation exchange capacity of pit soil was somewhat high compared to barren soil; however levels of these chemical properties were lower as compared to the rhizospheric soil samples of dominant plant species. Total and available nutrient content was high in rhizospheric soil as compared to the pit soil and barren. The bacteria, fungi, actinomycetes and nitrogen fixer namely *Rhizobium* and *Azotobacter* and vesicular arbuscular mycorrhizae (VAM) spores were found to be high in pit soil as compared to barren soil. However, the rhizospheric soil samples had comparatively high counts of tested microorganisms. Findings of this particular study will formed a basis for developing future plans required for development of forest cover with a diverse variety of plants of high economic value like timber, medicinal and ornamental including even the exotic and rare species by using the biotechnological approach of bioaugmentation.

Keywords: Gorewada forest, Rhizospheric Soils, *Rhizobium*, *Azotobacter*, VAM bioaugmentation.

Introduction

Many parts of country are facing problem of forest resources depletion, due to lack of proper physico-chemical and microbiological status of forest soils. This problem can be alleviated by site-specific soil augmentation process which has now been accepted as a tool for the sustainable management of forest resources. Sustainable management of forest practices are many. In forest management practices, high scientific skills are required. However, important steps are proposed herein as preliminary study is based on following aims and objectives. One of the main aim is to study native plant species in and around Gorewada forest. Another aim is to characterize rhizospheric soil of native plant species with respect to its physico-chemical and microbiological properties. This study will help to generate the basic data required for developing an appropriate bioaugmentation process that can be used to resolve the low nutrient status and less counts of useful micro flora present in forest soil for proper functioning of biogeochemical cycle. Another aim is to isolate site- specific biofertilizers *i.e.* strains of *Rhizobium, Azotobacter* and Vesicular Arbuscular Mycorrhiza (VAM) and to select a suitable organic ammendment for improving the physico-chemical and microbiological properties of soil. Results of above mentioned study will form a platform for future studies in order to define the most supportive and nutritive rhizosphere for the sustainable development of the forest canopy by using bioaugmentation process.

Materials and Method

The Gorewada forest is located in Nagpur at 21° 45'N latitude and 78° 15'E longitude. The annual rainfall is 750 to 1000 mm. Site surveys were conducted at Gorewada forest during December-2007 to March-2008. During every visit, as many specimens as possible were collected were brought to the college laboratory to identify all plants by using reference floras by Hooker (1872-1897), Ugemuge (1986), Almeida (1990), Joshi (2000) and Singh *et al.* (2001). Besides this representative soil samples from minimum five rhizospheres of same plant species were collected from each compartment with the help of hand auger and were mixed in equal proportions to get a composite sample on dry weight basis, air dried, homogenized, sieved through 2.0 mm mesh and stored in polythene bags. Analysis of composite barren, pit and rhizospheric soil samples of dominant plant species available in experimental site were conducted for different physico-chemical and microbiological properties to know the fertility status of the soil. Particle size analysis of soil samples was done by international pipette method (Piper, 1966). Bulk density was determined by using clod method while Particle density was determined by a pycnometer method as per Black *et al.* (1965), Water holding capacity was determined by using Keen-Roozkowski brass boxes as described by Piper (1965a). The pH of soil was determined by using 1:2.5 soil water suspension ratio and Electrical conductivity was determined with the help of conductivity meter using 1: 2.5 soil water suspension ratios (Jackson, 1967). Organic carbon was determined by following Walkly and Black wet digestion method, (Black; 1965b). Available N (was determined by following alkaline permanganate (0.31% $KMnO_4$) method (Subbaih and Asija, 1956). Available P_2O_5 determined colorimetrically by following Bray's II method at 660 nm wavelength, (Black; 1965b). Available K_2O determined air dried soil with 50.0 ml neutral N NH_4OAc. (Jackson; 1967). Total N: Samples were digested using conc. H_2SO_4 (wet digestion) and taken for nitrogen distillation by micro Kjeldahl method using Kjel Plus instrument (Tondon, 1993). Total phosphorous is estimated by digesting the soil with mixture of nitric acid and perchloric acid (9:4) and determines the phosphate by ammonium vanadomolybdate as described by Jackson (1967). Total potassium is estimated by digesting the soil with mixture of nitric acid and perchloric acid (9:4) and determines the potassium by using flame photometer (Hesse 1971). Microbes such as bacteria, fungi, actinomycetes and nitrogen fixing strains of *Rhizobium* and *Azotobacter* were analyzed by following standard procedure for soil microbial populations and were expressed in terms of colony forming units cfu/g (Black *et al.*,1965). Vesicular Arbuscular Mycorrhizal (VAM) spores were enumerated by Wet Sieving and decanting technique (Gerdmann and Nicolson, 1963).

Results and Discussion

Types of Flora present

During course of study a total of 42 different plant species classified as trees, shrubs, herbs, climbers and grasses were reported in compartment number 792 (Table 1). Among the tree species dominant type were *Acacia catechu*, *A. leucophloea. A. nilotica*, *Albizia odoratissima*, *Butea monosperma*, *Cassia siamea*, *Mitragyna perbifolia*, and *Ziziphus mauritiana.*

Textural classification

The textural classification *i.e.* sand, silt and clay percentages of soils collected from 792 compartment of Gorewada forest are presented in Table 2 and results showed that the textural class of all the soil samples analyzed were clayey in texture. The rhizospheric soil of the *Albizia odoratissima* had highest clay content of 62.12 percent and very low content of sand (17.57%) and silts (20.31%). Due to clayey textural class the clay content was high in all the rhizospheric soils of different species as compared to the sand and silt percentages. In this compartment it was reported that the majority of the soils have high amount of clay as compared to sand and silt. As basalt (rock) being the parent material for formation of these soils, therefore, it is known to impart higher amount of clay to the soil.

Table 1: List of Various Plant Species Present in Compartment Number 792

S. No.	*Botanical Name*	*Family*	*Local Name*
	Trees		
1	*Acacia catechu*	Leguminosae	Khair
2	*Acacia leucophloea*	Leguminosae	Hivar
3	*Acacia nilotica*	Leguminosae	Babhul
4	*Albizia odoratissima*	Leguminosae	Shiras
5	*Andropogon pumilus*	Poaceae/Gramineae	Diwartan
6	*Azadirachta indica*	Meliaceae	Kaduneem
7	*Bauhinia racemosa*	Leguminosae	Apta
8	*Butea monosperma*	Leguminosae	Palas
9	*Cassia siamea*	Leguminosae	Cassia siamea
10	*Dalbergia sissoo*	Leguminosae	Shisham
11	*Dandrocalamus strictus*	Poaceae	Bamboo
12	*Dolichandrone falcata*	Bignoniaceae	Mersing
13	*Maytenus emarginata*	Celastraceae	Bharati
14	*Mimosa hamata*	Leguminosae	Chilati
15	*Mitragyna perbifolia*	Rubiaceae	Kalamb
16	*Phoenix sylvestris*	Palmae	Sindi
17	*Semicarpus anacardium*	Anacardiaceae	Biba
18	*Tamarindus indica*	Leguminosae	Imli/Chinch
19	*Tectona grandis*	Verbenaceae	Teak
20	*Terminalia bellirica*	Combretaceae	Bihada
21	*Ziziphus glaberrima*	Rhamnaceae	Guti
22	*Ziziphus mauritiana*	Rhamnaceae	Bor

contd...

S.No.	*Botanical Name*	*Family*	*Local Name*
	Shrubs		
23	*Lantana camera*	Verbenaceae	Ghaneri
24	*Xanthium strumarium*	Asteraceae	Gokharu
	Climbers		
25	*Cocculus hirsutus*	Menispermaceae	Vasanvel
26	*Coix lacryma-jobi*	Poaceae/Gramineae	Kasai/Garu
27	*Hemidesmus indicus*	Periplocaceae	Khobarvel/Anantvel
28	*Ipomea quamoclit*	Convolvulaceae	Ganeshvell
	Herbs		
29	*Ageratum conyzoides*	Asteraceae	Osadi
30	*Alternanthera pungens*	Amaranthaceae	Galighosh
31	*Blumea eriantha*	Asteraceae	Nirmudi
32	*Cassia tora*	Leguminosae	Tarota
33	*Hyptis suaveolens*	Lamiaceae/Labiatea	Rantulas
34	*Parthenium*	Asteraceae	Gajarghas
35	*Sida acuta*	Malvaceae	Chikana/Tupkadi
36	*Sida cordata*	Malvaceae	Bhuichikana
37	*Tridax procumbens*	Asteraceae	Kambarmodi
38	*Vernonia cinerea*	Asteraceae	Sahadevi
39	*Vicoa indica*	Asteraceae	Sankuli (Sonuli)
	Grasses		
40	*Apluda mutica*	Poaceae/Gramineae	Ponai
41	*Hackelochloa granularis*	Poaceae/Gramineae	Phulwa
42	*Iseilema laxum*	Poaceae/Gramineae	Mushan

Physical properties

Results of different physical properties of the soils collected from compartment number 792 of Gorewada forest are presented in the Table 3 and results showed that the barren soil was although clayey in texture the porosity and WHC was very low (51.32% and 50.45%, respectively) and the bulk density was very high (1.29 mg m^{-3}). Due to presence of vegetation in the soil the rhizospheric soil showed increase in porosity and WHC percentages. This phenomenon was observed in case of rhizospheric soil of *Albizia odoratissima* which reported the highest WHC (61.45%)

and porosity (65.34%) with the lower BD of 1.10 mg m^{-3}. Similar findings were observed in case of the rhizospheric soils of other plant species namely *Acacia nilotica, Mitragyna perbifolia, Acacia catechu, Ziziphus mauritiana* and *Bauhinia racemosa* where the WHC ranged from 52.65 percent to 61.45 percent.

Table 2: Textural classification of barren and rhizospheric soils of dominant plant species available from compartment number 792 of Gorewada forest.

S. No	*Sample Name*	*Sand (%)*	*Silt (%)*	*Clay (%)*	*Textural class*
Compartment number 792					
1	Barren patch	19.29	28.61	52.10	Clayey
Rhizospheric soils of dominant plant species					
1	*Ziziphus mauritiana* (Bor)	22.34	26.29	51.37	Clayey
2	*Acacia catechu* (Khair)	28.39	27.45	44.16	Clayey
3	*Bauhinia racemosa* (Apta)	20.71	23.53	55.76	Clayey
4	*Acacia nilotica* (Babul)	20.39	27.68	51.93	Clayey
5	*Albizia odoratissima* (Shiras)	17.57	20.31	62.12	Clayey
6	*Tamarindus indica* (Imli/ Chinch)	24.02	23.84	52.14	Clayey
7	*Mitragyna perbifolia* (Kalamb)	23.87	25.69	50.44	Clayey

Chemical properties

The chemical properties of the different plant species found in compartment number 792 are presented in Table 4. The result shows that pH and the EC of this site were observed to be high as compared to the other compartments. The organic carbon status varied from 1.78 percent in barren soil to 2.34 percent in the *Mitragyna perbifolia.* In leguminous plant species like *Acacia catechu, Acacia nilotica, Mitragyna perbifolia* and *Tamarindus indica* high organic carbon, *viz.* 2.90, 2.22, 2.34 and 2.24 percent WOS reported. The CEC was obviously high in the rhizospheric soil of all the species of the compartment as compare to the barren soil.

The rhizospheric soil exhibits biological nitrogen fixation in leguminous plants; hence rhizosphere contains the higher level of nutrient availability and exchange of the cations and thus CEC is high in the rhizosphere. In this compartment the CEC was very high in the rhizospheric soil of the *Mitragyna perbifolia* (60.61CEC cmol (p^+) kg^{-1}) which was followed by the *Tamarindus indica* (58.67 cmol (p^+) kg^{-1}) and 58.20 cmol (p^+) kg^{-1} in the *Acacia catechu.*

Table 3: Physical properties of barren and rhizospheric soils of dominant plant species available from compartment number 792 of Gorewada forest

S. No.	*Sample Name*	*Bulk density (mg m^{-3})*	*Porosity (%)*	*Water holding capacity (%)*
Compartment number 792				
1	Barren patch	1.29	51.32	50.45
Rhizospheric soils of dominant plant species				
1	*Ziziphus mauritiana* (Bor)	1.14	58.92	56.39
2	*Acacia catechu* (Khair)	1.15	56.60	52.65
3	*Bauhinia racemosa* (Apta)	1.09	58.87	56.34
4	*Acacia nilotica* (Babul)	1.17	60.74	57.78
5	*Albizia odoratissima* (Shiras)	1.10	65.34	61.45
6	*Tamarindus indica* (Imli/Chinch)	1.13	57.36	54.45
7	*Mitragyna perbifolia* (Kalamb)	1.10	58.59	55.34

Total Nutrient status

The total nutrient status of the 792 compartment are given in the Table 5. The highest N (0.26 %), K (0.57%) and S (0.067%) content was found in the rhizospheric soil of *Acacia catechu,* because the *Acacia catechu* is the leguminous plant which harbors *Rhizobium* in its root nodules and its growth is preferentially stimulated in the rhizosphere of legume than in that of non legumes. Moreover, legumes excrete a large number of substances into the rhizosphere principally sugars, amino acids and vitamins such as biotin and pantothenic acid, therefore the total N was high in the rhizosphere of this plant. The K and S content are high due to the rhizospheric effect of soil microorganisms residing in the root zone of *Acacia catechu.* In rhizospheric soils of other plant species the total N ranged from 0.1 to 0.25, P: 0.045 to 0.068; K: 0.27 to 0.45 and S: 0.043 to 0.063 percent.

Table 4: The chemical properties of the barren and rhizospheric soils of dominant plant species available from compartment number 792 of Gorewada forest.

S. No.	*Name of Sample*	*pH*	*Electrical conductivity dS m^{-1}*	*Organic carbon (%)*	*Cation exchange capacity cmol (p^+) kg^{-1}*
Compartment number 792					
1	Barren soil	7.80	0.30	1.78	53.14
Rhizospheric soils of dominant plant species					
1	*Ziziphus mauritiana* (Bor)	7.44	0.32	1.89	56.57
2	*Acacia catechu* (Khair)	7.76	0.32	2.90	58.20
3	*Bauhinia racemosa* (Apta)	7.67	0.22	2.21	56.66
4	*Acacia nilotica* (Babhul)	7.12	0.29	2.22	55.34
5	*Albizia odoratissima* (Shiras)	7.32	0.26	1.90	57.59
6	*Tamarindus indica* (Imli/Chinch)	7.35	0.28	2.24	58.67
7	*Mitragyna perbifolia* (Kalamb)	7.65	0.20	2.34	60.61

Available nutrient status

The available nutrient status of the compartment number 792 is given in the Table 6. The nitrogen content of this site was low as compared to the other compartment. In all the rhizospheric soil of the different plant species of this compartment the highest N content was observed in the *Ziziphus mauritiana* (0.022%) followed by the *Mitragyna perbifolia* (0.021%). The P_2O_5 and K_2O content was high in the *Albizia odoratissima* 0.00129 and 0.00443 percent respectively. The *Acacia catechu* also showed the high P_2O_5 and K_2O *viz.* 0.00129 and 0.0435 percent respectively.

Microbiological counts

The microbiological count of the different plant species collected from the compartment number 792 is presented in the Table 7. The highest bacterial and fungal count was reported in the rhizospheric soil of *Acacia catechu* (53 × 10^5, 42 × 10^3 cfu g^{-1} respectively) due to high organic carbon and CEC of the soil. The actinomycetes were highest in the *Bauhinia racemosa* (41 × 10^3 cfu g^{-1}) followed by *Ziziphus mauritiana*. The nitrogen fixing bacteria *i.e. Rhizobium* was high in the *Acacia nilotica* (48 × 10^3 cfu g^{-1}) closely followed by *Acacia catechu* (46 × 10^3 cfu g^{-1}) being a leguminous plant. However, the non symbiotic nitrogen fixing bacteria *i.e. Azotobacter* was highest in the *Ziziphus mauritiana*. The count of VAM spores was high in the rhizosphere of *Bauhinia racemosa* (12 cfu 10^{-1} g). In the barren soil, the microbiological counts of useful microorganisms was very low owing to low total and available nutrient status of barren soil in terms of contents of carbon, N, P and K levels.

Table 5: Total nutrient status of the barren and rhizospheric soils of dominant plant species available from compartment number 792 of Gorewada forest.

Sample No.	*Total nutrient status*
	0.056
	0.047
	0.063
	0.043

Table 6:. The available nutrient status of barren and rhizospheric soils of dominant plant species available from compartment number 792 of Gorewada forest.

S. No.	*Name of Sample*	*Available nutrient status*		
		N (%)	*P_2O_5 (%)*	*K_2O_5 (%)*
Compartment number 792				
1	Barren soil	0.012	0.00111	0.0224
Rhizospheric soils of dominant plant species				
1	*Ziziphus mauritiana* (Bor)	0.022	0.00112	0.0345
2	*Acacia catechu* (Khair)	0.016	0.00129	0.0435
3	*Bauhinia racemosa* (Apta)	0.012	0.00063	0.0234
4	*Acacia nilotica* (Babul)	0.018	0.00110	0.0378
5	*Albizia odoratissima* (Shiras)	0.016	0.00129	0.0443
6	*Tamarindus indica* (Imli/ Chinch)	0.019	0.00122	0.0238
7	*Mitragyna perbifolia* (Kalamb)	0.021	0.00120	0.0367

Conclusion

Development of richness in the vegetation wealth and biodiversity of soil is of prime importance in the forest cover as well as on the non-forest area to conserve the bioresources and tree wealth. Many parts of country are facing problem of forest resources depletion, due to lack of proper physico-chemical and microbiological status of forests soils. Therefore, sustainable management of forest soil is required. This can be achieved by virtue of performing the proposed preliminary study of native plant species in and around Gorewada forest, characterization of rhizospheric soil of native plant species with respect to its physico-chemical and microbiological properties, isolation of site specific biofertilizer strains of *Rhizobium*, *Azotobacter* and Vesicular Arbuscular Mycorrhiza (VAM) and selection of a suitable organic ammendment to improve the physico-chemical and microbiological properties of soil from Gorewada forest. Findings of this particular study will form a basis for developing future plans required for development of forest cover with a diversed variety of plants of high economic value like timber, medicinal and ornamental including even the exotic and rare species by using the biotechnological approach of bioaugmentation. The bioaugmentation process can be used to resolve the low nutrient status and less counts of useful micro flora present in forest soil.

Table 7: The microbiological count of the barren and rhizospheric soil of dominant plant species available in the compartment number 792 of Gorewada forest.

S. No.	*Name of Sample*	*Counts of different types of useful microorganisms*					
		Bacteria cfu g^{-1}	*Fungi* cfu g^{-1}	*Actinomycetes* cfu g^{-1}	*Rhizobium* cfu g^{-1}	*Azotobacter* cfu g^{-1}	*VAM spores cfu* 10^{-1} g
Compartment number 792							
1	Barren patch	14×10^5	11×10^3	7×10^3	12×10^3	17×10^3	4
Rhizospheric soils of dominant plant species							
1	*Ziziphus mauritiana* (Bor)	49×10^5	34×10^3	40×10^3	36×10^3	43×10^3	9
2	*Acacia catechu* (Khair)	53×10^5	42×10^3	36×10^3	46×10^3	34×10^3	10
3	*Bauhinia racemosa* (Apta)	52×10^5	37×10^3	41×10^3	39×10^3	24×10^3	12
4	*Acacia nilotica* (Babul)	43×10^5	40×10^3	39×10^3	48×10^3	27×10^3	10
5	*Albizia odoratissima* (Shiras)	41×10^5	32×10^3	34×10^3	29×10^3	26×10^3	9
6	*Tamarindus indica* (Imli/Chinch)	32×10^5	41×10^3	27×10^3	26×10^3	21×10^3	10
7	*Mitragyna perbifolia* (Kalamb)	40×10^5	35×10^3	23×10^3	29×10^3	20×10^3	3

References

Almeida, S.M., 1990. *The Flora of Savantwadi.* Vol. I–II, *Scientific Publishers, Jodhpur - India.*

Black, C.A., D.D. Evans, J.L. White L.E. Ensminger and F.E. Clark, 1965. *Methods of Soil Analysis; Chemical and Microbiological, Propertie.* Part II, American Society of *Agronomy 9*, Part II, Medison, Wisconsin, USA.

Black, C. A. (1965b), Methods of Soil Analysis: Part – II Am. Soc. Agron., Madison, Wisconsin, U.S.A.

Gerdmann, J.W. and T.H. Nicolson, 1963. Spores of mycorrhizal endogone species extracted from soil by wet sieving and decanting. *Trans. Br. Mycol. Soc.* 46: 235-244.

Hesse, P.R., 1971. A text book of soil chemical analysis. John Murray, (Pub) Ltd. London. 377.

Hooker, J.K., 1872-1897. *The Flora of British India.* Vol. I – VII, L Reeve and Co., London.

Jackson, M.L., 1967. Soil chemical analysis. Prentice Hall of India (Pvt.) New Delhi.

Joshi, S.J., 2000. Medicinal Plants. *Oxford and IBH Publishing Co., Pvt. Ltd.,* New Delhi,

Piper, C.S., 1965a. Soil and Plant Analysis. (Indian Reprint) Hans Publishers, Bombay.

Piper, C.S., 1966. Soil and Plant Analysis. (Indian Reprint) Hans publishers, Bombay.

Singh, N.P., L. Narasimhan, S. Karthikeyen and P.V. Prasanna, 2001. *Flora of Maharashtra. BSI Publication, Kolkata.*

Subbaih, B.V. and G.L. Asija, 1956. A rapid procedure for estimation of available nitrogen in soil. Curr. Sci., 25 – 259-260.

Tondon, HLS Ed., 1993. Methods of Analysis of Soil, Plants, Water and Fertilizers, FDCO, New Dehli, India.

Ugemuge, N.R., 1986. *Flora of Nagpur District. Shree Prakashan, Nagpur,* India.

Biodiversity Conservation and Envir. Management (2012)
Editors: D.R. Khanna et al.
Pub. by Biotech Books. *ISBN: 978-81-7622-262-4*

Pages: 213-220

EVALUATION OF ACTIVE PRINCIPLES OF *OCIMUM SANCTUM*, *ALOE VERA*, *AZADIRACHTA INDICA* AGAINST CERTAIN STRAINS OF *STAPHYLOCOCCUS AUREUS*, *PSEUDOMONAS AERUGINOSA*, *SALMONELLA TYPHI*, *ESCHERICHIA COLI* AND *CANDIDA ALBICANS*

Shozeb Jawed[1], Fouzia Ishaq[2], Amir Khan[3]✉

[1]Department of Biotechnology /Microbiology, Beehive College of Advance Studies, Dehradun, Uttarakhand (INDIA)

[2]Department of Zoology and Environmental Science, Gurukula Kangri University, Haridwar, Uttarakhand (INDIA)

[3]Department of Biomedical Science and Biotechnology, Dolphin P.G. Institute of Biomedical and Natural Sciences, Dehradun, Uttarakhand (INDIA)

E-mail: amiramu@gmail.com

India is rich in medicinal plant diversity and is considered as a treasure house of valuable medicinal and aromatic plants. Due to problems like adverse effects, limiting life span &

✉ Corresponding author

misuse of traditional antibiotics efforts are currently under way to look for products of natural origin. In the present study three plants were tested against different strains of microbes in varying solvents. Antimicrobial activity of ethanol and chloroform extract of *Ocimum sanctum* was found to have maximum inhibitory activity against *S. aureus* followed by *E. coli* and the ethanol extract of *Aloe vera* inhibited the growth of *Staphylococcus aureus,* the gram negative *Salmonella typhi* and fungi *Candida albicans.* The methanol extract of the leaves of *Azadirachta indica* inhibited pronounced activity against the gram positive organism *Staphylococcus aureus* and the gram negative organisms like *Salmonella typhi* followed by *Pseudomonas aeruginosa.* Di Methyl Sulfoxide (DMSO) extract of *Azadirachta indica* inhibited the growth of *Staphylococcus aureus* and *Salmonella typhi* but was inactive against *Pseudomonas aeruginosa.* The ethanol extract of *Aloe vera* inhibited the growth of *S. aureus* and *Salmonella typhi,* followed by *Candida albicans.*

Keywords: Plants extract, DMSO, Antimicrobial activity, fungi, medicinal.

Introduction

The discovery of medicine is an effort of man over millions of years in search for eternal health, longevity and remedies to relief pain and discomfort with promotion of early man to explore his natural surroundings and tried many plant minerals to develop a variety of therapeutic agents. The systematic record and incorporation into a regular system of medicine which was refined and developed, became a part of metaria, medica of many external cultures including those of India, China & Arab world. There are many approaches for the search of new biologically active principles in higher plants (Farnsworth and Loub, 1983).The ancient civilization of India, China Greece & other countries of the world developed their system of medicine independent of each other. One of such resources is folk medicine and systematic screening of them may result in the discovery of novel effective compounds (Janovska *et al.*, 2003). The oldest of them were predominantly plant based. But the theoretical foundation and in-depth understanding of practice of medicine that is defined in ayurveda, is much superior among all the organized ancient system of medicines. It is perhaps the oldest among the organized traditional system of medicines which spreaded with Vedic, Hindu, & Buddhist cultures and reached as far as Indonesia in the east and to the west. It influenced the ancient Greek to develop similar forms of medicines.

About 80% of the world population still depends on traditional medicines for primary health care. Interest in traditional medicine is reserved and now a days the demand is more & more for these drugs. In developing countries, medicine based plant resources have been grown specially during the past decades. This is because of the wide belief that the green medicine is safe and more effective than the costly synthetic drugs which have adverse side effects. Due to interest in plant based drugs there is an increased demand for medicinal plants. India is rich in medicinal plant diversity and is considered as a house of valuable medicinal and aromatic plants. Out of 17,000 species of flowering plants, the classical system of medicines like ayurveda, siddha & unani make use of only 2000 plants in various formulations. The Indian system have identified 15000 medicinal plants of which 500 species are commonly used in the herbal formulation by the different pharmaceutical industries. A number of reports regarding the antibacterial screening of plant extracts of medicinal plants have appeared in the literature (Salvet *et al.*, 2001).

Herbal medicines are in great demand both in developed and developing countries in primary health care because of their great efficiency and less side effects. People of all continents & civilization used plants in one form or the other. Due to problems like adverse effects, limiting life span & misuse of traditional antibiotics, efforts are currently under process to look for products of natural origin. Since time immemorial, man has used various parts of plants in the treatment and prevention of various ailments (Tanaka *et al.*, 2002). As a result of indiscriminate use of antimicrobial drugs in the treatment of infection, microorganisms have developed resistance against many antibiotics which result in need for more developed alternative microbial drugs. Bacteria and fungi both are responsible in causing diseases not only in human beings but also in plant and animals. The increasing failure of chemotherapeutics and antibiotic resistance exhibited by pathogenic microbial infectious agents has led to the screening of several medicinal plants for their potential antimicrobial activity (Colombo and Bosisio, 1996 and Iwu *et al.*, 1999).

Materials and Method

a) Preparation of crude extracts

Hundred grams of each of the air dried and coarsely powdered plant material was meticulously extracted for 2 hours with petroleum ether (60-80°C) in soxhlet apparatus. The petroleum ether extract was filtered and evaporated under reduced pressure using Rota vapour. The extracted plant material was then air dried, repacked in the soxhlet apparatus and carefully extracted with methanol (98.8%) for 2 hours. The methanol was filtered and evaporated under reduced pressure using rota vapour.

The extracts were dissolved in (DMSO) di-methylsulphoxide to make the final concentration which were kept in refrigerator till used. Simultaneously water extract was prepared by adding (10 ml) of boiled distilled water to 5gm of coarsely powdered plant leaf material in a beaker on water bath with occasional stirring for 4 hours. The aqueous extract was then filtered and rewashed with small volume and used immediately.

b) Preparation of the tested organisms

i) Preparation of standard bacterial suspension: The average number of viable *Staphylococcus aureus*, *Salmonella typhi*, *Pseudomonas aeruginosa*, *Escherichia coli*, organisms per ml of the stock suspensions was determined by means of the surface viable counting technique. About (10^8 – 10^9) Colony Forming Unit per ml (CFU/ml) was used. Each time a fresh stock suspension was prepared, the experimental conditions were maintained constant so the suspension with close viable counts would be obtained.

ii) Preparation of standard fungal suspension : The fungal culture *Candida albicans* was maintained on Sabouraud Dextrose Agar incubated at 25 °C for 4 days. The fungal growth was harvested and washed with sterile normal saline solution and the suspension was stored in refrigerator till used.

c) *In vitro* testing of extracts for antimicrobial activity

i) Testing for antibacterial activity: The cup plate agar diffusion method was adopted accordingly to assess the antibacterial activity of the prepared extracts (Parekh *et al.*, 2005). 0.6ml standardized bacterial stock suspensions (10^8–10^9) Colony Forming Units per ml was thoroughly mixed with 60 ml of sterile nutrient agar. 20 ml of the inoculated nutrient agar was distributed into sterile petridishes. The agar was left to set and in each of these plates, 10mm in diameter were cut using a sterile cork borer and the agar disc were removed. Alternate cups were filled with 0.1ml of each extract using micro titer- pipette and allowed to diffuse at room temperature for 2 hours. The plates were then incubated in an upright position at 37°C for 18 hours. Two replicates were carried out for each extract against each of the test organism. Simultaneously addition of the respective solvents instead of extracts was sorted as control. After incubation the diameter of the results and growth inhibition zones were measured, averaged and the mean values were tabulated.

ii) Testing for antifungal activity: The same method as for bacteria was adopted. Instead of nutrient agar, yeast and mould agar was used. The inoculated medium was incubated at 25 °C for 2 days for *Candida albicans.*

Results and Discussion

Successful prediction of botanical compounds from plant material is largely dependent on the type of solvent used in the extraction procedure. The traditional healers or practitioners make use of water primarily as a solvent, but our study showed that chloroform extracts of these plants were certainly much better and powerful. These observations can be rationalized in terms of the polarity of the compounds being extracted by each solvent and in addition to their intrinsic bioactivity, by their ability to dissolve or diffuse in the different media used in the assay. The growth media also seem to play an important role in the determination of the antibacterial activity. (Lin *et al.* 1999) reported that Muller-Hinton agar appears to be the best medium to explicate the antibacterial activity and the same was used in the present study. Amongst the gram-positive and gram-negative bacteria, gram-positive bacterial strains were more susceptible to the extracts as compared to gram negative bacteria. This is in agreement with previous reports that plant extracts are more active against gram-positive bacteria than gram-negative bacteria (Vlietinck *et al.* 1995; Rabe and Van Staden, 1997).

A) *Azadirachta indica* (Neem)

(Zone of Inhibition in mm)	*DMSO*	*Methanol*
Staphylococcus aureus	22	18
Salmonella typhi	20	20
Pseudomonas aeruginosa	——	14

B) *Ocimum sanctum* (Tulsi)

(Zone of Inhibitions in mm)	*Petroleum Ether*	*Benzene*
Escherichia coli	10	5
Candida albicans	12	9

C) *Aloe vera* Linn (*Aloe vera*)

(Zone of Inhibitions in mm)	*Ethanol*
Staphylococcus aureus	18
Salmonella typhi	4
Candida albicans	17

The methanol extract of the leaves of *Azadirachta indica* inhibited pronounced activity against some gram positive and gram negative bacteria and fungi. It was highly active against gram positive organism *Staphylococcus aureus* (18 mm) and against the gram negative organism *Salmonella typhi*, but was inactive against *Pseudomonas aeruginosa.* The methanol extract of *Azadirachta indica* inhibited the growth of *Staphylococcus aureus* (18 mm), *Salmonella typhi* (20 mm), *Pseudomonas aeruginosa* (14 mm) and Dimethylsulfoxide extract of *Azadirachta indicia* inhibited the growth of *Staphylococcus aureus* (22 mm) and *Salmonella typhi* (20mm).

The petroleum ether and Benzene extract of *Ocimum sanctum* was highly active against *Candida albicans* and showed low activity against *Escherichia coli.* The Petroleum ether extract of *Ocimum sanctum* inhibited the growth of *Candida albicans* (12 mm) and *Escherchia coli* (10 mm). Similarly the extract of benzene obtained from *Ocimum sanctum* inhibited the growth of *Candida albicans* (9 mm) followed by *Escherchia coli* (5 mm).

The ethanol extract of *Aloe vera* showed highest activity against *Staphylococcus aureus,* very low activity against *Salmonella typhi* but high against *Candida albicans.* The ethanol extract of Aloe vera inhibited the growth of *Staphylococcus aureus* (18 mm), followed by *Salmonella typhi* (4 mm) and *Candida albicans* (17 mm).

The results of present study supports the traditional usage of the studied plants and suggests that some of the plant extracts possess compounds with antimicrobial properties that can be used as antimicrobial agents in new drugs for the therapy of infectious diseases caused by pathogens. The most active extracts can be subjected to isolation of the therapeutic antimicrobials and carry out further pharmacological evaluation. The higher resistance of Gram-negative bacteria to plant extracts has previously been documented and related to thick murein layer in their outer membrane, which prevents the entry of inhibitor substances into the cell (Brantner *et al.,* 1996; Palombo and Semple, 2001; Tortora *et al.,* 2001; Matu and Van Staden, 2003). Similarly, our results indicated that the antibacterial activities of the extracts were more pronounced on Gram positive than on Gram-negative bacteria.

References

Brantner, A., Males, Z., Pepeljnjak, S. and Antolic, A. (1996): Antimicrobial activity of Paliurus spina-christi Mill (Christ's thorn). *J.Ethnopharmacol.* 52: 119-122.

Colombo, M.L. and Bosisio, E. (1996): Pharmacological activities of Chelidonium majus L (Papaveraceae). *Pharmacol Res* 33: 127-134.

Farnsworth, N.R. and Loub, W.D. (1983): *Information gathering and data bases that are pertinent to the development of plant-derived drugs in Plants: The Potentials for Extracting Protein, Medicines, and Other Useful*

Chemicals. Workshop Proceedings. OTA-BP-F-23. U.S. Congress, Office of Technology Assessment, Washington, D.C., pp. 178-195.

Iwu, M.W., Duncan, A.R. and Okunji, C.O. (1999): *New antimicrobials of plant origin. In: Janick J. ed. Perspectives on New Crops and New Uses. Alexandria*, VA: ASHS Press: pp. 457-462.

Janovska., D., Kubikova, K. and Kokoska, L. (2003): Screening for antimicrobial activity of some medicinal plant species of traditional Chinese medicine. *Czech. J. Food Sci.* 21: 107-111.

Lin, J., Opoku, A.R., Geheeb-Keller, M., Hutchings, A.D., Terblanche, S.E., Jager, A.K. and Van Staden, J. (1999): Preliminary screening of some traditional zulu medicinal plants for anti-inflammatory and antimicrobial activities. *J. Ethnopharmacol.* 68: 267-274

Matu, E.N. and Van Staden, J. (2003): Antibacterial and anti-inflammatory activities of some plants used for medicinal purposes in Kenya. J. *Ethnopharmacol.* 87: 35-41.

Palombo, E.A. and Semple, S.J. (2001): Antibacterial activity of traditional Australian medicinal plants. *J. Ethnopharmacol.* 77: 151-157.

Parekh, J., Nair, R. and Chanda, S. (2005): Preliminary screening of some folklore medicinal plants from western India for potential antimicrobial activity. *Indian J Pharmacol* 37: 408-409.

Rabe, T. and Van Staden, J. (1997): Antibacterial activity of South African plants used for medicinal purposes. *J Ethnopharmacol*; 56: 81-7.

Salvet, A., Antonnacci, L., Fortunado, R.H., Suarez, E.Y. and Godoy, H.M. (2001): Screening of some plants from northern Argentina for their anti-microbial activity. *Journal of Ethnopharmacology* 32: 293-297.

Tortora, G.J., Funke, B.R. and Case, C.L. (2001): *Microbiology: An Introduction.San Francisco:* Benjamin Cummings.

Tanaka, H., Sato, M. and Fujiwara, S. (2002): Antibacterial activity of isoflavonoidsisolatedfromErythrinavariegataagainstmethicillinresistant *Staphylococcus aureus. Lett Appl Microbiol* 35: 494-498.

Vlietinck, A.J., Van Hoof, L. and Tott, J. (1995): Screening of hundred Rwandese medicinal plants for antimicrobial and antiviral properties. *J. Ethnopharmacol* 46: 31- 47.

Biodiversity Conservation and Envir. Management (2012)
Editors: D.R. Khanna et al.
Pub. by Biotech Books. *ISBN: 978-81-7622-262-4*

Pages: 221-235

ASSESSMENT OF LICHEN PATCHES AND MOSS COMMUNITIES IN THE VICINITY OF LARSEMANN HILLS, EAST ANTARCTICA

Pawan Kumar Bharti
30th Antarctica Expedition Member, Bharti Station,
Environment Protection Division,
Shriram Institute for Industrial Research, Delhi (INDIA)
E-mail: gurupawanbharti@rediffmail.com

India has conducted scientific experiments, both at Dakshin Gangotri and Maitri stations in Antarctica in various disciplines. It now intends to broaden the scope of its scientific research by complementing the existing studies from an additional location. The proposed location for the new research base is at Larsemann Hills, Prydz Bay area, East Antarctica, which is about seven days away by ship from Maitri. The Larsemann Hills (69°20'–69°30'S lat: 75°55'–76°30'E long), named after Larsemann

Christensen, is an ice-free coastal oasis with exposed rock and low rolling hills.

However, it is mandatory to have some background ecological information prior to the initiation of station activity in the proposed area. Hence, the lichens, one of the major biological elements of Antarctica and highly privileged environmental indicators in addition to Moss communities, are studied to generate baseline information for future biomonitoring studies in the area to assess anthropogenic activities in the area after the construction of the 3rd Indian research station in Larsemann Hills.

Rock is the major substratum in the island accommodating many lichen species followed by moss species. True soil is virtually absent in the studied area, but a thin soil may be accumulated in rocks crevices, base of the rocks or in moss beds. The closely packed soil grains form a hard crust, a suitable habitat for lichens, as different species have been collected from such habitats. Organic matter comprising dead birds was frequent in the island and *Caloplaca citrina* was found growing luxuriantly in such habitats. Most of the lichens are substrate-specific, while some were found growing on all available substrate. *Buellia frigida*, *Candeleriella flava* and *Rhizoplaca melanophthalma* were found most abundant and dominant lichen species in various islands/peninsulas of Larsemann Hills especially at Bharti Promontory and Fisher Island.

Introduction

Antarctica is the most precious asset on the earth and is the last heritage of human kind. Antarctica is the only area on earth planet which is strictly devoted to scientific research and the continents of extremes come to be known as the "Continent of Science". It is the nature biggest laboratory on earth where no outside anthropogenic (human activities) interference has taken place over the centuries till recent times. Being at a unique geographic location, it offers unique opportunities for Scientists to conduct number of scientific research experiments. Antarctica is attracting world attention because of the tremendous biological species in surrounding seas and likelihood of vast hydrocarbons. Even though it is difficult to survive at Antarctica, still Scientists all around the worlds have been engaged in pursing the exciting scientific research investigations. The investigations are essential not for the exploitation of natural resources buried under the region but for the preservation of environment and ecology on earth; especially in the light of climate change.

The north and south poles of Arctic and Antarctic polar region maintain the heat budget of the world in balance. The heat transported through the atmosphere and the oceans to the poles is dissipated in space in the form of long wave radiation. The cold air going from Antarctica, when meets the warm air in the atmosphere of the lower latitude, changes into moisture bearing clouds. Thus, Antarctica polar region regulates the global climate and more particularly southern atmosphere.

Antarctica provides a unique, unpolluted and stable pure environment for carrying out scientific observation. It is far away from all sources of environmental contamination and thus remains an unpolluted datum point from which global changes due to pollution could be monitored and is suitable for a wide range of scientific research.

Antarctic Ocean supports biological communities of few species with large population and short food chain magnification. It is among the richest biological provinces on the earth. It is far away from main earth. Further, life forms in the Antarctica are also concentrated primarily in coastal lakes, which are poor in species and low in productivity, or in the surrounding seas, which are relatively rich in species. Thus, this is an ideal location for study of new species undiscovered so far.

Antarctica polar region contains almost 90 % of the world ice which constitutes about 70 % of the world's store of fresh water reserves. Further, huge amounts of water from all the regions of the world are constantly mixing together making Antarctic Ocean absolutely essential in the earth climate and heat balance (Hodgson *et al.*, 2001).

Since Antarctica is virtually untouched by human civilization and has no higher plant and animal life.

Because of the influence of world weather, and climate change, Antarctica lies at the heart of the debate on climate change and has become the premier location to study the effects of global warming and climate change.

The Larsemann Hills (69°20'–69°30'S lat: 75°55'–76°30'E long.), named after Larsemann Christensen, is an ice-free coastal oasis with exposed rock and low rolling hills. The Larsemann Hills contain hundreds of freshwater lakes of varying sizes, depth and biology (Hodgson *et al.*, 2001).

Important information about Antarctica

Parameters	Value
• Total area, million sq km	14.2
• 98% of Ice sheets	
• 2% Ice free areas	
• Antarctica possess	
• 90% of world ice	
• 70% of world fresh water resources (Stored)	
• Highest peak (Vinson Massif), m	4,897
• Height of surface of South Pole, m	2,635
• Maximum Known Thickness of ice, m	4,776
• Shortest Distance Across Antarctica, km (*via* South Pole)	3250

Antarctica is the worlds coldest and most isolated continent on the earth. The Indian Ocean, Pacific Ocean and Atlantic Ocean surround the continent. Antarctic continent covers 10% of the earth surface and has a surface area of nearly 14 million square kilometres. It also has 70% of the world's fresh water resources in the form of ice sheets. Antarctica possesses 20 % of the world seas and ocean. Thick ice sheets cover the whole continent (almost) 98% of the surface. As a result of the environmental conditions, the remaining (~2%) portion without ice cover is basically the barren soil and rocks. The Antarctica is 990 km away from Cape horn-the southernmost tip of Argentina, which is the nearest land area to Antarctica.

Antarctica is the most inaccessible to human beings in comparison to the other continent on the earth. Antarctica continent is endowed with the harshest climate on earth. The coldest temperature recorded in the Antarctica was –90°C at Vostok Station (in the year 1983). The mean temperature in Antarctica lies between –20°C to –30°C on the coast of Antarctica and –40°C to –90°C in the interior of Antarctica (http://www.ncaor.gov.in/Draftcee/index.htm). The Antarctica region in addition to extreme cold is also full of ultraviolet rays during the summer period. It is also the stormiest, driest and windiest place on the earth. There is frequent blizzards, snow storms and snow drifting in Antarctica. These factors coupled with others make Antarctica conditions harsh stressful, which makes human life very difficult in the region and survival in such conditions for the mankind is a real struggle. It is also the remotest continent. Antarctica has very few living organisms such as lichens, mosses, algae, yeast *etc.* However, the Antarctica lacks higher forms of plants and wild life except Penguins, Skua, Snow Petrel *etc.*

Study area

Larsemann Hills is at the Ingrid Christensen Coast of east Antarctica (69° 21' 00" S to 69° 27' 00" S and 75° 57' 00" E to 76° 27' 00" E). It is an ice-free Antarctic coastal oasis, located between the Amery Ice Shelf and the Vestfold Hills. Isolated islands, Peninsulas and Nunataks occur along the coastal continental ice. The areas are mainly exposed hilly and rocky terrains. There are two main peninsulas in the Larsemann Hills, of east Antarctica (i) Broknes Peninsula and (ii) Stornes Peninsula. In between these two peninsulas, there are numbers of promontory, islands of varying shapes, sizes, heights and water bodies.

Bharti Island (69° 24' to 69° 26' S; 76° 10' to 76° 15' E), is situated over the Larsemann Hills of east Antarctica. The ice patches over the lake and the rocky regions are typical characteristics of the oasis regions (Hall and Waltson, 1992). The polar ice is also seen in the background of the Bharti island, which is one of the main feeders of fresh water to various lakes in this oasis region. This promontory is sandwiched in the northern direction by McLeod Island; eastern direction comprises the Quilty bay, north-eastern direction by Fisher Island, western direction is the Stornes peninsula and the southern direction by the Antarctic continent and polar ice caps.

The ecosystem of this area is simple and in the primary stage of ecological succession. The area is devoid of any higher organism and plants except of some sea birds, Seals, Penguins, Skuas, Algae, Lichen and Mosses.

Thick mosses, algae and lichens crusts were observed mainly near the lake banks. Most of the areas are ice/snow free.

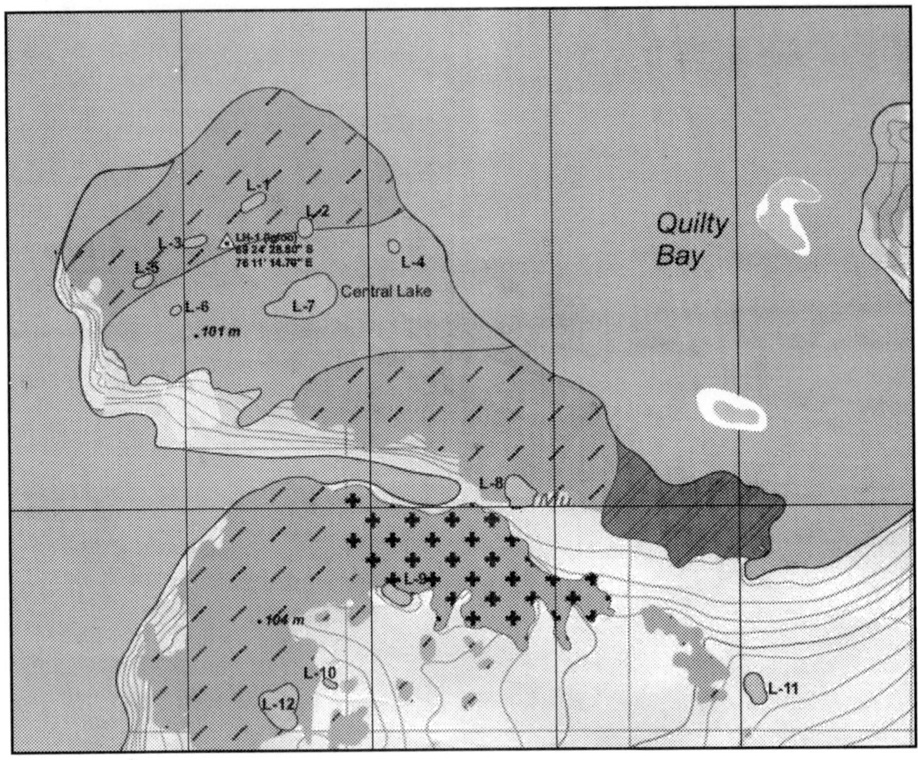

Fig.1: Site for Bharti station, Bharti Island, Larsemann hills.

A major feature of the climate of the Larsemann Hills is the existence of persistent, strong katabatic winds that blow from the northeast in summer days. Daytime air temperatures from December to February frequently may exceed 4°C, with the mean monthly temperature a little above 0°C (Draft Comprehensive Environmental Evaluation, 2004). Mean monthly winter temperatures are between –15°C and –40°C. Pack ice is extensive inshore throughout summer and the fjords and bays are rarely ice-free. Precipitation occurs as snow and is unlikely to exceed 250 mm water equivalent annually (Report of the Norwegian Antarctic Inspection, 2001). Snow cover is generally deeper and more persistent on Stornes than Broknes, due to north easterly prevailing winds and the perennial sea ice held in by the islands offshore from Stornes.

Biological study

Weak development of the weathering forms on the rock surfaces and fresh traces of glacial impact, indicate recent ice disappearance. The weathering of rocks releases minerals, an essential component for the survival and development of the ecosystem (Hall and Waltson, 1992). It may be noted that the weathering of rocks leads to the formation of sandy soil, which cannot support normal forms of plant and animal lives. At the same time, brown rocks absorb solar energy, leading to a much higher temperature of the rock surfaces, thus providing a better habitat for the unique micro flora and fauna of Antarctica.

Fig.2: (A-C): Lichen, Moss and Algae

Fig.3: Collection of Plankton from a lake in Antarctica

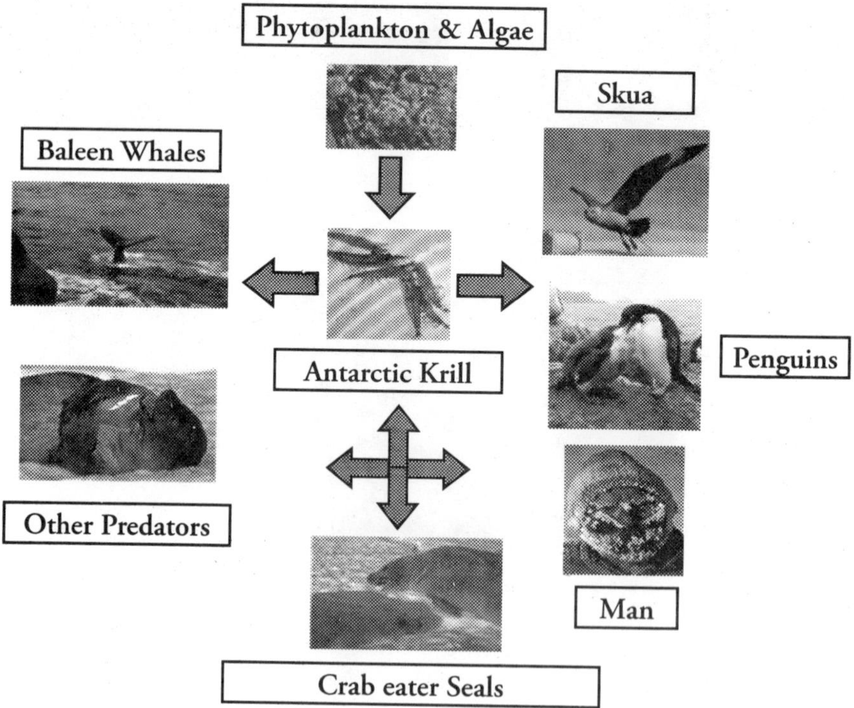

Fig.4: Antarctica Ecosystem

Lichens

Lichen is a combination of two organisms, an algae and a fungus, living together in symbiotic association. The algal component in the lichen is called phycobiont or photobiont while fungus as mycobiont. The phycobiont and the mycobiont loose their original identity during the association and the resulting entity (lichen) behave as a single organism, both morphologically and physiologically. Hence the lichen is called as a composite organism. In lichen thallus (body) the mycobiont predominates with 90% of the thallus volume and provides shape, structure and colour to the lichen with partial contribution from algae. Whatever visible from outside in a lichen thallus is fungal part, that holds algal cell inside. Hence the lichens are placed in the Kingdom – Mycota (Fungi). The fungi present in lichens are called as lichenized fungi. Among the 20,000 lichen species known in the world 95% belongs to the Ascomycetes group of fungi while Basidiomycetes and Deuteriomycetes groups are represented by only 3% and 2% of species respectively.

Lichens can grow in diverse climatic conditions and on diverse substrates. Lichens that are growing on tree trunk and bark are called corticolous lichens, twig inhabiting ones are ramicolous, on wood - legnicolous, on rocks and boulders – saxicolous (epilithic), on moss - muscicolous, on soil - terricolous and on evergreen leaves – foliicolous (epiphyllous). In general any lichen growing on other plant is called as epiphytic. The lichens can grow on underwater rocks, but not freely in water or on ice. The lichens are widely distributed in almost all the phytogeographical regions of the world. Sufficient moisture, light and altitude, unpolluted air and undisturbed, perennial substratum often favour the growth and abundance of lichens.

By their appearance the lichens can be grouped into three main categories of growth forms:

- **Crustose lichens:** The thallus in crustose lichen is closely attached to the substratum without leaving any free margin. The thallus usually lacks lower cortex and rhizines (root like structure). Such lichens are collected along with their substratum for the detailed study.
- **Foliose lichens:** They are also called as leafy lichens. The thallus in this case is loosely attached to the substratum at least at the margin. Lichens are collected by scraping them from the substratum.

- **Fruticose lichens:** Here the lichen thallus is attached to the substratum at one point and remaining major portion is either growing erect or hanging. The lichen usually appears as small shrub or bush and easy to collect with hand.

Table 1: List of lichens from McLeod Island, Larsemann Hills, East Antarctica (Singh *et al.*, 2007).

Lichen taxa	*Substratum*	*Growth form*
Acarospora gwynnii C. W. Dodge & E. D. Rudolph	Rock	Crustose
Arthonia lapidicola (Taylor) Branth & Rostr.*	Rock	Crustose
Buellia frigida Darb.	Rock	Crustose
Buellia grimmiae Filson# Moss,	Soil	Crustose
Caloplaca athallina Darb. #	Moss	Crustose
Caloplaca sp. A	Moss	Crustose
Caloplaca citrina (Hoffm.) Th. Fr.	Rock	Crustose
C. lewis-smithii Søchting & Øvstedal*	Moss	Crustose
C. saxicola (Hoffm.) Nordin* Rock,	Soil	Crustose
Candelariella flava (C. W. Dodge & Baker)	Soil, moss	Crustose
Castello & Nimis Rock,		
Carbonea vorticosa (Flörke) Hertel*	Rock	Crustose
Huea coralligera (Hue) C. W. Dodge & G. E. Baker*	Moss	Crustose
Lecanora expectans Darb. Soil,	Moss	Crustose
L. geophila (Th. Fr.) Poelt*	Moss	Crustose
Lecidea cancriformis C. W. Dodge & G. E. Baker Rock,	Soil	Crustose
Lecidella patavina (A. Massal.) Knoph & Leuckert*	Rock, soil	Crustose
L. siplei (C. W. Dodge & G. E. Baker) May. Inoue*	Rock	Crustose
Lepraria sp.# Rock, soil,	Moss	Crustose
Physcia sp.* Rock,	Soil	Foliose
P. caesia (Hoffm.) Furnr.	Rock, moss	Foliose
P. dubia (Hoffm.) Lettau*	Rock	Foliose
Pseudophebe minuscula (Nyl. ex Arnold)	Soil	Fruticose
Brodo & D. Hawksw. Rock,		
Pseudephebe minuscule (Nyl. ex Arnold)#	Rock	Fruticose
Rhizoplaca melanophthalma (Ram.)	Moss	Crustose
Leuckert & Poelt Rock, soil,		
Rinodina olivaceobrunea C. W. Dodge & G. E. Baker Soil,	Moss	Crustose
R. peloleuca (Nyl.) Mull. Arg.*	Rock	Crustose
Sarcogyne privigna (Ach.) A. Massal.*	Rock, Soil	Crustose
Umbilicaria decussata (Vill.) Zahlbr.#	Rock	Foliose
Usnea antarctica Du Rietz#	Rock	Foliose
Xanthoria elegans (Link) Th. Fr.	Rock	Foliose
Xanthoria mawsonii C. W. Dodge	Rock	Foliose

Moss

Mosses are small, soft plants that are typically 1–10 cm (0.4–4 in) tall, though some species are much larger. They commonly grow close together in clumps or mats in damp or shady locations. They do not have flowers or seeds, and their simple leaves cover the thin wiry stems. At certain times mosses produce spore capsules which may appear as beak-like capsules borne aloft on thin stalks. There are approximately 12,000 species of moss classified in the Bryophyta. The division Bryophyta formerly included not only mosses, but also liverworts and hornworts. These other two groups of bryophytes now are often placed in their own divisions. Botanically, mosses are bryophytes, or non-vascular plants. They can be distinguished from the apparently similar liverworts (Marchantiophyta or Hepaticae) by their multi-cellular rhizoids. Other differences are not universal for all mosses and all liverworts, but the presence of clearly differentiated "stem" and "leaves", the lack of deeply lobed or segmented leaves, and the absence of leaves arranged in three ranks, all point to the plant being a moss.

Benthos

Benthic macroinvertibrate are the animals inhabiting the sediment, or living on or in other available bottom substrates of freshwater, estuarine and marine ecosystems. During all or part of their life cycles, these organisms may construct attached cases, tubes or nets that they live on or in; roam free over rocks, organic debris and other substrates; or burrow freely in substrates. Although they vary in size from small form, difficult to see without magnification, to other individuals large enough to see without difficulty, macroinvertebrate are considered historically by definition to be visible to the unaided eye and retained on a US standard No. 30 sieve (0.595 mm or 0.600 mm openings) (Ellis-Evans *et al.*, 1998).

The major macroinvertebrate found in freshwater are flatworms, annelids, mollusks, crustaceans and insects. The major macroinvertebrate groups included in estuarine and marine waters are bryozoans, sponges, annelids, mollusks, roundworms, cnidarians (coelenterates) crustaceans, insects and echinoderms.

Periphyton

Biodiversity play a significant role in the structure and balance of an ecosystem. The biodiversity rich ecosystem is generally called as wealthy ecosystem. Primary producers, primary consumers, secondary consumers, ultimate consumers and decomposers all are the part of biodiversity and all have the specific role to sustain the ecosystem.

Microorganisms growing on stones, sticks, aquatic macrophytes and other submerged surfaces are useful in assessing the effect of pollutants on lakes, streams and estuaries.

Included in this group of organisms, here designated periphyton are the zoogleal and filamentous bacteria, attached protozoa, rotifers, algae and the free living microorganisms that swim, creep and lodge among the attached forms.

Plankton

The term plankton refers to those aquatic forms having little or no resistance to currents and living free floating and suspended in natural waters. Planktonic plants (Phytoplankton) and planktonic animals (Zooplankton) are covered in this section.

Plankton, particularly phytoplankton have long been used as indicators of water quality. Some species flourish in highly eutrophic waters while other is very sensitive to organic and chemical waste. The species assemblage of phytoplankton and zooplankton may be useful in assessing water quality.

Many researchers have already discovered many species of bacteria in Antarctica. There are lot of opportunities to find out the new species of plankton and bacteria in Larsemann Hills area.

Microbes

Microbiological study was also conducted for the Lake water ecosystem around Bharti Island. These few microbial parameters were selected for study-

1. Total Bacterial Count/ml
2. Psychrophillic count/ml
3. MPN coliform/100 ml
4. Yeast and Mould counts/ml

5. *Salmonella*/ 25 ml
6. *Staphylococcus aureus*/25 ml
7. *Pseudomonas aeruginosa*/ 10 ml and
8. Alga

Biodiversity in Antarctica

Many species of Antarctic wildlife are unique to the southern region. Each of the major species shares a variety of adaptations that enables them to survive in the harsh environment. They all take their food from the sea that surrounds the continent; indeed, most live at the shore, although some breed on land.

The flightless birds - Penguins are able to withstand the extreme cold because their short, densely-packed feathers forms a waterproof coat and provides insulation. A thick layer of fat or blubber also serves as an energy store. These adaptations, among others, enable them to minimize heat loss in icy cold waters so they can cope with the harsh Antarctic conditions.

Most of the seal pups can swim on the first day of their birth. Among the 35 varieties of the seals, 6 types live in Antarctica. They are Leopard seals, Crabeater seals, Ross seals, Weddell seals, Southern elephant seals & Antarctic fur seals.

Along with seals & penguins, whales also share the Antarctic waters. They belong to the family Cetacea, which includes about 75 species of huge whales, smaller dolphins & porpoises. The blue, fin, sei, humpback & minke whales migrate regularly in summers feeding on krill & small fish. These are the most dangerous of all the Antarctic animals.

At the center of the Antarctic marine ecosystem is the krill. These are underwater shrimp like creatures up to 8 cm in length. They live on phytoplankton, which grow in the warm sunlight in summer. Obviously these are the days when krill are found in unimaginable numbers. And why not? Imagine the female laying 2000 to 3000 eggs twice a year, which grow into adults in 2 or 3 years. One also may be surprised to notice that nearly 250 million tonnes of krill are consumed every year.

Antarctica also has a distinct variety of birds, which are distinguished by a thick layer of insulating fat under their skins & extensive fat deposits throughout their bodies. Their feathers provide excellent protection from the cold. There are 43 species of birds that breed within the limits of

Antarctic convergence. The giant Albatrosses, the colorful shags, ferocious skuas and the small, white, pigeon like sheathbills are some of them.

A variety of fresh water and saline lakes in Antarctica contain a limited range of aquatic life. The only vegetation observed here is mosses and lichens. They inhabit an exposed ground where moisture is available. Mosses grow rarely more than 100 mm deep even in most favorable conditions. Lichens are best adapted to survive at lower temperatures and with less light and water. More than 300 species of non-marine algae are found under stones. They may form spectacular red, yellow or green patches on the areas of permanent snow. Few species of mites, insects and invertebrates are also rarely found in the outer environment of Antarctica.

One must appreciate that even after the advent of man on this continent, the life here has remained as unspoilt & innocent as it was. And the credit for this mainly goes to the Antarctic treaty, which has played and is playing a pivotal role in the explorations of this continent.

Acknowledgement

I want to thank the Secretary, MoES, Director, Shriram Institute for Industrial Research and Director, NCAOR for providing me opportunity to execute the environmental monitoring in Antarctica. Thanks to Voyage Leader, members of XXX ISEA and all the crew members *M/V IVAN PAPANIN* who braved the tough challenges of the weather deserve special thanks.

References

Draft Comprehensive Environmental Evaluation for the concept of upgrading the Norwegian Summer Station Troll in Dronning Maud Land, Antarctica to permanent station, 2004, Norwegian Polar Institute, Polar Environmental Centre, Tromsa, Norway.

Ellis-Evans, J. C., Laybourn-Parry, J., Bayliss, P. R. and Perriss, S. J., *Arch. Hydrobiol.*, 1998, 141, 209–230.

Hall K.J. and Waltson D.W.H. (1992). "Rock weathering, soil development and colonization under a changing climate". Philosophical Transactions of the Royal Society of London. Series B, Biological Sciences 338 (14) 269-277.

Hodgson, D. A., Noon, P. E., Vyverman, W., Bryant, C. L., Gore, D. B., Appleby, P., Gilmour, M., Verleyen, E., Sabbe, K., Jones, V. J., Ellis-Evans, J. C. and Wood, P. B. (2001), Were The Larsemann Hills Ice-Free Through the Last Glacial Maximum? *Antarctic Science,* Vol. 13 (4): 440-454. http://www.ncaor.gov.in/Draftcee/index.htm

Report of the Norwegian Antarctic Inspection under Article VII of the Antarctica Treaty and Article 14 of the Protocol on Environmental Protection to the Antarctica Treaty, January 2001.

Singh, S.M.; Nayaka, S. and Upreti, D. K. (2007): Lichen communities in Larsemann Hills, East Antarctica, *Current Science*, 93 (12): 1670-72.

Biodiversity Conservation and Envir. Management (2012)
Editors: D.R. Khanna et al.
Pub. by Biotech Books. *ISBN: 978-81-7622-262-4*

Pages: 237-243

RHIZOBACTERIA: A BOON TO ENVIRONMENTAL CONCERN

Shailesh Joshi[1✉], Achleshwar Bohra[2], Amir Khan[1]
[1]Department of Biotechnology, Dolphin (P.G.) Institute of Biomedical and Natural Sciences, Dehradun, Uttarakhand (INDIA)
[2]Faculty of Sciences, J.N.V. Mahila (P.G.) Mahavidyalaya, Jodhpur, Rajasthan (INDIA)
E-mail: shailesh175@gmail.com

Extensive application of chemical fertilizers and pesticides have overwhelmed our aquatic bodies and pushed many species to the verge of extinction. Moreover many eco-friendly efforts have been put forward to resolve this issue, including the use of microorganisms for degradation of xenobiotic pesticides and as biofertilizers. Rhizospheric organisms are best suited to be used as biofertilizers as they are native to the niche. Rhizospheric bacteria that augment

✉ Corresponding author

the plant growth are referred as Plant Growth Promoting Rhizobacteria or PGPR, are sought as one of the best means to reduce the flooding aquatic bodies with chemicals. Phosphate solubilizing microorganisms (PSMs) are also a part of rhizobacterial population that can solubilize the insoluble forms of phosphate and make them available for plant uptake, thereby reducing the application of phosphate fertilizers. Another category of rhizobacteria can fix environmental nitrogen reducing the need of nitrogen fertilizers. Some of the PGPRs are also able to function as biocontroling agents, which could be a potential means of reducing application of pesticides. Rhizospheric population has an excellent potential to save the mother earth.

Keywords: Xenobiotic, Rhizobacteria, PGPR, PSM, Fertilizer.

Introduction

India is the seventh largest country in the world, with a total land area of 3,287,263 sq km (1,269,219 sq miles). A change in land use pattern implies variation in the proportion of area under different land uses at a point in two or more time periods. Over the past fifty years, while India's total population increased by about three times, the total area of land under cultivation increased by only 20.2 per cent (from 118.75 Mha. in 1951 to 141.89 Mha. in 2005-06). Most of this expansion has taken place at the expense of forest and grazing land. Despite fast expansion of the area under cultivation, less agricultural land is available on per capita basis. Direct consequences of agricultural development on the environment arise from intensive farming activities, which contribute to soil erosion, land salination and loss of nutrients. The introduction of Green Revolution in the country has been accompanied by over-exploitation of land and water resources and excessive usage of fertilizers and pesticides. (MOEF, 2009).

Moreover, intensive agriculture is an inevitable menace around the globe that has generated various negative side effects on the agro-ecosystem and the environment such as pollution and soil degradation. Various nations all around the world framed policies and laws to ensure sustainable agriculture and better environment. The policies aim to sustain productivity and conserve environmental quality of soil and water, reduce pollution and other environmentally harmful effects, recycle organic resources and produce safe foods. Rhizospheric bacteria can play an important role in proper implementation of eco-friendly policies of various nations.

Plant growth promoting rhizobacteria or PGPR benefit plants through different mechanisms of action including production of secondary metabolites such as antibiotic, cyanide and hormone like substance, production of siderophores, antagonism to soil born root pathogens, phosphate solubilization and dinitrogen fixation (Dubeikovsky *et al.*, 1993; Leong, 1986). Antagonistic attributes of PGPR could be exploited to develop biocontroling agent and direct plant growth promotion attributes could be used for developing fertilizers. Although efforts have been made to develop biocontroling agents and biofertilizers, need for better performers is always there.

Need of enhancing nitrogen bioavailability

Molecular nitrogen (N_2) is the major component (approximately 80%) of the earth's atmosphere. The element nitrogen is an essential part of many of the chemical compounds, such as proteins and nucleic acids, which are the basis of all life forms. However, N_2 cannot be used directly by biological systems to build the chemicals required for growth and reproduction. Before its incorporation into a living system, N_2 must first combine with the element hydrogen. This process of reduction of N_2, commonly referred to as "nitrogen fixation" (N-fixation) may be accomplished chemically or biologically (Hubbell and Kidder; 2003).

It is estimated that Bio Nitrogen Fertilizer (BNF) on a global scale may reach a value of 175 million metric tons of nitrogen fixed per year. The amount of nitrogen fixed in any given situation would depend upon the environmental conditions and the nature of biological system(s) present which are capable of nitrogen fixation (Zaharan, 1999) For particular situations, nitrogen fixation rates may vary from barely detectable to several hundred kilograms per hectare per year. The significance of the contribution of any BNF system to the nitrogen economy in any situation is a function of the supply and demand of the biological community for nitrogen (Hubbell and Kidder, 2003).

Along with BNF the Synthetic N fertilizer input into global agricultural systems increased by approximately 430% (~19 to ~82 Tg N) from 1965 to 1998 (Mosier, 2002). The forecast for world nitrogen fertilizer demand is estimated to increase at an annual rate of about 1.4% until 2011/2012, which is an overall increase of 7.3 million tonnes. About 69% tonnes of this growth will take place in Asia (Hubbell and Kidder, 2003).

The world's largest consumers of nitrogen are East Asia, South Asia, North America and West Europe. While their share of global consumption is modest, it is forecast that the relative contribution of Latin America and East Europe and Central Asia (EECA) to change in nitrogen use will be 10.4% and 5% respectively. The relative contribution to change in world nitrogen consumption by East Asia and South Asia is expected to be about 65% (Hubbell and Kidder, 2003).Globally fertilizer N use efficiency was approximately 50% in 1996. Inefficient use of N and energy is exacerbated by the global inequity of use distribution (Mosier, 2002).

Since fertilizer N is not used efficiently in most parts of the world, N use in excess of crop potential utilization leads to losses to the environment through volatilization and leaching. These N losses result in N fertilization of pristine terrestrial and aquatic systems through NH_x and NO_y deposition and contribute to global greenhouse gases through N_2O production and local elevated ozone concentrations due to NO_x emission. Inefficient use of Nitrogen fertilizers are likely to further degrade environment in South Asian countries including India, hence it became imperative for countries like us to employ biofertilizers to save natural environment of world.

Need of enhancing phosphorus bioavailability

Phosphorus (P) is the second most important plant nutrient after nitrogen. However, most Phosphorus in soil (up to 95-99%) is part of insoluble compounds, which makes P unavailable for plant nutrition (Corona *et al.*, 1996). In order to increase crop yields, mineral phosphate fertilizers are regularly incorporated into the soil. However, immediately after fertilizer application is done, most of the applied phosphorus transforms into an insoluble form (Pundarikakshudu, 1989). As a result, most P in the soil is found in poorly soluble, highly stable forms with limited availability to plants. Only 5% or less of the total amount of P in soil is available for plant nutrition (Boronin, 1998). The vicious cycle continues as such low bioavailability of P requires regular application of phosphate-based fertilizers (Omar, 1998). According to assessments made by experts from U.S. Geological Survey and the International Association of Fertilizer Producers, the demand for fertilizers over the next 5 years will increase by 2.5-3% annually (Gilbert, 2009). Given this rate of phosphate consumption, all global phosphate resources would be exhausted within 100-125 years (Gilbert, 2009). Taking into account the long-term increase in demand for P and phosphate production, peaking in 20 years, the importance of partial P recycling continues to grow. Recovering phosphates from livestock waste is one of the examples of reusing P for agriculture (Gilbert, 2009). Other ways to control the wastage of phosphate resources include reducing P run-off into the oceans.

Considering the anticipated food production crisis as it relates to phosphate deficit in the future, efforts to study and apply microbiological phosphate solubilization processes are well justified. Phosphate Solubilizing Microorganisms (PSM) play an important role in plant nutrition and growth promotion, especially when phosphate fertilizers are used extensively for long periods of time. It has been proven that agricultural application of PSM boosts crop yield (Khan *et al.*, 2007). On the other hand, soil activity depends on the activity of phosphate solubilizing bacteria. P solubilization mechanisms include acid formation, chelating metal ions and exchange reactions. The most active among PSM are genera: *Aspergillus*, *Penicillium*, *Curvularia* and phosphate-solubilizing yeast, which is more active in solubilizing phosphates than bacteria.

In soil with low P bioavailability, free-living phosphate-solubilizing bacteria may release phosphate ions from sparingly soluble inorganic and organic P compounds in soil (Kucey *et al.*, 1989), and thereby contribute with an increased soil phosphate pool (Smith and Read, 1997). The inorganic form of P may be held firmly in crystal lattices of largely insoluble forms, and may also be chemically bonded to the surface of clay minerals and unavailable to plants. Organic P is also largely unavailable to plants until it is converted to an inorganic form, by phosphate-solubilizing bacteria. Soluble P entering the soil after mineralization by such bacteria results in localized and short-term increases in the concentration of phosphate ions in the soil solution, which AM fungal hyphae and subsequently plants may benefit from. Organic P may be mineralized by bacteria that secrete phosphatases whereas inorganic P may be released by bacteria that excrete organic acids (Smith and Read, 1997).

Conclusion

The ever increasing demand for fertilizers is a manifestation of the increasing poplation and its demand. The indiscriminate use of fertilizers needs to be reduced with increasing use of biofertilizers. Research toward exploring or creating better bioinoculant needed to be accelerated.

References

Boronin, A.M. (1998). Rhizosphere bacteria of the genus Pseudomonas enabling plant growth and development. Sorovsky Educational Magazine, 10, 25-31.

Corona, M.E.P., Klundert, I.V.D. and Verhoeven, J.T.A. (1996) Availability of organic and inorganic phosphorus compounds as phosphorus sources for carex species. New Phytologist, 133(2), 225 231.

Dubeikovsky, A. N.; Mordukhova, E. A.; V. V. Kochetkov, F. Y.; Polikarpova, and A. M. Boronin. 1993. Growth promotion of blackeurrant softwood cuttings by recombinant strain Pseudomonas fluorescens BSP53a synthesizing in increased amount of indole-3-acetic acid. Soil. Biol. Biochem. 25: 1277-1281.

Gilbert, N. (2009) Environment: The disappearing nutrient. Nature, 461(7265), 716-718.

Hubbell, D.H. and Kidder, Gerald, 2003. Biological Nitrogen Fixation, series of fact sheets of the Soil and Water Science Department, Florida Cooperative Extension Service, Institute of Food and Agricultural Sciences, University of Florida.

Khan, M.S., Zaidi, A. and Wani, P.A. (2007). Role of phosphate-solubilizing microorganisms in sustainable agriculturer—A review. Agronomy for Sustainable Development, 27(1), 29-43.

Kucey, R.M.N., Janzen, H.H. and Leggett, M.E. 1989. Microbiologically mediated increases in plant-available phosphorus. In Advances in Agronomy, Edited by N.C. Brady. New York: Academic Press. 199 228. pp.

Leong, J. 1986. Siderophores: Their biochemistry and possible role in the biocontrol of plant pathogens. Annu. Rev. Phytopathol. 24: 187 209.

MoEF, (2009): State Trends of the Environment in State of Environment Report India, 2009. Environmental Information System (ENVIS) Ministry of Environment & Forests Government of India, 10-19.

Mosier, Arvin R. (2002). Environmental challenges associated with needed increases in global nitrogen fixation. Nutrient Cycling in Agroecosystems 63,(2-3), 101-116, DOI: 10.1023/A:1021101423341

Omar, S.A. (1998). The role of rock-phosphate-solubilizing fungi and vesicular-arbusular-mycorrhiza (VAM) in growth of wheat plants fertilized with rock phosphate. World Journal of Microbiology and Biotechnology, 14(2), 211-218.

Pundarikakshudu, R. (1989). Studies of the phosphate dynamics in a vertisol in relation to the yield and nutrient uptake of rainfed cotton. Experimental Agriculture, 25(4), 39-45.

Smith, S.E. and Read, D.J. 1997. Mycorrhizal symbiosis. Academic Press. San Diego pp.

Zahran, Hamdi Hussein, 1999. Rhizobium-legume symbiosis and nitrogen fixation under severe conditions and in an arid climate. Microbiology and Molecular Biology Reviews, 63 (4), 968–989.

Biodiversity Conservation and Envir. Management (2012)
Editors: D.R. Khanna et al.
Pub. by Biotech Books. *ISBN: 978-81-7622-262-4*

Pages: 245-254

ANTIMICROBIAL ACTIVITY AND SCREENING OF SOME MEDICINAL PLANTS (*AZADIRACHTA INDICA*, *MURRAYA KOENIGII* AND *NICOTIANA TOBACCUM*) AGAINST VARIOUS PATHOGENIC STRAINS

Amir Khan[✉1] **N. S. Rathore**[3] **and Fouzia Ishaq**[2]
[1]Department of Biomedical Science and Biotechnology, Dolphin P.G. Institute of Biomedical and Natural Sciences, Dehradun, Uttarakhand (INDIA)
[2]Department of Zoology and Environmental Science, Gurukula Kangri University, Haridwar, Uttarakhand (INDIA)
[3]Department of Biotechnology, Graphic Era University, Dehradun, Uttarakhand (INDIA)
✉E-mail: amiramu@gmail.com

Excessive uses of antibiotics have led to the prevalence of multidrug resistant microbial strains which adds

✉ Corresponding author

urgency to the search of new infection fighting strategies. Antimicrobial activity of plant extracts may be a good answer to multidrug resistance. In present study antimicrobial activity of *Azadirachta indica*, *Murraya koenigii* and *Nicotiana tobaccum* have been tested against five bacterial and five fungal strains. The present study reveals that Ethyl Acetate extract of *Azadirachta indica* showed increased antimicrobial activity against *Bacillus cereus*, *Shigella* species and *Pseudomonas aeruginosa* while Petroleum Ether extract showed maximum activity against *Klebsiella* species. No extract had shown antimicrobial activity against *Escherichia coli.* Petroleum Ether extract of *Murraya koenigii* showed increased antimicrobial activity against *Bacillus cereus*, *Shigella* species and *Pseudomonas aeruginosa* while no extract has shown antimicrobial activity against *Klebsiella* species and *Escherichia coli. Nicotiana tobaccum* had shown no antibacterial effect. None of the plant extracts have shown any antifungal activity. Finally it may be concluded that antibacterial activity of *Azadirachta indica* and *Murraya koenigii* were due to some bioactive compounds. The presence of these compounds in the extracts were checked by using thin layer chromatography. Four different compounds were separated, out of which one may be responsible for antibacterial activity. Thus these medicinal plants exhibit broad spectra of antibacterial activity which may serve as a better alternative for multidrug resistant pathogens.

Keywords: Antimicrobial activity, A-ntibiotics Multidrug resistant, Pathogens and Medicinal.

Introduction

People on all continents have long applied infusions of thousands of indigenous plants for treatment of several diseases. Chemical studies of Chinese medicinal plants provide a valuable material base for the discovery and development of new drugs of natural origin. Contrary to the synthetic drugs, antimicrobial agents of plant origin are not associated with many side effects and have an enormous therapeutic potential to heal many infectious diseases (Iwu, 1999). The increasing prevalence of multi drug resistant strains of bacteria and the recent appearance of strains with reduced susceptibility to antibiotics raises the specter of untreatable bacterial infections and adds urgency to the search of new infection fighting strategies. Systematic screening of folk medicines may result in discovery of novel effective compounds. Numerous studies have shown that aromatic and medicinal plants are sources of diverse nutrient and non nutrient molecules, many of which display antioxidant and antimicrobial properties which can protect the human body against both cellular oxidation reactions and pathogens. Thus it is important to characterize different types of medicinal plants for their antioxidant and antimicrobial potential (Mothana and Lindequist, 2004, Bajpai *et al.*, 2005; Wojdylo *et al.*, 2007). Many studies reported the activities of spices and herbs on food borne pathogenic microorganisms (Arora and Kaur, 1999; Yano *et al.*, 2006). There are many kinds of natural products that have been studied on plant protections, some of them are considered to have the ability of inducing plant defence reactions (Ghaouth *et al.*, 1997; Sticher *et al.*, 1997; Meir *et al.*, 1998; Wang *et al.*, 2009). Antimicrobial agents derived from plants include Phytoalexins, Allicin, Isothiocynate, Quinones, Flavons, Polyphenols, Coumarins, Alkaloids and Lectin compounds (Cutter, 1999). In present study *Azadirachta indica* and *Murraya koenigii* leaves were taken. *Azadirachta indica* (vernacular name "Neem") is found all over India and is being used as antimicrobial (Singh *et al.*, 1987) and antimalarial agent (Badam *et al.*, 1987) since time immemorial. Not only its seeds rather its flowers, fruits and bark are extensively used in traditional medicinal methodologies. It is reported not only to have antidiabetic (Alam *et al.*, 1989) rather antifertility properties also (Riar and Devakumar1991). The Indian plant *Murraya koenigii* commonly called "curry leaf" in English and locally known as Karivepu. The species is native of India and found everywhere in India (Bhattacharjee *et al.*, 2001). It commonly occurs in foothills of Himalaya, Assam, Sikkim, Kerala, Tamil Nadu, Andra Pradesh, Maharashtra, (Bhattacharjee, 2001). The leaves are pinnate, with 11-21cm broad and flower are small white and fragrant. On phytochemical investigation researcher claimed that leave of *Murraya koenigii* found to contain alkaloids, volatile oil, Gycozoline, Xanthotoxine and sesquiterpione (Pandey *et al.*, 2009). The leaf has been found to show antioxidant activity, hypoglycemic activity (Kesari *et al.*, 2005), antibacterial activity (Kesari *et al.*, 2005), antidysentery (Adebajo *et al.*, 2006) and also act as a hepetoprotective. The leaf extracts were used for assay of antibacterial activity against *Bacillus cereus, Shigella* species *Klebsiella* species, *Pseudomonas aeruginosa* and *Escherichia coli* and antifungal activity against *Aspergillus niger, Tricnoderma* species, *Aspergillus terrus, Candida albicans* and *Aspergillus flavus.*

Materials and Method

Leaves of *Azadirachta indica* was collected from village Sudhonwala and leaves of *Murraya koenigii* from village Manduwala, Distt. Dehradun (Uttarakhand).The leaves were washed with water to remove adhering particles and dried for one week in shady place, then in oven at 25-30 °C for few hours and powdered in mixer grinder. 50 g of dry powder of leaves was added in 700 ml of different solvents and packed in soxhlet apparatus for extraction of respective soluble bioactive molecules from the plant. The antimicrobial activity of plant extracts was assessed according to method of Liu *et al.* (2005), Cao *et al.* (2006) and Zhu *et al.* (2008) with slight modifications. Each plant extract with different solvents were incorporated onto the Whatman filter paper discs prepared and sterilized by dry heat at 140 °C in hot air oven for one hour. Test compound (10 µl) was loaded onto the filter paper disc (5mm). Different bacterial strains were swabbed in nutrient agar plates and fungal strains on Sabouraud's media plates. Discs with antimicrobial compounds of different plant extracts were placed in different media plates and zone of inhibition was measured after incubation under microaerophilic conditions. Further separation of bioactive compounds was carried out by thin layer chromatography using silica gel "G" Merck which contains 13% $CaSO_4$.

Results and Discussion

The present study reveals that Ethyl Acetate extract of *Azadirachta indica* showed increased antimicrobial activity against *Bacillus cereus, Shigella* species and *Pseudomonas aeruginosa* while Petroleum Ether extract showed maximum activity against *Klebsiella* species. No extract had shown antimicrobial activity against *Escherichia coli.* Petroleum Ether extract of *Murraya koenigii* showed increased antimicrobial activity against *Bacillus cereus*, *Shigella* species, and *Pseudomonas aeruginosa* while no extract has shown antimicrobial activity against *Klebsiella* species and *Escherichia coli.* The standard antibiotic Streptomycin in concentration (10μg/ml) is used for comparison.

Table 1: Zone of inhibition of different leaf extracts obtained from *Azadirachta indica* against various pathogenic bacterial strains.

Strain	*Solvents*	*Zone of inhibition(mm.)*			*Streptomycin control (μg/ml)*
		25%	*50%*	*100%*	
Bacillus cereus	Petroleum Ether	7	8	12	17
	Ethyl Acetate	12	13	14	25
	Ethanol	6	7	8	20
	Methanol	--	--	--	20
Shigella species	Petroleum Ether	9	11	17	20
	Ethyl Acetate	14	15	16	17
	Ethanol	8	8	10	23
	Methanol	--	--	--	20
Klebsiella species	Petroleum Ether	7	9	11	22
	Ethyl Acetate	7	8	10	21
	Ethanol	7	8	8	22
	Methanol	--	--	--	21
Pseudomonas aeruginosa	Petroleum Ether	6	7	10	23
	Ethyl Acetate	12	13	14	25
	Ethanol	--	--	--	20
	Methanol	--	--	--	19
Escherichia coli	Petroleum Ether	--	--	--	20
	Ethyl Acetate	--	--	--	19
	Ethanol	--	--	--	21
	Methanol	--	--	--	20

Table 2: Zone of inhibition of different leaf extracts obtained from *Murraya koenigii* against various pathogenic bacterial strains.

Strain	*Solvent*	*Zone of inhibition (mm.)*			*Streptomycin control*
		25%	*50%*	*100%*	
Bacillus cereus	Petroleum Ether	7	8	8	20
	Ethyl Acetate	--	--	--	20
	Ethanol	--	--	--	19
	Methanol	--	--	--	18
Shigella species	Petroleum Ether	7	7	8	18
	Ethyl Acetate	--	--	--	20
	Ethanol	--	--	--	17
	Methanol	--	--	--	21
Klebsiella species	Petroleum Ether	--	--	--	22
	Ethyl Acetate	--	--	--	21
	Ethanol	--	--	--	22
	Methanol	--	--	--	21
Pseudomonas aeruginosa	Petroleum Ether	6	7	8	22
	Ethyl Acetate	--	--	--	25
	Ethanol	--	--	--	19
	Methanol	--	--	--	18
Escherichia coli	Petroleum Ether	--	--	--	16
	Ethyl Acetate	--	--	--	18
	Ethanol	--	--	--	21
	Methanol	--	--	--	18

Thus from the results it is evident that 100% Petroleum Ether, 100% Ethyl Acetate extract of *Azadirachta indica* and 100% Petroleum Ether extract of *Murraya koenigii* showed strong antibacterial activity, especially against gram negative bacteria *Shigella* species and *Pseudomonas aeruginosa* and gram positive bacteria *Bacillus cereus.* No plant extract showed antifungal activity against any fungal strain taken.

Table 3: Minimum inhibitory concentration of extracts of *Azadiracta indica* with Petroleum Ether and Ethyl Acetate and *Murraya koenigii* with Petroleum Ether on different strains using disc diffusion method, (---) =no activity, (+) =less activity, (++) =more activity, (+++) =most activity

Concentration (mg/ml)	*Extract of Azadirachta indica with*		*Extract of Murraya koenigii with*
	Ethyl Acetate on Pseudomonas aeruginosa	*Petroleum Ether on Shigella species*	*Petroleum Ether on Pseudomonas aeruginosa*
0	---	---	---
2	---	---	---
4	+	---	---
6	++	---	---
8	++	+	---
10	++	++	+
12	++	++	++
14	+++	++	++
16	+++	++	++
18	+++	++	++
20	+++	++	++
22	+++	++	++

Table 4: Rf value of separated compounds of 100% Ethyl Acetate extract of *Azadirachta indica* with Ethyl Acetate : Hexane (15:85) solvent.

Compounds	**100%**	**50%**	**25%**
Compound 1	0.89	0.87	0.76
Compound 2	0.62	0.67	0.58
Compound 3	0.44	0.39	0.35
Compound 4	0.23	0.32	0.21

Finally it may be concluded that antibacterial activity of *Azadirachta indica* and *Murraya koenigii* is due to some bioactive compounds. The presence of these compounds in the extracts was checked by using thin layer chromatography. Four different compounds were separated, out of which one may be responsible for antibacterial activity. Thus these medicinal plants exhibit broad spectra of antibacterial activity which may serve as a better alternative for multidrug resistant pathogens.

References

Adebajo A.C., Ayoola O. F., Iwalewa E.O., Akindahunsi A.A., Omisore N.O., Adewunmi C.O. and Adenowo T.K., (2006). phytomedicine; 13(4): 246.

Alam, M.M., Siddiqui, M.B. and Hussain W. (1989). Treatment of diabetes through herbal drugs in rural India. Fitotherapia.Vol. LXI, No.3. 240-242.

Arora D.S., and Kaur J. (1999). Antimicrobial activity of spices. Int J of Antimicrobial Agents, 12: 257-262.

Badam, L., Deolankar, R.P., Kulkararni, M.M., Nagsanpgi, B.A. and Wagh, U.V. (1987). In vitro antimalarial activity of neem (Azadirachta indica) leaf and seed extracts. Indian Journal of Malariology. 24:111 117.

Bajpai M, Pande A, Tewari S.K., and Prakash D. (2005). Phenolic contents and antioxidant activity of some food and medicinal plants. International Journal of Food Sciences and Nutrition, 56(4): 287-291.

Bhattacharjee S K; Hand Book of Medicinal Plant. 3rd, pointer publishers, Jaipur, 2001, 230.

Cao J, Zeng K, Jiang W (2006). Enhancement of postharvest disease resistance in YaLi pear (Pyrus bretschneideri) fruit by salicylic acid sprays on the trees during fruit growth. Eur. J. Plant Pathol., 114:363 370.

Cutter, N. et al., (1999). Bioactivity of selected plant essential oils against L. monocytogenes. Journal of food microbiology, 7:23-30.

Ghaouth, El.A., Arul J., Wilson C., Behamou N. (1997). Biochemical and cytochemical aspects of the interaction of chitosan and Botrytis cinerea in bell pepper fruit. Postharvest Biol. Tec., 12: 183-194.

Iwu, M.W., (1999). New antimicrobials of plant origin. Perspectives on New Crops and New Uses. 457-462.

Kesari A.N., Gupta R.K. and Watal G. (2005). Hypoglycemic effect of Murraya Koenigii on normal and alloxan-diabetic rabbits. Journal of Ethnopharmacology; 97(2):247-251.

Liu H., Jiang W., Bi Y., Luo Y. (2005). Postharvest BTH treatment induces resistance of peach (Prunus persica L. cv. Jiubao) fruit to infection by Penicillium expansum and enhances activity of fruit defense mechanisms. Postharvest Biol. Tec., 35: 263-269.

Meir S., Droby S., Davidson H., Alsevia S., Cohen L., Horev B., Pilosophadas S. (1998). Suppression of Botrytis rot in cut rose flowers by postharvest application of methyl jasmonate. Postharvest Biol.Tec., 13: 235-243.

Mothana RAA and Lindequist U. (2005). Antimicrobial activity of some medicinal plants of the island Soqotra. J. of Ethnopharmacology 96: 177-181.

Pandey M.S., Gupta S.P.B.N., Pathak A. (2009). Hepatoprotective Activity of Murraya koenigii Linn Bark. Journal of Herbal Medicine and Toxicology; 3(1).

Riar, S.S. and Devakumar, C., (1991). Antifertility activity of volatile fraction of neem oil. Contraception, Sep; 44 (3):319-326.

Singh, P.P., Junnarkar, A.Y., Reddi, G.S. and Singh, K.V. (1987). Azadirachta indica neuro-psychopharmacological and antimicrobial studies. Fitotherpia 58, 235-238.

Sticher L., Mauch-Mani B., Métraux J.P. (1997). Systemic acquired resistance. Annu. Rev. Phytopathol., 35: 235-270.

Wang F., Feng G., Chen K. (2009). Defense responses of harvested tomato fruit to burdock fructooligosaccharide, a novel potential elicitor. Postharvest Biol. Tec., 1: 110-116.

Wilking J.E., Mathias J.K. Das K, Nidhaya I.S.R. and Sudhakar G. (2006). Comparative Hepatoprotective Activity of Leaf Extracts of Murraya koenigii from Indian Subtropics. Indian Journal of Natural Product; 23: 13-17.

Wojdylo A, Oszmianski J and Czemerys R. (2007). Antioxidant activity and phenolic compounds in 32 selected herbs. Food Chemistry, 105: 940-949.

Yano Y., Satomi M. and Oikawa H. (2006). Antimicrobial effect of spices and herbs on Vibrio parahaemolyticus. Int. J. of Food Microbiol., 111: 6-11

Zhu X., Cao J, Wang Q., Jiang W. (2008). Postharvest Infiltration of BTH Reduces Infection of Mango Fruits (Mangifera indica L. cv. Tainong) by Colletotrichum gloeosporioides and Enhances Resistance Inducing Compounds. J. Phytopathol., 156: 68-74.

Biodiversity Conservation and Envir. Management (2012)
Editors: D.R. Khanna et al.
Pub. by Biotech Books. *ISBN: 978-81-7622-262-4*

Pages: 255-282

MEDICINAL PLANTS AND THEIR IMPORTANCE

Shweta Sharma[1], Amir Khan[2]✉, Ashok Munjal[1]
[1]Department of Biotechnology, Banasthali Vidhya Peeth, Rajasthan, (INDIA)
[2]Department of Biotechnology and Biomedical Science, Dolphin (PG) Institute of Biomedical & Natural Sciences, Dehradun, Uttarakhand (INDIA)

The bioactive compounds and the role of medicinal plants in Ayurvedic systems of medicine in India and their earlier investigation have been studied. There has been an increase in demand for the Phyto-pharmaceutical products of Ayurveda in Western countries, because of the fact that the allopathic drugs have more side effects. Many pharmaceutical companies are now concentrating on manufacturing of Ayurvedic Phyto-pharmaceutical products. Ayurveda is the Indian traditional system of

✉ Corresponding author

medicine, which also deals about Pharmaceutical Science. Different type of plant parts are used for the Ayurvedic formulation; overall out line of those herbal scenario and its future prospects for the scientific evaluation of medicinal plants used by traditional healers is also discussed. Various plant parts have been used for different types of medicinal preparation. These medicines are very cost effective with no side effects.

Keywords: Ayurvedic system, medicinal plants, antibiotics, allopathy, traditional healers, plant extract.

Introduction

Plants have been used as medicines throughout history. Indeed, studies of wild animals showed that they also instinctively eat certain plants to treat themselves for certain illnesses.

Medicinal plants are widely and successfully used on every continent. In Asia, the practice of herbal medicine is extremely well established and documented. As a result, most of the medicinal plants that have international recognition come from this region, particularly from China and India.

In Europe and North America, the use of herbal medicine is increasing fast, especially for correcting imbalances caused by modern diets and lifestyles. Many people now take medicinal plant products on a daily basis, to maintain good health as much as to treat illness.

In Africa, attitudes towards traditional, herbal medicines vary strongly. One reason for this is the confusion between herbal medicine and witchcraft. The use of medicinal plant is sometimes associated with superstition and therefore rejected by some people in favour of western medicine. On the other hand, there are millions of Africans who prefer traditional methods of treatment. The valuable medicinal properties contained in certain plants are not, however, in doubt.

In recent years, for example, the Chinese plant *Artemisia annua*, has become the essential ingredient in a new generation of anti-malarial drugs. The plant is now being grown in East African countries to supply pharmaceutical manufacturers in Europe. The bark of the tree *Prunus africana* is used in making treatments for prostate cancer. *Sutherlandia*, a native plant of South Africa is being increasingly recognised for its value to HIV/AIDS sufferers. Other African plants, such as Devil's Claw and African Geranium, are also gaining popularity as herbal medicines, particularly in Europe.

Medicinal plants therefore represent an important opportunity to rural communities in Africa, as a source of affordable medicine and as a source of income. Governments too need to be thinking about how to promote the benefits that medicinal plants have to offer, which may involve integrating herbal medicine into conventional healthcare systems. This raises important issues, such as regulation of traditional healers and ensuring certain standards are met (Rural resource Pack, 2007).

There is a growing focus on the importance of medicinal plants and traditional health systems in solving the health care problems of the world. Because of this awareness, the international trade in plants of medical importance is growing phenomenally, often to the detriment of natural habitats and mother populations in the countries of origin. Most developing countries have viewed traditional medical practice as an integral part of their culture. In spite of this traditional health care systems suffered a setback during colonial times and lost patronage particularly in urban areas. Unfortunately, post-independent India still suffering from the colonial hang up continued to favor western allopathic medicine as the system of choice for India's health care system. The results are there for all to see. We have next to no primary health care and practically non-existent veterinary care in our rural areas because the expensive western system is too slow (it takes 5 years to train a doctor) to meet the critical health care needs of our exploding population.

These problems can be solved by a strong revival of the Indian systems of medicine like Ayurveda and Siddha and a thoughtful, high level investment in developing the medicinal plant base and manpower needed to translate our traditional skills and resources to a functional, modern system. Our role model should be China which has a tradition as old as ours but has done a far better job of maintaining that tradition and making it relevant to China's health needs. In China traditional medicine still retains 40 % share of the medicine market nationwide. In rural areas, 90 % of the medicines used are of traditional origin. The sale of traditional medicines in China has more than doubled in the last five years. Today, the majority of China's factory processed drugs is of plant origin and medicinal plant preparations are almost as important as others like antibiotics. A highly developed and well organized industry delivers quality drugs to the Chinese people and exports plant based drugs to markets where the Chinese system of medicine is practiced. Although most of this goes to Asian countries including India and China is developing markets in Europe, Canada and the USA.

India should develop its domestic market as well as exploit the international market which is growing very large, very quickly. With the interest in phyto or plant based medicines growing in Europe and the US, it is estimated that trade in herbal drugs and cosmetics is of the order of a few billion dollars annually and it is a growing market. One only has to see the instant success and rapid growth of a small level entrepreneur like Shahnaz Hussain to understand what a well organized industry backed by research and standardization could achieve. India should not export medicinal plants but go for value addition of the highest order and manufacture and export phyto medicines to international specifications. There is more than enough expertise available in the country and short training programs can bring local manufacturers up to international standards. Unfortunately, at present most Asian countries export raw materials even though they have the technical knowledge and pharmaceutical industries to add value to their resources. Very often, the financial resources are lacking or industry is not organized to exploit this branch. This results in the situation that almost 90 percent of the finished health products is still being manufactured in Europe and North America. At present a lot of raw material is being imported by Europe and America, often in violation of the Convention on International Trade in Endangered Species (CITES). Germany is the largest importer of unprocessed medicinal plants (and a significant offender of CITES regulations) since it has a strong tradition of alternative medicine. Along with the formal allopathic system, other, alternative forms of medicine are used and prescribed by doctors in conservative institutions like University clinics and Government hospitals. Von Ardennes' oxygen enrichment therapy is as widely used as ayurveda and acupuncture. Germany, which gave the world homeopathy, is comfortable with the use of plants as medicine and along with its own tradition of using European plants like lavender, rosemary and gentian for therapeutic purposes; it has willingly included alternative and traditional healing methods from other cultures. Policy and use regulation is one of the most sensitive aspects of developing and using plant based medicines and health products. This means that use must be sustainable and the harvesting and gathering of medicinal plants should be very strictly regulated so as not to kill off the goose that lays the golden egg. Indiscriminate harvesting will lead to the extinction of natural populations which are still the only source of bioresources. Already, world markets experience wild fluctuations in the price of herbals. Such fluctuations usually come in cycles of six to nine years since the availability of many wild plants goes from oversupply to scarcity very quickly and then stabilizes again. These swings reflect the stages when the plants are over harvested are therefore in short supply and command a high price. At this time natural populations are under extreme stress and some are threatened with possible extinction. This price swing would be a good indicator for the government to gauge the threat to distinct plant types by overexploitation and could help to identify the habitats that must be put under strict regulation to foster conservation.

At present there is almost no policy worth its name to regulate the procurement and sale of medicinal plants in India. At the other end of the spectrum, products derived from medicinal plants are subject to no controls either. Infact, herbal products do not even have to pass the scrutiny of the Drug Controller of India, like allopathic drugs have to. The Drug Controllers office has no authority to examine or challenge the contents of herbal products. This has led to stories about spurious contents and blatant cheating on the part of the better known herbal drug companies. One story, apocryphal or not, about a very well known company says that they produce more Chyavanprash than would be possible if they had bought every single Amla (*Emblica officinalis*) grown in the country. Amla is an essential ingredient of Chyavanprash. The allegation against this company is that it uses the humble pumpkin instead of the expensive Amla, thus inflating its profits several fold, even as it puts a spurious product on the market.

If India is to become a major player in the international phyto-medicine and herbal cosmetic market, it will have to enforce stringent quality control and transparency like China which has very strict criteria for regulating the sale of traditional plant based medicines. In a paper prepared for the World Bank, Lambert and coauthors describe the Chinese treatment and handling of medicinal plants and herbal drugs. There are several lessons that India can learn from China, like how to maintain the high standard of quality control that gives Chinese products their excellent reputation. All traditional Chinese medicines that are produced by the pharmaceutical industry for export or for local use, are subject to quality control tests before being released. Each factory has its own quality control unit that makes random quality checks.

The important thing about the Chinese quality control system is the attitude. Rather than wanting to cut corners and boost production (and increase profits) by putting in pumpkin instead of Amla, the attitude is that high standards must be maintained and the quality and reputation of Chinese traditional medicines produced in China must be ensured. The government has laid down rigid criteria for traditional medicines. The manufacturer must list all ingredients. Reviewing authorities check on the veracity of these listings. It is only when they are satisfied that the product conforms to the specifications described in Chinese traditional medicine, and that it is safe and effective, will the medicine be allowed to be released in the market or for export. The review and assessment is not carried out by some silly bureaucratic committee but by persons trained in Chinese medicine. The Chinese are aware of the scepticism with which new and alien products can be treated by the global community. In their efforts to integrate Chinese medicines into the modern medicine system and conquer a share of the international pharmaceutical trade, they have laid a very strong emphasis on transparency, standardization and quality control. In order to achieve this, they have developed pragmatic, workable methods to upgrade and modernize the production of Chinese medicines. Instead of pontificating and theorizing as we tend to do in India, the Chinese have made their system viable by a hands on approach. This includes establishing a close working relationship between field scientists, pharmacologists and clinicians so that an all round integration is achieved.

The Indian systems of Ayurveda, Unani, Siddha and the host of other lesser known medicine practices in tribal areas are a fund of knowledge and wealth. We have not yet developed a comprehensive, overall strategy for forward and backward integration between collectors, vaids and hakims on the one hand and the scientists, modern market place and export arena on the other. Until we do that with rigid quality control, we will not be able to translate our assets into export earnings or into desperately needed, affordable health care for our people.

List of Medicinally Important Plants

NB: (Fam - Family, T – Tree, H – Herb, C – Climber, S- Shrub)

Plant	*Common name / Maturity period*	*Botanical Name or Family*	*Parts Used*	*Average Price (Rs. / kg)*	*Medicinal Use*
	Amla (T) After 4th year	*Emblica officinalis* Fam: euphorbiaceac	Fruit	Rs 15 – 45/kg	Vitamin – C, Cough, Diabetes, cold, Laxativ, hyper acidity.
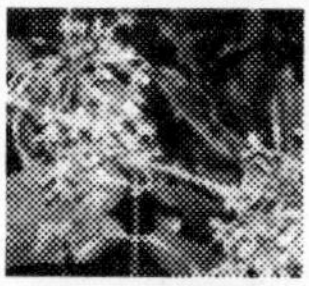	Ashok (T)10 years onward	*Saraca asoca* Fam: Caesalpinanceac	Bark Flower	Dry Bark Rs 125/ kg	Menstrual pain, uterine, disorder, Diabetes.
	Aswagandha (H), One year	*Withania somnifera* Fam: Solanacae	Root, Leafs	Rs 140/ Kg	Restorative Tonic, stress, nerves disorder, aphrodisiaic.
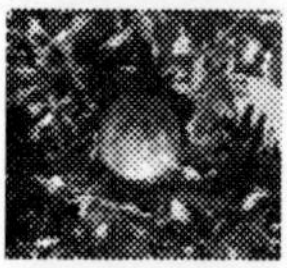	Bael/Bilva (T) After 4-5 year	*Aegle marmelous* Fam: Rutacace	Fruit, Bark	Fruit – Rs 125 / kg Pulp – Rs 60 / Kg	Diarrrhoea, Dysentry, Constipation.
	Bhumi Amla (H), with in one year	*Phyllanthous amarus* Fam : Euphorbiacae	Whole Plant	Rs 40 / kg	Aenimic, Jaundice, Dropsy.
	Brahmi (H) Indian penny worth/one year	*Bacopa, monnieri* Fam: Scrophulariacae	Whole plant	Rs 20 per kg	Nervous, Memory enhancer, mental disorder.
	Chiraita (high altituted) with in one year (H)	*Swertia chiraita* Fam: Gentianacace	Whole Plant	Rs 300-350 / per kg	Skin Desease, Burning, censation, fever.

contd...

Plant	*Common name / Maturity period*	*Botanical Name or Family*	*Parts Used*	*Average Price (Rs. / kg)*	*Medicinal Use*
	Gudmar/ madhunasini, after Four year (C)	*Gymnema sylvestre* Fam: Asclepiadaccac	Leaves	Rs 50 –75 per kg	Diabetes, hydrocil, Asthama.
	Guggul (T) after 8 years	*Commiphora wightii* Fam: burseracace	Gum resin	Rs 80 – 100 per kg	Rheumatised, arthritis, paralysis, laxative.
	Guluchi /Giloe (C)With in one year	*Tinospora* Fam: Cordifolia	Stem	Rs 20 –25 per kg	Gout, Pile, general debility, fever, Jaundice.
	Calihari / panchangulia Glori Lily Five years	*Gloriosa superba* Fam: Liliacace	Seed, tuber	Rs 60	Skin Desease, Labour pain, Abortion, General debility.
	Kalmegh/Bhui neem (H) with in one year	*Andrographis* Fam :Paniculata scanthacace	Whole Plant	Rs 12 - 20	Fever, weekness, release of gas.
	Long pepper / Pippali (C) after two to three years	*Peeper longum* Fam : Piperacace	Fruit, Root	Rs 100 – 150 per kg Root – 150 per kg	Appetizer, enlarged spleen, Bronchities, Cold, antidote.
	Makoi(H) Kakamachi/ With in one year	*Solanum nigrum* Fam: Solanacace	Fruit/ whole plant	Rs 40 per kg Seed – 200 per kg	Dropsy, General debility, Diuretic, anti dysenteric.
	Pashan Bheda / Pathar Chur (H) One year	*Coleus barbatus* Fam: Lamiacace	Root	Rs 40-50 per kg	Kidny stone, Calculus.
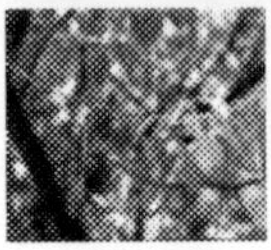	Sandal Wood (T)Thirty years onward	*Santalum Album* Fam: Santalinacace	Heart wood, oil	Rs 350 per kg	Skin disorder, Burning, sensation, Jaundice, Cough.

contd...

Plant	*Common name / Maturity period*	*Botanical Name or Family*	*Parts Used*	*Average Price (Rs. / kg)*	*Medicinal Use*
	Sarpa Gandha (H) After 2 year	*Rauwolfia serpentina* Fam: Apocynacace	Root	Root – Rs 60 per kg Seed – Rs 300 per kg	Hyper tension, insomnia.
	Satavari (C) After 2-3 year	*Asparagus racemosus* Family: Liliacace	Tuber, root	Rs 20 –50 per kg	Enhance lactation, general weekness, fatigue, cough.
	Senna (S)With in 1 year	*Casia augustifolia* Fam: Liliaceae	Dry Tubers	Rs 500/ kg seed Rs1200/ kg dry	Rheumatism, general debility tonic, aphrodisiac.
	Tulsi (perennial) Each 3 months	*Ocimum sanctum* Fam: Lamiaccac	Leaves/ Seed	Leaves Rs 10/kg	Cough, Cold, bronchitis, expectorand.
	Vai Vidanka (C), 2nd year onward	*Embelia ribes* Fam: Myrsinacace	Root, Fruit, Leaves	Rs 40-50 per kg	Skin disease, Snake Bite, Helminthiasis.
	Pippermint (h) Perennial	*Mentha pipertia* Fam:Lamiacace	Leaves, Flower, Oil	-	Digestive, Pain killer.
	Henna/Mehdi (S) 1/25 years	*Lawsennia iermis* Fam: Lytharaceae	Leaf, Flower, Seed	L – 50 /kg Powder- Rs.75 per kg	Burning, Steam, Anti Imflamatary.
	Gritkumari (H) 2nd-5th yr	*Aloe vera* Fam: Liliaceae	Leaves	Fresh L- Rs 5 kg Juice 90 Per kg	Laxative, Wound healing, Skin burns & care, Ulcer.
	Sada Bahar (H) Periwinkle/ Nyantara	*Vincea roseal* Catharanthus roseus Fam: Apocyanace	Whole Plant	R-Rs50 per kg L- Rs 25 S- Rs 10kg	Leaukmia, Hypotensiv, Antispasmodic, Antidot.

contd...

Plant	*Common name / Maturity period*	*Botanical Name or Family*	*Parts Used*	*Average Price (Rs. / kg)*	*Medicinal Use*
	Vringraj (H)	*Eclipta alba* Fam: Compositae	Seed/ whole	Powder- Rs 60/kg	Anti-inflamatory, Digestive, hairtonic.
	Swet chitrak Perennial (h)	*Plumbago zxeylanica* Fam: Plumbaginaceae	Root, Rootbar	-	Appetiser, Antibacterial, Aticancer.
	Rakta Chitrak (H)	*Plumbago indica* Fam : Plumbaginaceae	Root, Root bar	-	Indyspeipsia, colic, imflammation, cough.
	Kochila (T)15 yrs	*Strychinos nuxvomica* Fam: loganiaceae	Seed	-	Nervous, Paralysis, healing wound.
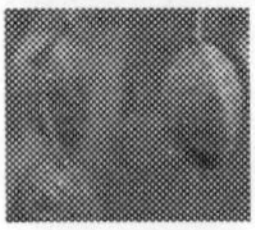	Harida (T)	*Terminalia chebula* Fam: Combretaceae	Seed	Rs. 80 per K Powder	Trifala, wound ulcer, leprosy, inflammation, Cough.
	Bahada (T)	*Terminalia bellerica* Fam: comretaceae	Seed, Bark	Fruit – Rs 20/k Powder- Rs 100/k	Cough, Insomnia, Dropsy, Vomiting, Ulcer, Trifala.
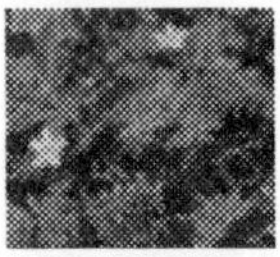	Gokhur (H) Crawling Puncture Vine/1 yr	*Tribulus terrestris* Fam: Lygophyllaceae	Whole Plant	Plant-Rs 10/K Fruit – Rs 15/k	Sweet cooling, Aphrodisiac, Appetizer, Digestive, Urinary.
	Neem (T)	*Azadirachta indica* Fam : Mahaceae	Rhizome	Rs 45/k	Sedative, analgesic, epilepsy, hypertensive.
	Anantamool/ sariva (S) Indian Sarap sarilla	*Hemibismus Indicus* Fam: Asclepiadaceae	Root/ Leaf	Rs 45/k root Rs 90/ kPowder	Appetiser, Carminative, Aphrodisiac, Astringent.

contd...

Plant	*Common name / Maturity period*	*Botanical Name or Family*	*Parts Used*	*Average Price (Rs. / kg)*	*Medicinal Use*
	Bach (H) Sweet Flag/1 yr	*Acorus calamus* Fam : Araceae	Rhizome	Rs 45/K	Sedative, analgesic, tpilepsy, hypertensive.
	Vasa (S)	*Adhatoda vesica* Fam : Sacanthaceae	Whole Plant	Leaf – Rs 25/ k	Antispasmodic, respiratory, Stimulant.
	Nageswar (T) Nag Champa	*Mesua ferrea* Fam : Guttiferae	Bark, Leaf, Flower	Flower – Rs 120/k Powder Rs 175/k	Asthma, Skin, Burning, Vomiting, Dysentry, Piles.
	Benachar (S) Khus/khus	*Vetiveria ziziinoides* Fam: Toaceae / Graminae	Root	Flower – Rs 120/k Powder Rs 175/k	Hyperdisia, Burning, ulcer, Skin, Vomiting.
	Mandukparni (H) Indian pennywort	*Centella asiatica* Fam: Umdelliferae	Whole plant	-	Antiinflamatory, Jaundice, Diuretic, Diarrhoea.
	Kaincha/ Creeper Baidanka	*Mucuna truriens* Fam: Fabaceae	Root, Hair, Seed, Leaf	-	Nervous, Disorder, Constipation, Nephroaphy, Strangury, Dropsy.
	Dalchini Perenial Shrub	*Cinnamomum zeylanicum* Fam: Lauraceae	Bark, Oil	-	Bronchitis, Asthma, Cardiac, Disorder, Fever.

Utilization of medicinal plants in traditional systems of medicine

Traditional medicine is widespread through the world and it comprises of those practices based on beliefs that were in existence, often for hundreds of years, before the development and spread of modern scientific medicines and which are still in use today. As its name implies, it is the part of tradition of each country which employes practices that have been handed down from generation to generation. Its acceptance by a population is largely conditioned by cultural factors and much of traditional medicine, therefore, may not be easily transferable from one culture to another.

An important feature of traditional therapy is the preference of practitioner for compound prescriptions over single substance/drug as it is being held that some constituents are effective only in the presence of others. This renders assessment of efficacy and eventually identification of active principles as required in international standards much difficult than for simple preparation. In India, earlier the medicines used in indigenous systems of medicines were generally prepared by the practicing physicians by themselves, but now this practice has been largely replaced by the establishment of organized indigenous drug industries. It is estimated that at present there are more than 1,00,000 licensed registered practitioners of Ayurveda, Siddha, Unani medicine or Homeopathy. As far as the Ayurveda system of medicine is concerned, it does not rule out any substances being used as potential source of medicine. Presently about 1000 single drugs and 8000 compound drug formulations of recognized merit are in vogue. In fact reliable data on availability in different regions of country as well as supply and demand of medicinal plants used in production of indigenous medicines are not available.

However, annual herbal drug market has been estimated around 2200 crores and is expected to reach up to 4000 crores by the year 2000.

Plant-parts, extracts and galenicals

The direct utilization of plant material is not only a feature of ISM in the developing world but also in developed countries like USA, UK, Germany *etc.*, the various herbal formulations are sold on health food shops. Preparation of decoctions, tinctures, galenicals and total extracts of plants also form a part of many pharmacopoeias of the world. The current trend of medicinal plants based drug industry is to procedure standard extracts of plants as raw material.

Essential Oils from plants

The essential oil industry was traditionally a cottage industry in India. Since 1947, a number of industrial companies have been established for large scale production of essential oils, oleoresins and perfumes. The essential oil from plants includes Ajowan oil, Eucalyptus oil, Geranium oil, Lavender oil, Palmarosa oil, Patchouli oil, Rose oil, Sandalwood oil, urpentine oil and Vetiver oil.

Parts of plant use for extraction

Leaves: For maximum value of most species, pick clean, dry undamaged leaves or sprigs at midmorning just before flowering and after the dew is off. Dry in bunches in warm, dust-free, circulating air, out of the sun, until brittle (4-10) days. Store in dark, airtight jars.

Flowers: Spread out small flowers or thick petals of large flowers on paper or gauge, in warm, dust free, circulating air for 1-3 weeks. Dry roses or other large flowers head upright in mesh. Hang loose bundles of Lavender stems and remove the flowers later.

Seeds: Pick seeds when ripe from healthy plants on a warm, dry day. Shake into a paper bag or cut whole stalks. Lay seeds or stalks on paper or hang above an open box in a warm place for two weeks to ensure no moisture remains. Rub seeds from their stalks or pods; store in airtight jars.

Roots: Shake or rub off the soil, remove fibrous roots, and scrub clean. Chop, then spread to dry in a warm oven (120 – 150 degrees F, 50 – 60 degrees C), for 2-6 hours until brittle. Store in dark, airtight jars and label. Most roots prepared in this way will keep for years without absorbing moisture.

Resin, Gum and Latex Extraction: Resin, gum, and latex are harvested by puncturing or cutting the bark in diagonal groves, avoiding the cambium layer. It is collected later. Pine, copal, dragons blood (from Dracaema species), dammar (from Shorea species), balsams, mastic, and storax are collected in this way. Resin is also collected as naturally exuded "teas" from Frankincense, Myrrh, and Gum trees.

Essential Oils: It's easier to buy essential oils, as it takes a lot of plant material to get an ounce of essential oil. It takes something like a ton of rose petals to get an ounce of rose essential oil. Buy oils that have been tested for purity and extracted from organically grown plants. They should be sold in dark glass bottles with a dropper, and labeled with the botanic name, country of origin, and safety advice. Keep oils in a cool dark place. They may be fatal if ingested, so store securely away from children. Users of essential oils must be extremely safety conscious. Essential oils should not be taken internally except when prescribed by a qualified person, and training is necessary to learn how to use oils safely. Some should-be avoided during pregnancy, for example, or by suffers of epilepsy or high blood pressure, or by people with sensitive skin. Essential oils should not be confused with pressed oils, usually from seeds, used in cooking.

Type of herbal preparations

Infusions: Some herbalists use the word tea when talking about either an infusion or a decoction. Using these more specific words helps remind you, first, that you are drinking herbs as medicine, not as thirst-quenchers, and second, how each one is made. You make an infusion with the softer parts of an herb that grow above ground stems, leaves and/or flowers. Because they are soft, you don't need to cook them over heat. Proportions of herb to water vary slightly depending on the herb you are using, but usually 1 to 3 teaspoons per cup of water is recommended. Because of the short storage time of infusions, you will usually make only one to six cups at a time. For most herbs, unless otherwise directed, you can drink one to three cups a day to relieve discomfort. To make an infusion; place the fresh or dried leaf, flower, crushed seed, bark, or root into a teapot; adding boiling water and brew for 5 minutes and strain and serve. Use a teaspoon of dried herb, or a fresh sprig of about 6-9 leaves per cup.

Decoctions: Because decoctions use the tough parts of an herb (bark, roots, or dry berries), you treat them in a slightly different manner from the way you would treat delicate flowers. A decoction is similar to an infusion except that you simmer or boil the herbs instead of simply steeping them. The proportions of the herb to water vary slightly depending on the herb you are using; usually 1 to 2 teaspoons of dried herbs per 2 cups of water. Like infusions, decoctions have a short storage time, so you will usually make only one to six cups at a time. Bruise the root, bark, or seed and put in a pan of cold water; cover, bring to a boil; and simmer until reduced to ¼ of the volume and then strain and use.

Tincture: Tinctures are made with alcohol. Most people use vodka, but you can also use grain alcohol (198 proof, compared to vodka's 40 – 100 proof). 100 proof alcohol is 50% alcohol; 80 proof is 40% alcohol. Some people use rum or brandy when they make tinctures with bitter herbs, to help disguise the taste. If you're preparing a tincture for use by children, recovered alcoholics, or others who should avoid alcohol, you can substitute cider vinegar for the alcohol. Unfortunately, vinegar doesn't extract some herbs as efficiently as alcohol does. Tinctures are more concentrated and keep longer than infusions and decoctions. Never use denatured or rubbing alcohol for this purpose. It is highly toxic and should never be taken internally.

Syrups: The bitterness of some herbs serves natural purposes; to keep us from overdosing, as well as to stimulate digestive juices. Unfortunately, the bitterness is no help to us if we turn up our noses at a medicine's taste and refuse to take it. Syrups are a good alternative way to prepare some mixtures and one that can extend the storage life of the herb. And for colds or the flu, syrups can also soothe a sore throat. Honey should never be given to children younger than a year old. Honey sometimes contains botulism spores and although the count is so low that it doesn't affect older children and adults, the digestive systems of very young children aren't mature enough to handle the spores.

Capsules: If you'd rather not drink your medicines, you can buy capsules. One advantage of capsules is that they are easier than liquids to take along to work or when traveling.

Poultices: If you are hiking in the woods and lose a fight with a briar patch, you can help stop the bleeding by pressing a handful of wild geranium (*Geranium maculatum*) or yarrow leaves (*Achillea millefolium*) against the cut. That bunch of leaves is actually a poultice. At home and with more time, you can make a more sophisticated version. You can use fresh, dry, or powdered herbs to make poultices. The goal is to chop them finely and get them damp enough that they release their volatile oils. Fresh herbs will supply some of their own moisture. The amount you make will depend on the size of the injured area. Depending on the type of injury, you might want to make enough so you can change the poultice several times, with two or three hours between changes. If possible, you can wrap the area with a piece of gauze to hold the poultice in place.

Compresses: The approach is similar to a poultice, but you use a liquid solution instead of whole herbs. Traditionally, a compress is hot, but sometimes, especially for a headache, a cold compress may fell better.

Herbal Oils: To make homemade oils (not essential oils), is a lot like making sun tea, only slower and most are made in the dark, with the exception of St. John's Wort oil which has to be made in the sun for it to work correctly. Herbal oil infusions are useful for massage to relieve pain or tension and also as a treatment for skin problems. None are for internal use. You need a large, clear glass container and a high quality cooking oil, such as extra virgin olive oil, sanflower oil, or grape seed oil. Fill the jar ¾ full (or ½ full) of the herb and then put enough oil in to cover it to about 1 inch above the herb. Let it sit in a darkened area for around 14 days. Then strain and bottle in darkened bottles.

Ointments: Ointments are used to protect the skin from air and moisture. Non-penetrating, they form a barrier on the surface of the skin, shielding raw, irritated, or wounded areas while providing antibacterial and/or antifungal healing benefits.

Salves: Although you can find a wide array of salves and creams at any health food store, you may enjoy making your own, tailored to you personal need and aromatic preferences. Salves are thicker than ointments. Although recipes vary considerably, most of them call for beeswax which you can find at health food stores in easy-to measure 1 ounce cubes. Some recipes call for lanolin or glycerin, which can also be found in most stores that sell herbal products. You can make salves using either an infusion or a decoction, using fresh or dried herbs.

Medicinal plants frequently used in ayurvedic formulation

Ayurveda, whose history goes back to 5000 B.C., is one of the ancient health care systems. The Ayurveda was developed through daily life experiences with the mutual relationship between mankind and nature. The ancient text of Ayurveda reports more than 2000 plant species for their therapeutic potentials. Besides Ayurveda, other traditional and folklore systems of health care were developed in the different time periods in Indian subcontinent, where more than 7500 plant species were used. According to a WHO estimate, about 80 % of the world population relies on traditional systems of medicines for primary health care, where plants form the dominant component over other natural resources (Ram, 1997).

Ayurveda, pañcavidhakaşayakalpana are the two basic pharmaceutical preparations, from which all the other preparations are formulated. Sarangdhara mentioned detailed information about various formulations with respect to their methods of preparation as well as basic standards and are documented in Sărangdhara Samhita (Mukherjee and Wahile, 2006). The different drug is prepared by percentage dry weight of the plant parts for the Ayurvedic formulations.

Anemia: *Asparagus racemosus* (roots) 20%, *Withania somnifera* (roots) 20%, *Phyllanthus emblica* (fruits) 15%, *P. amarus* (leaves) 10%, *Tephrosia purpurea* (leaves) 10%, *Plumbago zeylanica* (roots) 5%, *Glycyrrhiza glabra* (roots) 15% and *Piper longum* (fruits) 5%. 4 gm of powder is given to the patient, twice daily with water.

Asthma bronchitis: *Solanum xanthocarpum* (whole plant) 25%, *Piper longum* (fruits) 10%, *Adhatoda vasica* (leaves) 25%, *Zingiber officinale* (roots) 10%, *Curcuma zedoaria* (roots) 10%, *Ocimum sanctum* (leaves) 10% and *Phyllanthus emblica* (fruits) 10%. 4 gm (one teaspoonful) of mixed powder should be given to the patient, twice a day (morning and at bedtime) with water (Narayana and Subhose, 2005).

Arthritis: *Piper longum* (fruits) 10%, *S. xanthocarpum* (whole plant) 15%, *Withania somnifera* (roots) 10%, *Terminalia chebula* (fruits) 10%, *T. bellerica* (fruits) 10%, *Curcuma zedoaria* (roots) 15%, *Phyllanthus emblica* (fruits) 15% and *Ricinus communis* (roots) 15%. 4 gm of mixed powder should be given to the patient, twice daily (morning and evening, one hour before meals) with ginger juice for rheumatic problems (Narayana and Subhose, 2005).

Blood circulation: *Zingiber officinale* (roots) 20%, *Piper longum* (roots) 10%, *Withania somnifera* (roots) 10%, *Phyllanthus emblica* (fruits) 10%, *Curcuma longa* (roots) 10%, *Terminalia bellerica* (fruits) 10%, *T. chebula* (fruits) 10%, *Ocimum sanctum* (leaves) 10% and *Tephrosia purpurea* (leaves) 10%. 4 gm of powdered mixer is given to the patient, twice daily with water (Narayana and Subhose, 2005).

Chronic constipation: *Holarrhena antidysenterica* (bark) 10%, *Plumbago ovata* (husk) 20%, *Terminalia bellerica* (fruits) 10%, *T. chebula* (fruits) 15%, *Phyllanthus emblica* (fruits) 15%, *Cassia angustifolia* (leaves) 20% and *Glycyrrhiza glabra* (roots) 10%. 4 gm of powdered mixer is given to the patient at night before going to bed with water (Narayana and Subhose, 2005).

Cancer: *Azadirachta indica* (bark) 20%, *Bauhinia variegata* (bark) 15%, *Crataeva nurvala* (bark) 15%, *Terminalia chebula* (fruits) 15%, *T. bellerica* (fruits) 10%, *Holarrhena antidysenterica* (bark) 10% and *Tinospora cordifolia* (stems) 15%. 4 gm of mixed powder should be given to the patient, twice a day (morning and night) with lukewarm honey for cancer cure (Narayana and Subhose, 2005).

Chronic fever: *Tinospora cordifolia* (stems) 15%, *Ocimum sanctum* (leaves) 15%, *Adhatoda vasica* (leaves) 15%, *Azadirachta indica* (leaves) 15%, *Holarrhena antidysenterica* (bark) 10%, *Piper longum* (fruits) 10%, *Zingiber officinale* (roots) 10% and *Terminalia bellerica* (fruits) 10%. 4 gm of mixed powder is given to the patient, twice daily before meals with water (Narayana and Subhose, 2005).

Cough: *Phyllanthus emblica* (fruits) 25%, *Adhatoda vasica* (leaves) 20%, *Ocimum sanctum* (leaves) 10%, *Piper longum* (fruits) 10%, *Zingiber officinale* (roots) 10%, *Glycyrrhiza glabra* (roots) 15% and *Solanum xanthocarpum* (whole plant) 10%. 3 gm of mixed powder should be given to the patient twice daily (morning and at night before going to bed) with lukewarm mixed with honey to cure cold (Narayana and Subhose, 2005).

Cysts: *Terminalia chebula* (fruits) 20%, *Azadirachta indica* (bark) 20%, *Holarrhena antidysenterica* (bark) 10%, *Terminalia bellerica* (fruits) 10%, *Withania somnifera* (roots) 20% and *Tinospora cordifolia* (stems) 20%. 4 gm of mixed (one teaspoonful) powder is given to the patient, twice a day (morning and evening) with water (Narayana and Subhose, 2005).

Dental diseases: *Azadirachta indica* (leaves) 15%, *A.arabia* (bark) 15%, *Areca catechu* (bark) 15%, *Achyranthus aspera* (leaves) 10%, *Ficus benghalensis* (bark) 15%, *Quercus infectoria* (fruits) 15% and *Symlocos racemosa* (bark) 15%. The powder is applied to the gums and teeth, two times a day. Additionally a gargle of the decoction (3 gm of powder mixed in 150 ml of water) (Narayana and Subhose, 2005).

Diarrhoea: *Holarrhena antidysenterica* (bark) 25%, *Aegle marmelos* (fruits) 25%, *Zingiber officinale* (roots) 10%, *Terminalia chebula* (fruits) 10%, *Cyperus rotundus* (roots) 10%, *Syzygium cumini* (seeds) 10% and *Phyllanthus emblica* (fruits) 10%. 3 gm of mixed powder is given to the patient, three times a day, with curd for dysentery and diarrhoea (Narayana and Subhose, 2005).

Dislocation of bones: *Asparagus racemosus* (roots) 15%, *Withania somnifera* (roots) 15%, *Azadirachta arabica* (bark) 20%, *Terminalia arjuna* (bark) 20%, *T. chebula* (fruits) 10%, *T. bellerica* (fruits) 10% and *Phyllanthus emblica* (fruits) 10%. 3 gm of powdered mixer is given to the patient, twice a day with water for dislocation of bones and fractures (Narayana and Subhose, 2005).

Diabetes: *Gymnema sylvestre* (leaves) 30%, *Tinospora cordifolia* (stems) 15%, *Azadirachta indica* (leaves) 10%, *Phyllanthus emblica* (fruits) 20%, *Curcuma longa* (roots) 10% and *Aegle marmelos* (leaves) 15%. 4 gm of mixed powder should be given to the patient, twice a day with water (Narayana and Subhose, 2005).

Fistula: *Glycyrrhiza glabra* (roots) 20%, *Terminalia chebula* (fruits) 20%, *T. bellerica* (fruits) 15%, *Tinospora cordifolia* (stems) 15%, *Azadirachta indica* (leaves) 15%, and *Withania somnifera* (roots) 15%. 3 gm of mixed powder should be given to the patient, twice daily with water to treat fistula (Narayana and Subhose, 2005).

Female sterility: *Asparagus racemosus* (roots) 20%, *Withania somnifera* (roots) 20%, *Glycyrrhiza glabra* (roots) 20%, *Phyllanthus emblica* (fruits) 10%, *Ficus glomerata* (bark) 10% and *F. religiosa* (bark) 10%. 3 gm of powdered mixer is given to the patient twice daily, half an hour before meals with milk.

General health tonic: *Withania somnifera* (roots) 20%, *Asparagus racemosus* (roots) 10%, *Glycyrrhiza glabra* (roots) 10%, *Tribulus terrestris* (fruits) 10%, *Phyllanthus emblica* (fruits) 15%, *Terminalai arjuna* (bark) 15% and *Centella asiatica* (leaves) 10%. 4 gm of powder is given to the patient, twice daily (morning and evening) with milk (Narayana and Subhose, 2005).

Gastritis: *Zingiber officinale* (roots) 10%, *Piper longum* (fruits) 10%, *Mentha piperata* (leaves) 10%, *Terminalia chebula* (fruits) 15%, *T. bellerica* (fruits) 15%, *Phyllanthus emblica* (fruits) 15%, *Plumbago zeylanica* (roots) 10% and *Tinospora cordifolia* (stems) 15%. 4 gm of (one teaspoonful) powdered mixer is given to the patient twice daily, half an hour before meals with water (Narayana and Subhose, 2005).

Hair problems: *Eclipta alba* (leaves) 15%, *Centella asiatica* (leaves) 15%, *Terminalia chebula* (fruits) 10%, *T. bellerica* (fruits) 10%, *Phyllanthus emblica* (fruits) 15%, *Glycyrrhiza glabra* (roots) 15%, *Tinospora cordifolia* (stems) 10% and *Tribulus terrestris* (fruits) 10%. 4 gm of mixed powder is given to the patient, twice a daily with honey (Narayana and Subhose, 2005).

High blood pressure: *Terminalia arjuna* (bark) 35%, *T. chebula* (fruits) 15%, *Asparagus racemosus* (roots) 15%, *Zingiber officinale* (roots) 10% and *Withania somnifera* (roots) 25%. 4 gm of powder is given to the patient, twice a day (morning and night) with honey (Govindan, 1999).

Heart tonic: *Withania somnifera* (roots) 10%, *Terminalia arjuna* (bark) 30%, *T. bellerica* (fruits) 10%, *T. chebula* (fruits) 10%, *Cyperus rotundus* (roots) 10%, *Phyllanthus emblica* (fruits) 10% and *Ocimum sanctum* (leaves) 10%. 3 gm of powdered mixer is given to the patient, twice a day with water (Narayana and Subhose, 2005).

Intestinal worms: *Holarrhena antidysenterica* (bark) 10%, *Mentha piperata* (leaves) 10%, *Tinospora cordifolia* (stems) 20%, *Butea monosperma* (seeds) 20%, *Azadirachta indica* (leaves) 10%, *Pyllanthus emblica* (fruits) 20% and *Tribulus terrestris* (fruits) 10%. 3 gm of mixed powder is given to the patient, twice daily (morning and night) with water (Narayana and Subhose, 2005).

Epilepsy: *Centella asiatica* (leaves) 30%, *Withania somnifera* (roots) 20%, *Tribulus terrestris* (fruits) 15%, *Piper longum* (roots) 10%, *Achyranthus aspera* (leaves) 15% and *Plumbago zeylanica* (roots) 10%. 3 gm mixed powder is given to the patient, twice daily (morning & evening) with fruit juice to treat Hysteria (Narayana and Subhose, 2005).

Leucorrhoea: *Symplocos racemosa* (bark) 35%, *Asparagus racemosus* (roots) 15%, *Adhatoda vasica* (leaves) 10%, *Aegle marmelos* (fruits) 10%, *Phyllanthus emblica* (fruits) 10% and *Azadirachta indica* (bark) 10%. 3 gm of mixed powder is given to the patient, twice daily with water (Narayana and Subhose, 2005).

Leucoderma : *Psoralea cordifolia* (seeds) 20%, *Termins chebula* (fruits) 10%, *Phyllanthus emblica* (fruits) 20%, *Azadirachta indica* (bark) 20%, *Areca catechu* (bark) 10%, *Tinospora cordifolia* (stems) 10% and *Eclipta alba* (leaves) 10%. 3 gm of mixed powder should be given to the patient, twice a day before meals with water (Narayana and Subhose, 2005).

Liver tonic: *Holarrhena antidysenterica* (bark) 10%, *Eclipta alba* (leaves) 20%, *Tephrosia purpurea* (leaves) 20%, *Tinospora cordifolia* (stems) 10% *Azadirachta indica* (bark) 10%, *Phyllanthus amarus* (whole plant) 20% and *Plumbago zeylanica* (roots) 10%. 4 gm of powdered mixer is given to the patient twice daily, half an hour before meals with water (Narayana and Subhose, 2005).

Lack of appetite: *Zingiber officinale* (roots) 10%, *Piper longum* (fruits) 10%, *Pyllanthus emblica* (fruits) 30%, *Terminalia chebula* (fruits) 15%, *Tinospora cordifolia* (stems) 15%, *Cassia angustifolia* (leaves) 10% and *Mentha piperata* (leaves) 10%. 4 gm of mixed powder is given to the patient, two times a day after meals with water for indigestion (Narayana and Subhose, 2005).

Male sterility: *Withania somnifera* (roots) 15%, *Mucuna pruriens* (seeds) 25%, *Tribulus terrestris* (fruits) 20%, *Glycyrrhiza glabra* (roots) 10%, *Terminalia arjuna* (bark) 10%, *Phyllanthus emblica* (fruits) 10%, *Zingiber officinale* (roots) 5% and *Piper longum* (fruits) 5%. 4 gm of powdered mixer is given to the patient, twice a day with honey (Narayana and Subhose, 2005).

Migraine: *Curcuma longa* (roots) 15%, *Glycyrrhiza glabra* (roots) 15%, *Azadirachta indica* (bark) 15%, *Tinosporacordifolia* (stems) 15%, *Terminalia chebula* (fruits) 10%, *Ocimum sanctum* (leaves) 15% and *Eclipta alba* (leaves) 15%. 4 gm of mixed powder is given to the patient, twice a day with honey (Narayana and Subhose, 2005).

Obesity: *Terminalia chebula* (fruits) 15%, *Terminalia bellerica* (fruits) 15%, *Phyllanthus emblica* (fruits) 10%, *Crateva nurvala* (bark) 25%, *Tribulus terrestris* (fruits) 25% and *Zingiber officinale* (roots) 10%. 4 gm of powder is given to the patient, twice a day with warm water.

Paralysis: *Curcuma zedoaria* (roots) 20%, *Withania somnifera* (roots) 20%, *Tribulus terrestris* (fruits) 20%, *Zingiber officinale* (roots) 20%, *Piper longum* (fruits) 5%, *Crataeva nurvala* (leaves) 10% and *Plumbago zeylanica* (roots) 5%. 3 gm of powdered mixer is given to the patient, three times a day with honey (Narayana and Subhose, 2005).

Prostate enlargement: *Tinospora cordifolia* (stems) 15%, *Tribulus terrestris* (fruits) 15%, *Phyllanthus emblica* (fruits) 15%, *Zingiber officinale* (roots) 10%, *Butea monosperma* (seeds) 10%, *Adhatoda vasica* (leaves) 5%, *Terminalia chebula* (fruits) 10%, *T. bellerica* (fruits) 10% and *Glycyrrhiza glabra* (roots) 10%. 4 gm of powdered mixer is given to the patient twice a day, morning and evening before meals with water (Narayana and Subhose, 2005).

Piles: *Eclipta alba* (leaves) 35%, *Terminalia chebula* (fruits) 15%, *Terminalia bellerica* (fruits) 10%, *Phyllanthus emblica* (fruits) 10%, *Adhatoda* *vasica* (leaves) 10%, *Plumbago zeylanica* (roots) 5%, *Piper longum* (fruits) 5% and *Aegle marmelos* (fruits) 10%. 4 gm of mixed powder is given to the patient, twice daily (morning and at bedtime) with water (Narayana and Subhose, 2005).

Sleeplessness: *Withania somnifera* (roots) 20%, *Centella asiatica* (leaves) 30%, *Piper longum* (roots) 20%, *Glycyrrhiza glabra* (roots) 10% and *Terminalia bellerica* (fruits) 10%. 3 gm mixed powder is given to the patient, at night before going to bed, with milk (Narayana and Subhose, 2005).

Skin diseases: *Cyperus rotundus* (roots) 10%, *Tinospora cordifolia* (stems) 20%, *Azadirachta indica* (bark) 20%, *Terminalia chebula* (fruits) 10%, *T. bellerica* (fruits) 10%, *Curcuma longa* (roots) 10%, *Phyllanthus emblica* (fruits) 10% and *Centella asiatica* (leaves) 10%. 3 gm of powder is given to the patient, twice a day before meals with water to cure allergy problems (Narayana and Subhose, 2005).

Sexual debility: *Withania somnifera* (roots) 10%, *Mucuna pruriens* (seeds) 20%, *Asparagus racemosus* (roots) 10%, *Sida cordifolia* (seeds) 10%, *Tribulus terrestris* (fruits) 20% and *Glycyrrhiza glabra* (roots) 10%. About 4 gm of mixed powder should be given to the patient, twice daily (morning and at night before going to bed) with milk (Narayana and Subhose, 2005).

Throat diseases: *Glycyrrhiza glabra* (roots) 30%, *Terminalia chebula* (fruits) 10%, *T. bellerica* (fruits) 10%, *Solanum xanthocarpum* (whole plant) 20%, *Piper longum* (fruits) 10%, *Sida cordifolia* (roots) 10% and *Phyllanthus emblica* (fruits) 10%. 4 gm of powdered mixer is given to the patient twice daily, morning and at bedtime with honey (Narayana and Subhose, 2005).

Thyroid problems: *Crataeva nurvala* (bark) 20%, *Bauhinia variegata* (bark) 20%, *Sida cordifolia* (leaves) 15%, *Terminalia chebula* (fruits) 10%, *T. bellerica* (fruits) 10%, *Glycyrrhiza glabra* (roots) 15% and *Zingiber officinale* (roots) 10%. 3 gm of powdered mixer is given to the patient, twice daily with lukewarm water (Narayana and Subhose, 2005).

Urinary tract: *Tribulus terrestris* (fruits) 25%, *Zingiber officinale* (roots) 10%, *Solanum xanthocarpum* (whole plant) 10%, *Crataeva nurvala* (bark) 25%, *Tinospora cordifolia* (stems) 10%, *Asparagus racemosus* (roots) 10% and *Tephrosia purpurea* (leaves) 10%. 4 gm of powdered mixer is given to the patient, twice a day with water (Narayana and Subhose, 2005).

Economic value of medicinal plants

Plant species are used for medicinal purposes in two ways:

a) as traditional medicines, singly or in formulations, such as those prepared and dispensed by traditional medical practitioners, which may or may not attract a market price; and

b) Commercial products, dispensed by prescription or over the counter sales, such as patented/licensed medical products of allopathy or traditional systems of medicine.

Both these uses have economic value. For the lack of adequate and appropriate data, it is near impossible to evaluate the returns from the first category. The economic value of plant based drugs, therefore, largely rests on the second category uses. It is estimated that in the rich world, 25 per cent of all medical drugs are based on plants and their derivatives (Principe, 1991). In the poor world, this is closer to 75 percent.

The economic value of a particular species of plant in medicinal use depends upon number of factors, among which the following are important:

- Certain plant species are used in a large number of formulations. The use of a particular species with reference to the number therapeutic effects it exerts or the number of formulations in which it is an ingredient, is expressed as the therapeutic index and frequency index, respectively. A higher index reflects a higher economic value attributable to a particular species. Such species are often referred to as the 'elite species'. For example, neem (*Azadirachta indica*) is indicated for use against 10 out of 18 symptoms in gastro-intestinal disorders (Sharon, 1994) and against eight out of 11 symptoms in dental care (Shubharani, 1995). Neem has several other medicinal and non-medicinal uses as well. Consequently, neem is one of the elite species of Indian medicinal plants.
- Certain species are of great importance in the treatment of a particular disease as they happen to be the only (or one of very few) species with that therapeutic potential, as the alkaloids of *Catharanthus roseus* in the treatment of leukemia. The importance of the disease also is a factor. Such species attract high market rate.

- Some species have a narrow distribution and/or occur in small populations. Some may be difficult to cultivate. Such species also command a higher price. For example, *Trichopus zeylanica*, occurring in south India, Sri Lanka and the Malay peninsula, is recently projected as the Indian equivalent of ginseng. This species now attracts high market value.
- Certain species of medicinal plants like *Rauwolfia serpentina* and *Saraca asoca* have been over exploited and so now occur rather scarcely in nature. It is difficult to cultivate *Rauwolfia serpentina* on a large scale while *Saraca asoca* is rather easier to propagate. The market value of some species thus depends upon such criteria.
- There are synthetic substitutes for several originally plant derived products, as for example clove oil. If the synthetic substitute is cheaper to produce than the plant based product and/or if the natural products have no other uses, the economic value of the natural product falls. However, certain therapeutically active constituents produced by plants like digoxin and digitoxin have not been produced synthetically. Some like vincristine, vinblastin, opiates *etc.*, that have been synthesized have proven to be less efficient than the natural products. The economic value of a species of medicinal plants depends upon this situation as well.
- The economic value of a particular species varies with time. An effective synthetic substitute or the discovery of a better natural alternative or the disuse of the species/product over a period of time may deplete the species of its market value. For example, till sulphonamides came into use, sandalwood oil was the most widely used effective antiseptic. Subsequently, sandalwood oil (*Saantalum album*) has fallen into disuse as an antiseptic. Its economic value should have come down but did not as sandalwood oil has other uses with higher economic returns.
- The costs involved in isolation and purification of an active principle involve several considerations. It requires about a tones of leaves of *Catharanthus roseus* to obtain one gram of the alkaloid vincristine, essentially needed to treat leukaemia. Vincristine is one of the expensive plant products costing about US$ 24,000/g.

Vinblastin, another alkaloid from the same species, used to treat Hodgkin disease is present in quantity 1,000 more times than vincristine. One gram of vinblastine costs about US$ 6,800. It is now possible to convert vinblastine into vincristine through biotransformation. There is also a growing interest in the other alkaloids present in *Catharanthus roseus*. Thus, several factors govern the cost of the raw material and the final product of a medicinal plant from time to time.

The potential of plants as sources of medicine is often taken in support of identification and preservation of the world's most species rich ecosystems. Such assessments are speculative and totally based on chance. Screening the vast vegetable world for potential sources of medicine and their use in the traditional way or through the application of biotechnology for a large scale industrial production of the active constituents or by chemical synthesis is a very uncertain and a long term proposition that involves heavy financial investment with no certainty of the economic returns. There is no guarantee that the future drugs will all be derived from plants.

Principe (1989) reported that pharmaceutical companies have shown decreasing interest in the development of new plant products in favor of molecular biology and biotechnology applications to micro-organisms. However, some others in the industry believe that plant based resources will regain their importance. This hope has been bolstered by the much publicized prospecting agreement between pharmaceutical giant Merck and the Costa Rican National Institute for Biodiversity (Laird, 1993).

Although research based on micro-organisms to synthesize therapeutically active chemical compounds is promising, several limitations have to be borne in mind (Principe, 1989). The steps of identifying the chemical structure required to achieve a given effect and creating a proper genetic code for this structure are the most difficult stages of drug development for which plant based genetic material appears to be better than micro-organisms. Genetically engineered micro-organisms can substitute only for some of the plant based chemicals. The vast majority of plant based chemicals have not been successfully synthesized (Principe, 1989).

Many consider that plant based drug resources have been exhausted in the early part of this century (Findeisen, 1991). However, the discovery of vincristine in *Catharanthus roseus*, reserpine in *Rauwolfia serpentina*, taxol in *Taxus baccata* and *Taxus wallichiana*, camptothecin in *Camptotheca acuminata* and *Nothopodytes foetida etc.* have changed the direction and tempo of screening plants for drugs. This phase almost halted by the 1970s with a large scale failure to discover any 'blockbuster' drug from plant sources. Currently, once again there is some considerable revival of interest in plant based drugs. Interest in plant based drugs has thus been going in cycles or more precisely in spirals, reaching a similar situation at a different point in time.

Added to the problems outlined above, there are legal difficulties related to land rights, plant rights, patenting *etc.* Patenting has to relate to the process of manufacture (and not the product) or to some uncertain anticipated value because biological organisms or natural compounds cannot be patented. At the present moment, material and intellectual property rights are in a fluid state all over the world, but more particularly in India. Indications are that, in due course of time a clear cut policy and rules would emerge, when there would be more incentive for personal initiative and effort.

Notwithstanding all the difficulties, there has been a fresh revival of interest in plant based drugs in the last ten years or so. This is mostly based on improved techniques of purifying, analyzing and assaying samples and use of robots for continuous assay of material. Another factor, the consumers' demand for natural products has also contributed to the new herbal culture. As a result, internationally, the US National Cancer Institute, Monsanto, Smith Kline, Merck and Glaxo have resumed plant screening on a fairly large scale (Reid, 1993). Affymax and Shaman are new US companies which deal with drugs solely from natural products with emphasis on traditional medicine.

The significant point regarding medicinal plants is that they are used to support arguments for conserving entire biological resources in the third world, though the actual relevance is difficult to justify. On the one hand it is argued that if genetic engineering procedures become more successful, they would replace plant based research. On the other hand it is said that since the knowledge of medicinal plants is very limited, not paying adequate attention to them the soonest, may prove disastrous in the long run, as they may be lost for ever.

Evenson (1991) distinguishes between a) general strategic research for new resources that justifies the maintenance of most biological materials, and b) specialized research for genetic material to meet specific needs, which justifies the collection and preservation of genetic resources.

Ascribing economic value to medicinal plants can be done on two bases (Pearce and Moran, 1994). The first relates to the existing values which, in turn, are for commercial drugs and for traditional medicine. The other relates to the option value of plants that is the extent to which conservation is required to protect the future use values. Unfortunately, option values depend on a critical judgments and the correctness of this judgment, on which a species is to be conserved.

References

CTA Technical Centre for Agricultural and Rural Co-operation. 2007. Medicinal plants, RRRP: 3.

Evenson, R.E. 1991. Genetic resources: assessing economic value. In valuing environmental benefits in developing economies. (eds.) Vincent, J., Crawford, E. and Hoehn, J. Proceedings of Seminar, Michigan State University, special report no. 29.

Findeisen, C. 1991. Natural product research and the potential role of the pharmaceutical industry in tropical forest conservation. Rain Forest Alliance, New York.

Govindan, S. 1999. J Ethnopharmacol., 66: 205.

Laird, S. 1993. Contracts for biodiversity prospecting. In Biodiversity prospecting:using genetic resources. (ed.) Reid, W. World Resources Institute, Washington, D.C.

Mukherjee, P.K. and Wahile, A. 2006. J Ethnopharmacol. 103: 25.

Narayana, A. and Subhose, V. 2005. Bull Indian Inst Hist Med Hyderabad, 35: 21.

Pearce, D. and Moran, D. 1994. The economic value of biodiversity. Earthscan Publications, London.

Principe, P. 1991. Monetising the pharmacological benefits of plants. US Environmental protection Agency, Washington, D.C.

Principe, P. 1989. The economic significance of plants and their constituents as drugs. In Economic and medical plant research. (eds.) Wagner, H., Hakino, H. and Farnsworth, N. Academic Press, London.

Ram, A. 1997. J. Ethnopharmacol. 55: 165.

Reid, W. 1993. Biodiversity prospecting: using genetic resources for sustainable development. World Resources Institute, Washingron, D.C.

Sharon, A. 1994. Database of plants used in gastrointestinal disorders and effects of plant extracts on pathogenic bacteria. Ph.D. thesis, two vols. Bangalore University, India.

Shubharani, R. 1995. Plants used in dental care in India. Ph.D. thesis, two vols. Bangalore University, India.

Biodiversity Conservation and Envir. Management (2012)
Editors: D.R. Khanna et al.
Pub. by Biotech Books. *ISBN: 978-81-7622-262-4*

Pages: 283-295

PHYSICO-CHEMICAL CHARACTERISTICS IN RELATION TO MACRO-BENTHIC ORGANISMS OF SAHASTRADHARA STREAM AT DEHRADUN (UTTARAKHAND)

D. S. Malik, Umesh Bharti✉ and Swati Nagar
Department of Zoology & Environmental Sciences
Gurukula Kangri University, Haridwar, Uttarakhand
✉E-mail: saraswati_umesh@yahoo.co.in

Sahastradhara hill stream, a natural perennial stream originated from the hills of Mussoorie is well known for sulphur springs as cold sulphur water gush out of mountain cavities in a spectacular way, attracts maximum tourists for bathing and other adventurous activities. In the present study, the hydro-biological changes were recorded as annual average air temperature (3.4 °C – 38.3 °C),

✉ Corresponding author

wind velocity (0.9 km - 2.5 km), rainfall (225mm - 371mm), precipitation and sedimentation rate just double from two decade in the stream. The physico-chemical characteristics were recorded as, water temperature (15.6-23.5 °C), Velocity (0.34-0.74 m/s), pH (7.5-8.5), DO (7.01-10.87 mg/l), BOD (0.84-3.12 mg/l), Free CO_2 (0.01-2.96 mg/l), Alkalinity (30.1-44.8 mg/l), Sodium (7.66-11.53mg/l), Potassium (14.92-17.72 mg/l), Calcium (83.62-88.64 mg/l) Magnesium (50.51-58.97 mg/l) and Chloride (14.43-16.50 mg/l). Sahastradhara hill-stream ecosystem has moderate trophic conditions for different biotic communities and its water quality is quite rich in natural nutrients but presently, eco-biological characteristics of Sahastradhara stream exhibited continuous degradation nature in and around catchments basin ecosystem for suitability of aquatic biodiversity and maintaining the optimum biological productivity.

Introduction

It is well known that during last few decades, human activities have dramatically altered natural ecosystem due to increasing demand for food, clean water, energy and other resources. Aquatic system or aquatic harbours are the important base of the life on the earth along with the air and land. Hill streams, generally the important source of natural water to fulfil the basic requirement of peoples of the mountainous region. The process of economic growth and development, virtually have inverse relationship with the hill stream resources and quality of the environment. A hill stream, Sahastradhara situated in the foothills of Garhwal Himalayas is a major tributary of River Song, which flows downwards through Dehradun Valley. Sahastradhara stream is situated in the globe on 29°57'- 31°20' Latitude and 77°35' - 79°20' Longitude in Garhwal Himalaya.

Impact of climate change on river system is the subject of active research as reviewed by Arnell *et al.* (2001), because of the importance of water for human activities, in terms of resource (quantitatively and qualitatively) and risk factor, impact on hydrology (include changes in runoff, river flow and groundwater storage). The impact of climate change on river water quality is also heavily dependent on the future evolution of human activities (pollutions, withdrawals *etc.*), so that the direct effect of climate change may end up being small in relative terms (Hanratty and Stefan, 1998; Ducharne *et al.*, 2007). The main focus of this study was describe the variation in the hydrology, meteorology and changes in the stream water variables that shows the future based prediction about degradation of the water quality due to results of eutrophic condition, less productivity, shrinkage of stream width *etc.* Sahastradhara stream have lot of sulphide caves, contributed to minimize the risk of skin disease by bathing activities. So many anthropogenic activities and soil erosion of surrounding hilly carbonate rocks lead to the nutrient load in the aquatic ecosystem of the stream.

Materials and Method

The sampling site was selected in Sahastradhara stream flowing about 15 km away from Dehradun city. The three different stretches were selected for eco-biological study in the stream. The water samples were collected seasonally from different sampling station during May, 2009 to April, 2010 in morning period (9:00 AM). The samples were examined for physical, chemical and biological analysis on site and in laboratory by following standard methods by APHA (1995) & Trivedi and Goel (1984). All benthic samples were collected with the Ekman's Dredge sampler, sieved by size US No.60 cms.The specimen were preserved in 4.0 % formalin and identified by the keys mentioned in Edmondson (1992).

Results and Discussion

Hill streams are fresh water resources and naturally occurring water source in mountainous regions. The observations recorded during different seasons at different stretches of stream *viz.* upper stretch, middle stretch and lower stretch during the year May, 2009 – April, 2010 are shown in Table 1. In the present study, the maximum temperature was recorded during summer season at middle stretch (23.5 °C) and minimum (15.6 °C) at upper stretch during winter. Lower temperature recorded during winter and higher during summer that may be due to extreme cold and extreme sunshine period. Ward and Stanford (1979) suggested that temperature pattern influence the life cycle phenomenon of insects, which leads to increase in density.

Stream velocity increases as the volume of the water rises in the stream which influences the species richness and distributional patterns of organisms existed in the stream. The maximum value of velocity was reported (0.74 m/s) during monsoon at upper stretch and minimum (0.34 m/s) was recorded at middle stretch in summar season, where anthropogenic activities were examined at great extent by making small ponds for bathing & washing activities. The flow of a stream is directly related to the amount of water flowing off watershed into the stream channel. The pH was maximum (8.5) during monsoon season at upper stretch of the stream and minimum (7.5) during summer season at middle stretch. Here the study revealed that water was almost alkaline in nature throughout the study period. Sharma (1986) observed the alkaline water characteristics of River Bhagirathi and Joshi (1996) observed the similar range of pH in the riverine system of western Himalaya.

Alkalinity of stream water was in correlation with pH values, which was recorded as maximum (44.8 mg/l) during summer at middle stretch and minimum (30.1 mg/l) during monsoon at upper stretch of the stream due to excess runoff from the adjacent catchments area drained into the stream which causes such critical condition in riverine system. It may be due to increase in bicarbonates concentration through domestic waste discharge from near by villages and other tourist activities. Streams with limestone created high alkalinity and good buffering capacity. Direct sewage from near by villages into stream water contributed maximum alkalinity in middle stretch of the stream water.

The maximum free CO_2 (2.96 mg/l) was recorded during monsoon season at middle stretch of stream and minimum (0.01 mg/l) during summer season at upper stretch of stream. However, CO_2 was found nil during the few months of study period. Higher free CO_2 concentration occurred in water samples in monsoon season due to discharge of domestic water, inflow of sewage and mostly decomposition of organic wastes. Similar observation was reported by Sharma *et al.* (2008) in Garhwal Himalayas. Dissolved oxygen is a significant factor to determine the ecological health of any aquatic ecosystem, which brings the various bio-chemical changes amongst aquatic environment and organisms. The maximum DO (10.87 mg/l) was recorded during winter season at upper stretch of stream due to continuous flow of water with low temperature and minimum (7.01 mg/l) recorded during monsoon at middle stretch. Low values of DO, due to the increasing the temperature after the rainfall and decaying of macro-vegetation in the stream water. Mishra and Yadav (1978), Adebisi (1981) and Mitra (1982) also discussed seasonal changes and fluctuations in DO in different hill streams and riverine ecosystem of Garhwal region. Sharma *et al.* (2009) discussed the DO level for the macro-invertebrate existence in the Tons river at Doon Valley, Uttarakhand.

Biological Oxygen demand is an important parameter to estimate the pollution level and determines the present aquatic quality of water body. In the present investigation, BOD values clearly showed higher concentration (3.12 mg/l) during monsoon at middle stretch of the stream. The value of BOD recorded comparatively low (0.84 mg/l) during winter at upper stretch. Similar seasonal trends were obtained by John (1952) Robert (1969) and Richard (1966) in different hill streams.

Calcium is an essential component referred as nutrient parameter and plays important role in biological activities of living organisms. The maximum value (83.62 mg/l) of calcium was observed during monsoon at middle stretch and minimum (86.8 mg/l) during winter season at upper stretch of stream. The maximum concentration may be due to the erosion and landslides of lime stone into the stream with rain water and heavy soil erosion during monsoon and minimum due to low flow and less erosion possibilities in hilly areas. Similarly, Khanna and Singh (2000) stated the variation in calcium and magnesium concentration in Suswa river in Garhwal region.

Magnesium is useful with low concentration and essential element for phytoplankton and other aquatic organism in water but it becomes toxic at higher concentration. In the present study, the concentration of magnesium was higher (58.97 mg/l) at middle stretch during monsoon and lower (50.51 mg/l) in the same stretch during summer season. Jenkins *et al.* (1995) recorded similar findings on magnesium concentration in the streams of middle hills and high mountainous region at Nepal. Sodium was reported maximum (11.53 mg/l) at middle stretch during monsoon and minimum (7.66 mg/l) at upper stretch during winter season. The findings of Bond (1979) supported such nutrient concentration pattern in a stream of mountain ecosystem in Utah. Sharma *et al.* (2008) also observed the concentration of sodium ion in Chandrabhaga river in Garhwal Himalaya.

Potassium is naturally occurring element originated by the weathering and leaching from growing vegetation and decomposition of organic matter (Berndtsson, 1990).The maximum concentration of potassium was reported (17.72 mg/l) at middle stretch during monsoon season and minimum (14.92 mg/l) at upper stretch during winter season that may be due to runoff, increasing the potassium ratio in stream and minimum during winter due to low flow of water and less rocks erosion in the hills of Sahastradhara. Sharma *et al.* (2007 and 2009) pointed out the maximum and minimum concentration of potassium throughout the year in Tons river. Chloride was reported as chloride ion and inorganic anion that generally occurred in natural condition in fresh water. It was recorded maximum (16.50 mg/l) at lower stretch during monsoon and minimum (14.43 mg/l) at middle stretch of stream during winter season. Mishra and Joshi (2003) observed the fluctuation in chloride concentration in Ganga river at Haridwar. Pande and Mishra (2000) also observed the fluctuation in the chloride concentration in Sahastradhara stream.

The correlation was calculated by the mean values of water variables of different stretches of Sahastradhara stream. Correlation generally, draws between temperature, velocity, pH, alkalinity, CO_2, DO, BOD, calcium, magnesium, sodium, potassium and chloride. Temperature showed the positive correlation with pH (0.75), alkalinity (0.54), DO (0.98), BOD (0.88), calcium (0.73), sodium (0.76) and potassium (0.37) respectively. The pH showed positive correlation with DO (0.65), BOD (0.35), calcium (0.12), sodium (0.15) and chloride (0.25). Alkalinity correlated with Free CO_2 (0.67), DO (0.65), BOD (0.88), calcium (0.97), sodium (0.96) and potassium (0.98) (Table 2). Yeragi and Shaikh (2003) calculated correlation pattern between DO and net production efficiency in Tansa river. The velocity showed negative correlation (-0.94) with temperature; Free CO_2 (-0.26) with temperature, magnesium (-0.97) and chloride (-0.44). pH showed with alkalinity (-0.14), Free CO_2. (-0.83) with temperature, magnesium (-0.56) and potassium (-0.32) with temperature. DO correlate negative relationship with BOD (- 0.92) and magnesium (-0.99) and chloride (-0.56). BOD also showed negative relationship with magnesium (-0.97) and chloride (-0.82), calcium with magnesium (-0.88) and chloride (-0.93). Magnesium also showed negative correlation with sodium (-0.90)

Table 1: Physico-chemical characteristics (n=5, mean values) of Sahastradhara Stream during May, 09-April, 10.

Parameters	*Upper stretche S_1*				*Middle stretche S_2*				*Lower stretche S_3*			
	Summer	*Monsoon*	*Winter*	*Mean*	*Summer*	*Monsoon*	*Winter*	*Mean*	*Summer*	*Monsoon*	*Winter*	*Mean*
Temperature (°C)	21.9	19.8	15.6	19.1	23.5	20.2	16.2	19.9	21.8	18.5	15.7	18.7
Velocity (m/s)	0.37	0.74	0.66	0.59	0.34	0.60	0.69	0.55	0.39	0.71	0.65	0.59
pH	8.1	8.5	7.8	8.1	7.5	8.2	8.1	8.1	7.9	8.3	7.8	7.9
Alkalinity (mg/l)	42.5	30.1	39.4	37.3	44.8	31.7	41.9	38.9	43.6	30.8	40.4	38.3
Free CO_2 (mg/l)	0.01	2.10	1.90	1.99	0.25	2.96	1.86	2.18	0.17	2.46	1.86	2.33
DO (mg/l)	7.76	7.28	10.87	8.09	8.25	7.01	8.45	8.13	7.81	7.29	9.13	8.08
BOD (mg/l)	1.41	2.39	0.84	1.55	1.65	3.12	1.45	1.89	1.45	2.42	0.95	1.61
Calcium (mg/l)	84.48	85.15	83.62	84.42	84.88	88.64	84.07	84.74	84.67	85.18	83.8	84.55
Magnesium (mg/l)	50.92	53.81	51.51	52.11	50.51	58.97	50.86	51.77	50.70	54.3	51.41	52.14
Sodium (mg/l)	8.62	10.97	7.66	9.09	9.37	11.53	8.81	9.903	8.87	11.30	7.90	9.39
Potassium (mg/l)	15.67	16.97	14.92	15.85	16.17	17.72	15.16	16.34	16.22	17.31	15.21	16.24
Chloride (mg/l)	15.79	16.23	15.76	15.92	15.16	16.42	14.43	15.33	15.34	16.50	14.6	15.49

Table 2: Co-relation matrix in-between physico-chemical characteristics (mean value) of Sahastradhara stream.

Parameters	*Temperature*	*Velocity*	*pH*	*Alkalinity*	*Free CO_2*	*DO*	*BOD*	*Calcium*	*Magnesium*	*Sodium*	*Potassium*
Temperature											
Velocity	-0.94										
pH	0.75	-0.5									
Alkalinity	0.54	-0.78	-0.14								
Free CO_2	-0.26	-0.07	-0.83	0.67							
DO	0.98	-0.98	0.65	0.65	-0.12						
BOD	0.88	-0.98	0.35	0.88	0.23	-0.92					
Calcium	0.73	-0.91	0.12	0.97	0.46	0.82	0.97				
Magnesium	-0.97	0.99	-0.56	-0.74	0.005	-0.99	-0.97	-0.88			
Sodium	0.76	-0.93	0.15	0.96	0.43	0.84	0.98	0.99	-0.90		
Potassium	0.37	-0.66	-0.32	0.98	0.79	0.50	0.77	0.90	-0.60	0.89	
Chloride	-0.44	0.71	0.25	-0.99	-0.75	-0.56	-0.82	-0.93	0.66	-0.92	-0.99

and potassium (-0.60). Sodium and potassium showed negative relationship with chloride (-0.92) and (-0.99) respectively (Table 2). Dobriyal (1985) and Khanna *et al.* (1993) reported such relationship of temperature with DO and other parameters in streams and rivers of Garhwal region.

During the study period, a total of 23 genera belonging to 7 orders of benthic fauna were recorded. The most abundant order having maximum number of genera recorded was *Ephemeroptera* at each stretches of stream followed by *Oligocheata, Plecoptera, Trichoptera, Diptera, Ephemeroptera, Odonata* and *Hemiptera.* Table 3, showed the variations in the abundance of benthos at each site in the stream. *Chironomous* sp. and *Tubifex* sp. was observed at middle stretch that may be considered as pollution indicator species in this stretch and *Ephemeralla indica* and *Aquarius remigis* were reported dominated at upper and lower stretches of stream, indicating most suitable quality of water in the stream. The observations inferred that maximum number of genera was recorded at middle stretch, where the anthropogenic activities were highest and minimum at upper stretch due to less interference of human activities. Malik *et al.* (2010) reported the different species distribution and structure of macro-benthos in Song river in doon valley of Garhwal region.

Conclusion

Sahastradhara stream has a good potential of high trophic level for maintaining the aquatic biodiversity. In spite of the differences in substrate composition & other anthropogenic activities, the benthic distribution at all three different stretches remains stable and supported that water quality of stream has good nutrients to maintain the diverse groups of macro-benthic communities in stream. The present monitoring of the stream exhibited that a moderate change occurred in the water variables due to the anthropogenic activities like washing clothes, drainage of hotels and domestic liquid wastes at middle stretch of stream. Many remedial measures for solid waste dumping, control of point sources of sewage, liquid waste and other unhygienic activities should be properly managed for the static equilibrium of aquatic ecosystem in Sahastradhara stream at Dehradun in Garhwal Himalayan region.

Table 3: The occurrences of benthic organisms in Sahastradhara stream at Dehradun.

OLIGOCHAETA	*Upper stretches*	*Middle stretches*	*Lower stretches*
Tubifex	+	+++	++
Branchiora	++	++	+
Limnodrillus	+++	++	+++
PLECOPTERA			
Pteronarcys	-	++	++
Acroneuria	-	+	+++
Isoperia	-	+	++
TRICHOPTERA			
Hydrosyche	+++	++	++
Leptocella	+++	+	++
Ochrotrichia	++	++	++
DIPTERA			
Chironomous	++	+++	+
Tendipestentans	+	++	++
Culicoides	+	++	++
Alabesmyia	+	++	++
EPHEMEROPERA			
Adult Mayfly	+++	++	++
Stenononema	++	+	++
Leptophlebia	++	+	++
Ephemeralla	+++	++	+++
Cinygma	++	+	++
ODONATA			
Epicordulia (Dragonfly)	-	+++	++
Macromia	++	+	+++
HEMIPTERA			
Aquarius (Water striders)	+++	++	+++
Sigara (Water boatmen)	++	-	+++
Notonecta (Backswimmers)	++	+	++

+++: Abundant; ++: Common; +: Rare, -: Absent.

References

Adebisi, A.A., 1981. The physico-chemical hydrology of a tropical seasonal river upper Ogun river Nigeria. Hydrobiologia, 79(2): 157 165.

APHA, 1995. Standard methods for examination of water and waste water. American Public Health Association, 19th ed. Inc. New York, pp: 1150.

Arnell, N. W., Liu, C., Compagnucci, R., da Cunha, L., Hanaki, K., Howe, C., Mailu, G., Shiklomanov,I., and Stakhiv, E., 2001: Hydrology and water resources, in: Climate Change. Impacts, Adaptation, and Vulnerability, edited by: McCarthy J. J., Canziani, O., Leary, N. A., Dokken, D. J., White, K. S., Contribution of Working Group II to the Third Assessment Report of the Intergovernmental Panel on Climate Change, Cambridge University Press, Cambridge, 191-233.

Berndtsson, R., 1990. Transport and sedimentation of pollutants in a river: A chemical mass balance approach. Wat. Resour. Res. 26(7): 1549-1558.

Bond, H. B., 1979. Nutrient concentrations patterns in a stream draining a montane ecosystem in Utah. Ecology, 60(6): 1184 – 1196.

Dobriyal, A.K., 1985. Ecology of Limnofauna in the small streams and their importance to the village life in Garhwal Himalaya. Uttar Pradesh J. Zool., 5(2):139-144.

Ducharne, A., Baubion, C., Beaudoin, N., Benoit, M., Billen, G., Brisson, N., Garnier, J., Kieken, H., Lebonvallet, S., Ledoux, E., Mary, B., Mignolet, C., Poux, X., Sauboua, E., Schott, C., Thoery, S., and Viennot, P., 2007. Long term prospective of the Seine River system: Confronting climatic and direct anthropogenic changes, Sci. Total Environ., 375, 292-311.

Edmondson, W.T., 1992. Fresh water biology Second Edition, pp: 1248.

Hanratty, M. P. and Stefan, H.G., 1998. Simulating climate change effects in a Minnesota agricultural watershed, J. Environ. Quality, 27:1524-1532.

Jenkins, A., Sloan W. T. and Cosby, B. J., 1995. Stream chemistry in the middle hills and high mountain of the Himalaya, Nepal. Journal of Hydrology. Vol – 166, No-1-4: 61-79.

John, D. P., 1952. Water pollution, its effects on the public health. Proc. Fish Ohio Water Clinic, Ohio State Univ. Eng. Series Bull. 147: 34-9.

Joshi, B. D., 1996. Hydro-biological profile of river Sutlej in its middle stretch in Western Himalayas. U.P. J. Zool., 16(2): 9-103.

Khanna, D. R and Singh, R. K., 2000. Seasonal fluctuations in the plankton of Suswa river at Raiwala, Dehradun. Env. Conser. J. 1(2&3): 89-92.

Khanna, D.R., Badola, S.P. and Dobriyal, A.K. (1993). Plankton ecology of the river Ganga at Chandighat, Haridwar, Advances in limnology, Ed. By H.R. Singh, Narendra publishin house New Delhi. 171-174.

Malik, D. S., Bharti, U. and Kumar, P., 2010. Macro-benthic diversity in relation to biotic indices in Song river at Dehradun, India, Env. Conser. J., 11 (1 & 2): 99-104.

Mishra, G. P. and Yadav, A. K., 1978. A comparative study of physicochemical characteristics of river and lake water in Central India. Hydrobiologia. 59(30) 275-278.

Mishra, S. and Joshi, B. D. 2003. Assessment of water quality with few selected parameters of river Ganga at Haridwar. Him. J. Zool., 17(2): 113-122.

Mitra, A.K., 1982. Chemical characters of surface water at a selected gauging station in the river Godavari, Krishna and Tungbhadra. Indl. J. Environ, Hlth., 24 (2): 165-179.

Pande, R. and Mishra, Asha, 2000. Water quality study of freshwaters of Dehradun (Sahastradhara stream and Mussoorie lake). Aquacult. 1: 57-62.

Richard, L. W. 1966. Environmental hazard of water pollution. New England. J. Medicine. pp: 819-25.

Robert, D.H. 1969. Water Pollution. Bioscience, 19: 976.

Sharma, A., Sharma R. C. and Anthwal, A., 2007. Monitoring phytoplankton diversity in the hill stream Chandrabhaga of Garhwal Himalaya. Life Science Journal, Vol. 4, No.-1: 80-84.

Sharma, A., Sharma R. C. and Anthwal, A., 2008. Surveying of aquatic insect diversity of Chandrabhaga river, Garhwal Himalayas. Environmentalist, 28: 395-404. DOI 10, 1007/s10669-0079155-z.

Sharma, R. C., 1986. Effect of physico-chemical factors on benthic fauna of Bhagirathi river Garhwal Himalaya. Indian J. Ecol. 13(1): 133 – 137.

Sharma, R. C., Arambam R. and Sharma R., 2009. Surveying macro-invertebrate diversity in the Tons river Doon Valley, India. Environmentalist, 29: 241-254. DOI 10, 1007/s10669-008 9187 z.

Trivedi, R.K. and Goel, P.K., 1984. Chemical and biological methods for water pollution studies. Karad. Environmental publication. pp: 1 25.

Ward, J.V., and Stanford, J. A. 1979. Ecological factors controlling stream zoobenthos with emphasis on thermal modification of regulated streams. Ecology of regulated streams, Plenom Publishing Corporation, pp: 35-53.

Yeragi, S. G. and Shaikh, N. (2003). Studies on primary productivity of Tansa river. J. Natcon 15(1): 125-130.

landmark in the development of this subject. However, this monumental work could not draw wide attention of mycologists and as a result a wide lacunae remained till a comprehensive account of the systematic morphology and general biology of thermophilic fungi came into picture with the monograph of Cooney and Emerson (1964) which served as the major understanding of thermophilic fungi (Rawat and Johri, 2002).

The inherent ability of this group to occupy a temperature niche that would preclude most other fungi results in their ubiquitous distribution. The community structure of this small group of Eumycota is very heterogenous exhibiting a wide structural distributional pattern. They occur both in natural and man-made habitats like soil, compost, wood chip piles, coal spoil tips, stored grains and though limited to few species, encompasses various taxonomic groups. The most interesting feature of this group is that they have been recovered not only from hot climatic regions but also from cold and stressed environments (Ellis, 1980a, b; Mouchacca *et al.*, 1995).

Their ubiquitous presence in different temperature zones and habitats raises many question regarding their survival and dispersal. The presence of self-heating masses of organic debris all around the globe lends sustainability to this community. It is therefore imperative to explore out their geographically diverse reservoir in order to have a fair picture of their species spectrum and as well as to understand their functionality in a given habitat as this group is not only structurally diverse but also functionally. The survival and dominance exhibited by thermophilic fungi in a wide array of habitats is attributed to their inherently strong extracellular enzymatic machinery, release of volatile sporostatic and fungistatic compounds and colonization potential (Satyanarayana and Johri, 1981).

The use of molecular tools in recent years has however added considerable information to extend our knowledge of this group. The availability of molecular and immuno probes have opened up newer possibilities, which would permit retrieval of information concerning the role of thermophilic fungi in colonization, succession and *in situ* functioning in various habitats.

Compost is an important habitat of thermophilic fungi which serves as their rich reservoir. The concentration of thermophilic propagules is approximately 10^6 times higher in compost than in soil and therefore thermophilic fungi are considered primarily as compost fungi. Optimization of compost quality is directly linked to the composition and succession of microbial communities. This means tools are required to monitor and characterize microbial communities during composting and to relate them to compost quality. Although, analysis of the presence and distribution of different operational taxonomic units (OTU) within a population provides insights into the ecological functioning of communities, the analysis of taxonomic structure of communities alone, however, limits insight into the ecological relevance of community structure. The exclusion or addition of different microflora does not necessarily change the resultant function of the community (White and Findlay, 1988). Analysis of phenetic characteristics which directly relate to important processes in the environment under investigation and correlation of such characteristics to environmental parameters allow for greater insight into factors which regulate the community structure (Garland and Mills, 1991). The changes in individual abundances may not equate to meaningful shifts in the community function. Cultivation approaches inevitably favour growth of some community members due to the selective nature of the media and can thus exclude majority of the endogenous microbes (Troussellier and Legendre, 1981) and thus do not give a full description of the microbial diversity. Moreover, the time consuming nature of isolate-based methods severely limits the spatial and temporal intensity to sampling (Garland and Mills, 1991).

Cultivation-independent methods have been in recent use to characterize the microbial community succession and structure. These include assessment of the diversity based on the directly extracted phospholipids (Boggs *et al.*, 1998; Herrmann and Shann, 1997; Klamer and Baath, 1998), measurement of carbon source utilization by substrate extracted microbial cell consortia (Boggs *et al.*, 1998; Inssam *et al.*, 1996) and nucleic acid-based techniques (Kowalchuk *et al.*, 1999; Peters *et al.*, 2000).

This chapter deals with the physico-chemical aspects of composting along with microbial community structure and functionality in compost ecosystem. While information on some aspects is still scanty due to only horizontal advancements in that area as compost has a very complex ecology, but an effort to compile all the available information has been made. However, readers may find a *bias* towards diversity of thermophilic mycoflora in mushroom compost, an intersting ecosystem, due to our active engagement during the past many years with the thermophiles in this habitat.

Composting

Composting represents an astonishing example of solid-state fermentation (SSF) wherein a crude variety of waste such as sewage sludge, refuse, animal manure, industrial wastes, food wastes, leaves, tree bark, agriculture residues, abattoir residues *etc.* can be treated through microbial route irrespective of their suitability as feed-stock for compost production (Satyanarayana and Grajek, 1999). Refuse (municipal solid waste) is partly compostable but poses problems due to its extreme heterogeneity as it consists of food scrapes (garbage), paper, glass, plastic, metal, sweepings, yard waste, ash *etc.* The organic rich fraction after separation can however be composted; the whole municipal solid waste can also be passed through the composting stage (mass composting), possibly with subsequent segregation (Satyanaryana and Grajek, 1999).

The basic aims of composting according to Miller (1994) are: (1) achievement of a suitable bulk density (compost makes a more physically stable landfill and can be easily stored, transported and disposed off than the original material as bulk density of former is higher), (2) modification of complex polysaccharides and plant materials, (3) biological removal of readily available nutrients to avoid overheating, (4) building up of an appropriate biomass and a variety of microbial products, (5) establishment of selectivity, (6) conversion of nitrogen into stable organic form, and (7) sanitation *i.e.*, killing of pathogenic microbes, larvae and weeds.

The various methods of composting are: Sheet composting, Trench composting, In-vessel composting, Bin composting and Aerated static pile composting. It involves air for the operation by placing the heap on holed piping that allows circulation (Hultman, 2009). Composting is done at small scale like decomposition of domestic wastes as well as large scales like decomposition of industrial wastes.

The composting is carried out *via* either batch or a continuous mode. Batch mode is a long process comprising of four sequential phases, the mesophilic phase, thermophilic phase, cooling phase and the curing phase. The initial stage of decomposition of mass of organic matter, initiated by mixing and wetting the substrates, the mesophilic stage is governed by the mesophilic microflora, which uses up the readily available nutrients.

The aerobic fermentation (composting) commences as a result of growth and activity of microorganisms resulting in release of heat, ammonia and CO_2 as byproducts along with other unpleasant smelling compounds. The metabolic activity of the mesophilic microorganisms that results in rise in temperature and paves way for development of thermophilic microflora which initiates the second phase of composting. This phase of composting starts very rapidly and may last days or weeks or even months. It is the thermophilic stage that results in maximum decomposition of organic matter besides sanitation. During the cooling stage, mesophilic microflora recolonizes the compost and it is during this phase that more resistant organic matter is degraded (Tiquia *et al.*, 2002; Hiraishi *et al.*, 2003). The final phase of composting is called curing, aging or maturing stage, which is long and an important one since it provides a safety net for destruction of the pathogens. Uncured compost can produce phytotoxins, besides depriving soil of oxygen and nitrogen and can contain high levels of organic acids. In the continuous mode of composting, similar phases occur but they are not as apparent as they are in the batch mode and these could occur concurrently rather than sequentially. A well-designed continuous system can eliminate the need for a mesophilic stage and operate continuously at thermophilic temperatures. This mode offers a mean of decomposition of putrescible materials quickly under close process control.

Municipal compost is prepared by batch mode while garden composting is performed in a continuous mode. Mushroom compost and vermicompost, though prepared by batch mode, are significantly different from general municipal compost. Municipal compost is prepared by batch mode while garden composting is performed in a continuous mode. Mushroom compost and Vermicompost, though prepared by batch mode, are significantly different from general municipal compost. Mushroom compost is prepared very rapidly (18-24 d) and does not involve curing stage. It is prepared, either by long method (LMC) or short method (SMC), from various agro-residues *viz.*, wheat straw/ paddy straw/ sugarcane bagasse as a base material along with other additives *viz.*, chicken manure, calcium ammonium nitrate, urea, superphosphate, muriate of potash, wheat bran, gypsum as additives. LMC is the primitive, cheap method involving only one phase (without pasteurization) (Mantel *et al.*, 1972). SMC is a quick method constituting of a general advance in controlled composting (Sinden and Hauser, 1950) and involves two sequential phases: phase I (an uncontrolled self-heating process initiated by mixing and wetting the ingredients as they are stacked in windrows which are periodically turned and watered at approximately two days interval) and Phase II which is the indoor process of pasteurization, carried out in tunnels.

Vermicomposting is a method of decomposing organic materials *viz.*, straw, shredded newspaper, saw dust and horse manure using surface feeding worms especially *Eisenia foetida* (redworm). The worms primarily feed on the microbes that are actively involved in the break down of the organic matter.

Physico-chemical aspects

The quickly changing physico-chemical conditions during composting select for a succession of microbial communities and thus exerts a profound effect on the entire process. Amongst the various parameters, composting is affected mainly by temperature, ammonia, carbon dioxide, moisture and C: N ratio of the substrate. The mass of decomposing organic materials is an exception to most ecosystems as it results in not only intensive heat production due to the metabolic activity of microorganisms but also acts as an effective retention system resulting in a significant rise in temperature. Generally, self-heating occurs when organic materials are assembled provided there is sufficient mass, atleast one ton, for insulation, and that moisture, aeration and nutrition level are adequate (Satyanarayana and Grajek, 1999).

Temperature is directly proportional to the biological activity within the composting system. As the metabolic rate of the microbes accelerates, the temperature within the system increases while with the decrease in the metabolic rate, the system temperature decreases (Namkoong *et al.*, 2002; Antizar-Ladislao *et al.*, 2005). The temperature in the composting stack usually reaches as high as 70 °C within 2 or more days (Finstein and Morris, 1975). A peak temperature of 70 °C can be attained within 4 days in metro-waste composting system meant for processing 150 tons of general municipal refuse daily (Kane and Mullins, 1973). A maximum temperature of 67 °C was observed in wheat straw composting after 8 days, remained above 50 °C for about 3 weeks (Chang and Hudson, 1967); on the other hand, composting of grass cuttings can attain 66 °C within 16 h, reaching a maximum of 76 °C after 48 h before a decline. The difference in temperature recorded is attributed to the type of material and other factors, such as age of the plant material (Satyanarayana and Grajek, 1999). In pig manure and chicken manure compost too, maximum temperature can reach upto 75 °C (Kowalchuk *et al.*, 1999).

The size of the compost pile is crucial for not only temperature build up also for maintenance of appropriate microbial equilibrium and successional pattern. In phase I of mushroom compost, a gradient of 20 °C (outer region of stack) to 70 °C (center) exists while phase II runs at lower temperature of 50 to 60 °C (Johri and Rajni, 1999; Peters *et al.*, 2000). The high temperature (75-80 °C) of compost pile is necessary to induce Millard reactions *i.e.*, fixing of free ammonia through reactions with carbohydrates and lignatious polymers. Millard reaction transforms carbohydrates into chemical from which is accessible to the mushroom crop but not to competitors and hence, provides for compost selectivity. High temperature is also necessary for the chemical incorporation of nitrogen into stable form within the compost (Johri and Rajni, 1999).

The initial pH of the compost varies between 6 and 7 and rises by 0.9-1.8 units during the first week due to the release of ammonia but later drops off (Chang and Hudson, 1967). During vermicomposting the pH changes from acidic to neutral while in municipal waste (Stutzenberger *et al.*, 1970), poulty manure-saw dust mixtures (Galler and Davey, 1971) and grass composts (Forsyth and Webley, 1948), initial pH of 4.5-6.0, stabilizes within 7 to 9, possibly due to the loss of organic acids through volatilization and microbial decomposition and the release of ammonia through mineralization of organic nitrogen; Peters *et al.* (2000) have reported the decrease in pH from 5.6 to 4.5 within the first two days of mushroom composting followed by an increase to 7.0 (Peters *et al.*, 2000) while Rawat (2004) reported variation in pH from 6.7 to 8.6 (end of phase I) with phase II composting pH stabilizing at 7.8. During the early stages of composting, a rapid growth of microbes reduces the pH value due to the formation of short chain organic acids, mainly lactic acid and acetic acid, and thereafter, the pH value of the composting materials rises gradually due to the increment in the amount of ammonia generated by the biochemical reactions of nitrogen-containing materials (Khiyami *et al.*, 2008).

Moisture content has significant effects on enzyme activities and microbial respiration of the composting process (Horng, 2003; Margesin *et al.*, 2006). A moisture content of 50-60 % is ideal for composting and microbial activity (Horng, 2003). The ratio between dry matter, water and air is an important factor in composting because raw materials have different moisture contents. Straw compost has high organic matter and can therefore retain more water. In mushroom compost, the optimum moisture is generally 75 % at the time of filling, 69 % at the time of spawning and 66 % after spawn run (Johri and Rajni, 1999). Rawat (2004) reported a variation of 60 % (zero day) to 56 % (filling) in moisture content during phase I and 62 % to 56% during phase II. An increase in electrical conductivity (8.28 to 33.86 mmhos. cm^{-1}) occurs from zero day to end of phase I of mushroom compost with a sharp decline to 23.11 at peak-heat stage, and slight increase (25.60) at the end of phase II (Rawat, 2004).

The changes in oxygen concentration in the compost atmosphere between turnings represent net result of O_2 utilized by microorganism and that replenished by convection and diffusion through compost. The change in O_2 concentration is related with temperature. In the center of compost pile where the temperature is highest, anaerobic conditions exist. When the stack is turned, heat loss occurs and the resulting temperature of about 50 °C stimulates microbial activity and depletes oxygen. Turning of the stack also results in a change in physico environment whereby the existing temperature gradients are disturbed. The loss and recovery of oxygen in the compost atmosphere is thus largely due to combined influence of temperature, which affects microbial activity and changes the ventilation pattern (Johri and Rajni, 1999). CO_2 evolution on the surface of compost pile shows good correlation with microbial activity compared to the number of propagules, for example it has been shown that whereas the total number of propagules is low during peak heating, CO_2 evolution is high due to higher rate of respiration of the abundant thermophilic microflora (Johri and Satyanarayana, 1984). Respiratory CO_2 of *Scytalidium thermophilum* was documented to be the likely reason for the growth promotory effect of this fungus on mushroom yield (Weigant, 1992).

Johri and Rajni (1999) reported that decomposition during composting was dependent upon carbon, nitrogen, lignin and carbohydrate composition of the organic material and expressed this relationship as:

$$CO_2 \text{ evolved} = \sqrt{\frac{\text{C/N residues} \times \text{\% lignin}}{\text{\% carbohydrate}}}$$

The rate of decomposition in compost pile is maximum during the first few days followed by a decline and ceases after sometime although large quantities of cellulose and hemicellulose are still present. In mushroom compost, about 60-70% polysaccharides are consumed by compost microorganisms, however, only 15-25% of total polysaccharides are used during mushroom production. The absolute amount of lignin remains unaltered though changes occur in the degree and condensation of lignins during composting because general microflora is devoid of basidiomycetes. During spawn running and fruiting about 15% of wall polysaccharides are utilized from compost and thus, considerable amounts (17-31%) remain at the end of mushroom production (Iiyama *et al.*, 1996).

The rate of decomposition is markedly influenced by size as well as C: N ratio of the composting material. The variation in composition (% by dry weight basis) of municipal compost to garden composts ranges from 25.0-80.0 (organic matter), 8.0-50.0 (carbon), 0.4-3.5 (nitrogen); a C: N ratio of 30: 1 is considered ideal for the activity of most microbes (Biddlestone and Garay, 1985).

Microbial Ecology

Compost is an interesting example of man-made ecosystem which harbours a complete spectrum of microbial diversity. The microbial abundance, composition and activity changes substantially during the composting process and is correlated with high microbial diversity and low activity in matured compost.

Microbial community succession during composting is a classical example of how the growth and activity of one group of organisms can create conditions necessary for the growth of others. Several generations of microorganisms succeed each other during composting wherein each crop of microbial form utilizes the available material in the substrate as also the cellular components of its predecessors for growth, spread and sustenance. The study of community structure and diversity by various workers (Bilai, 1984; Beffa *et al.*, 1996; Straatsma *et al.*, 1994 a, b; Peters *et al.*, 2000; Rawat *et al.*, 2005) has been instrumental in manipulating the compost environment in order to quicken the composting process and to improve the compost quality.

(a) Structural diversity

The compost is a rich reservoir of microbial types, comprising of mesophilic and thermophilic bacteria, fungi and actinomycetes. Fungi are the most predominant component. *Aspergillus, Chaetomium, Humicola, Mucor, Penicillium* and *Thermomyces* are the dominant fungi of compost ecosystems. Species of *Aspergillus* and *Mucor* are predominant in composting of biowaste (Ryckeboer *et al.*, 2003). *Aspergillus fumigatus* and *Humicola grisea* var. *thermoidea* have been reported to be the dominant member of the spent mushroom compost. Other fungi reported from spent mushroom compost are *Aspergillus flavus, Aspergillus nidulans, Aspergillus terreus, Aspergillus versicolor, Chrysosporium luteum, Mucor* spp., *Nigorospora* spp., *Oidiodendron* spp., *Paecilomyces* spp., *Penicillum chromogenum, Penicillum expansum, Trichoderma viride* and *Trichurus* spp. (Kleyn and Wetzler, 1981). The mesophilic microflora forms the pioneer community while thermophiles form the climax community.

In paddy straw compost, *Chaetomium thermophile, Humicola* spp. and *Sporotrichum thermophile* have been reported to be abundantly present in paddy straw compost (Satyanarayana, 1978). Antagonism appears to play a significant role in determining the population structure. The volatiles of *Chaetomium thermophile* and *Sporotrichum thermophile* can inhibit conidial germination of *Humicola lanuginosa* by impairing essential metabolic processes whereas *Chaetomium thermophile* suppresses mycelial growth of *Humicola lanuginosa* and *Torula thermophila.* However, effect of the fungistatic volatile factors in compost ecosystem is only marginal in view of high temperature at which they grow (Johri and Rajni, 1999).

Satyanarayana and Grajek (1999) observed that the colonizing ability of thermophilic fungi on paddy straw was directly proportional to the inoculum concentration. For example, colonization by *Humicola lanuginosa, Sporotrichum thermophile* and *Torula thermophila* increased with higher inoculum dose. During peak heating period, only a few thermophilic fungal propagules were present exhibiting high rate of respiration. However, Johri and Rajni (1999) reported that thermophilic fungi were not present at peak high temperature in wheat and broadbean straw composts. When it cooled down to 51.5°C, *Myriococcum albomyces Penicillium dupontii* and *Sporotrichum thermophile* were found in abundance.

Chang and Hudson (1967) studied the diversity of mycoflora of wheat straw compost. *Absidia ramosa, Aspergillus fumigatus, Chaetomium thermophile, Humicola grisea* var. *thermoidea, H. insolens, H. lanuginosa, Mlabranchea pulchella* var. *sulfurea, Mucor pusillus* and *Stilbella thermophila* occurred abundantly in the compost. The mesophilic fungi were soon killed off by the high temperature developed during the first few days; *A. fumigatus* and *M. pusillus* were also suppressed by the high temperature. *Mucor pusillus* did not recur even when the temperature of the compost later become suitable presumably due to lack of easily available carbon sources in the later stages. In contrast *A. fumigatus* persisted, with its ability to utilize cellulose and hemicellulose. *Humicola lanuginosa* developed early and persisted throughout composting due to its ability to lead commensal life with other forms (Rawat and Johri, 2002). Besides, this eukaryotic thermophile can tolerate wide temperature fluctuations on either side of optima and elaborates a variety of hydrolytic enzymes that help in continued presence. *Chaetomium thermophile, H. insolens, H. lanuginosa* and *Talaromyces dupontii* develop abundantly in the plateau period and rapidly utilise cellulose and hemicellulose. *Sporotrichum thermophile* alongwith some mesophilic fungi succeed when the temperature of the compost dropped. However, Moubasher *et al.* (1982) reported that thermophilic fungi were not present at peak high temperature in wheat and broadbean straw composts. When it cooled down to 51.5 °C, *Myriococcum albomyces, Penicillium dupontii* and *Sporotrichum thermophile* were found in abundance.

In paddy straw compost, *Chaetomium thermophile, Humicola* spp. and *Sporotrichum thermophile* are abundantly present (Satyanarayana, 1978). Satyanarayana and Johri (1984) observed that the colonizing ability of thermophilic fungi on paddy straw was directly proportional to the inoculum concentration. The colonization by *Humicola lanuginosa, Sporotrichum thermophile* and *Torula thermophila* increased with higher inoculum dose. During peak heating period, only a few thermophilic propagules were present exhibiting a higher rate of respiration.

Klamer *et al.* (2001) studied the succession of mycoflora during eight months of composting of miscanthus straw and pig slurry in well insulated containers. Before peak heating, *Aspergillus fumigatus* and *Rhizomucor pusillus* were dominant. Forms developing after peak heating could be divided into two groups: those appearing from day 15 to 27 and others developing from day 50 to 225. The first group was dominated by *Paecilomyces variotii, Scytalidium thermophilum* and *Thermomyces lanuginosus* and the second by *Acremonium* spp. and *Thermomyces lanuginosus.* The Brillouin diversity index changed with temperature; diversity was high before peak-heating, low during elevated temperature, and increased again during the third phase of composting. Temperature was the main controlling parameter that changed fungal community during the first month of composting. Based on principal component analysis (PCA), it could be concluded that when the temperature reached the ambient level, only minor change in fungal community was detected although some changes in certain species were still discernible.

Municipal wastes generally contain substrates rich in lignohemicellulose, among other substrates. Thermophilic fungi play a significant role in the conversion of these materials into farmyard manure or single cell protein (Kane and Mullins, 1973; Eriksson and Larson, 1975). This aroused interest in studying the diversity of mycoflora in municipal compost. The unique ability of some species to degrade plastic substances envisaged special interest in their study (Eggins and Mills, 1971; Brown *et al.,* 1974).

Subrahamanyam *et al.* (1977) isolated a new genus *Thermomcor* besides other fungi from municipal waste compost in India. *Thermoascus aurantiacus* and *Myceliophthora thermophila* were the two major thermophilic fungi recovered from municipal waste compost (Sen *et al.,* 1980; Subrahmanyam, 1980). Kane and Mullins (1973) reported *Thermoascus aurantiacus* and other thermophilic fungi at the beginning of the digestion period, which persisted throughout composting. These workers were of the opinion that high temperature, acidity and anaerobic conditions could limit fungal growth in the interior of the compost and thus restrict the role of these moulds. They recorded the presence of thermophilic fungi at all times during composting of municipal refuse and did not find any apparent succession within the species.

Thermophilic fungi grow extensively during the last phase of composting in mushroom compost from the spores that survive the pasteurization temperature (Straatsma *et al.,* 1989). Thus, they contribute significantly towards the quality of compost. However, their presence throughout the course of composting is largely responsible for the maintenance of biological equillibrium that ultimately leads to unique selectivity wherein *A. bisporus* multiplies without competition. These fungi influence growth of *A. bisporus* at three distinct levels (Weigant, 1992): First, they decrease concentration of ammonia in compost which otherwise would counteract the growth of the mycelium. Second, they immobilize nutrients in a form, which improves apparent availability to the mushroom mycelium. Third, they exert direct growth promotory influence on the mushroom mycelium *viz.*, *S. thermophilum*. The course of fungal succession is partially dependent on the ecophysiological conditions in compost (Satyanarayana *et al.*, 1992).

In mushroom compost maximal diversity amongst thermophilic fungal morphotypes is observed with fourth turning of phase I compost (H'=2.14; E1=0.89 and D=8.88); and least by peak heat stage morphotypes (H'=0.75 and D=1.80). End of Phase I is most rich in species make up (R1=4.07 and R2=3.37). Fourth turning of phase I compost is maximally diverse for thermophilic community. The least fungal diversity is observed at peak-heat stage (Rawat, 2004). This is not unusual since only limited fungal species spectrum has been reported from this stage of composting (Straatsma *et al.*, 1994a).

The pioneer thermophilic mycoflora of mushroom compost comprises of fast growing and rapidly sporulating fungi such as *Aspergillus fumigatus* and *Rhizomucor* spp. with a pH optima below 7.0 and temperature optima of about 40°C. When self-heating and ammonification starts and pH reaches 9.0, the pioneer flora disappears and paves way for *Talaromyces thermophilus* and *Thermomyces lanuginosus*; during massive heat production these fungi possess moderate growth rate, as they exhibit high thermal death point and pH tolerance, but do not degrade cellulose. At the end of the composting process, about 50-70% of the compost biomass is constituted by thermophilic fungi (Sparling *et al.*, 1982; Weigant, 1992). While most of the species are eliminated, *S. thermophilum* appears as near exclusive species after phase II composting and constitutes a climax species in the mushroom compost along with thermophilic actinomycetes (Straatsma *et al.*, 1994b). The number of CFU of *S. thermophilum* in fresh matter of phase II is about 10^6 g^{-1} compost (Bilai, 1984), however, actinomycetes and bacteria appear to play a decisive role in successful colonization by this thermophile. The presence of *S. thermophilum* throughout the composting period *i.e.*, from zero day, dominance during phase II and at the end of phase II is supported by its relative abundance (0.68) as observed by Rawat (2004) and earlier observations of Straatsma *et al.* (1994a) on the subject.

In the beginning of phase II of mushroom compost, thermophilic fungi and actinomycetes extensively colonize the plant matter until temperature reaches 60°C, as an out come of slow peak-heating for about two days (Straatsma *et al.*, 1994a). The high temperature of the first indoor period of phase II kills most of the pathogenic and non-pathogenic microorganisms, except the spores of actinomycetes and thermophilic fungi such as *Scytalidium thermophilum* (Straatsma *et al.*, 1991); the later was most abundant at the end of phase II compost (abundance = 0.68) (Rawat, 2004). Klamer *et al.* (1998) reported *A. fumigatus* and *Rhizomucor pusillus* as predominant species before peak heating and *P. variotii*, *S. thermophilum* and *Thermomyces lanuginosus* as dominant forms after peak heating. Tewari (2000) reported the presence of *H. lanuginosa*, and *S. thermophilum* during peak heat stage of phase II composting.

Thermophilic fungi of the *Torula-Humicola* complex are a necessary and dominant component of the community in mushroom compost during Phase II. *S. thermophilum* is a natural inhabitant of compost ingredients including drainage from compost, and has been documented to be present throughout composting. Dominance of *S. thermophilum* has been reported by several workers (Straatsma *et al.*, 1991; Vijay, 1996; Klamer *et al.*, 1998; Rajni *et al.*, 1998) while *H. grisea* var. *thermoidea* and *H. insolens* have been described by others (Fergus, 1964). They are inherently close partners in the degradation processes in compost and provide selectivity to compost (Straatsma *et al.*, 1989; Opden Camp *et al.*, 1990). Rajni *et al.* (1998) and Rawat (2004) observed nearly similar microbial distribution pattern in compost as reported by Straatsma *et al.* (1991) with predominance of *S. thermophilum* although inputs in the European and Indian composts are substantially different. In twenty days schedule of compost preparation, *S. thermophilum* was detected from the first turning (*i.e.*, after 5th day of composting) till fifth turning (*i.e.* 17th day) whereas species of *Paecilomyces* were present only during the last turning i.e., 20th day. Two isolates of *Malbranchea cinnamomea* were recorded between the second and fifth turning stage (8th to 17th day). The presence of *Paecilomyces* sp. and *M. cinnamomea*, which are normally slow growers, provided an opportunity to evaluate their influence on *in situ* mycelial extension of *A. bisporus*. At 24th

day of composting the population of *S. thermophilum* was 10^8 propagules g^{-1} of compost.

Thermophilic fungi of the *Torula-Humicola* complex are a necessary and dominant component of the community during Phase II. Dominance of *S. thermophilum* has been reported by several workers (Straatsma *et al.*, 1991; Vijay, 1996; Klamer *et al.*, 1998; Rajni *et al.*, 1998) while *H. grisea* var. *thermoidea* and *H. insolens* have been described by others (Fergus, 1964). They are inherently close partners in the degradation processes in compost and provide selectivity to compost (Straatsma *et al.*, 1989; Opden Camp *et al.*, 1990). The density of *S. thermophilum* was found to be positively correlated with mushroom yield (Straatsma *et al.*, 1989). Improved growth of *A. bisporus* mycelium in composts treated with *S. thermophilum* has been extensively reported in literature (Ross and Harris 1983; Straatsma *et al.*, 1989; 1991; Weigant, 1992; Straatsma and Samson 1993; Straatsma *et al.*, 1994 a, b). The causal relationship between the presence of *S. thermophilum* and the crop yield of mushroom remains still obscure. Straatsma *et al.* (1993) observed that this fungal species merely affects the radial extension rate rather than having a positive influence on the surface growth rate of *A. bisporus* mycelium. It reduces the growth of pathogenic microorganisms by virtue of inhibitory influence (Straatsma *et al.*, 1991). Respiratory CO_2 of this species may play a stimulatory role (Weigant *et al.*, 1992) but under different experimental conditions, neither volatiles nor CO_2 were stimulatory (Straatsma *et al.*, 1993; Straatsma *et al.*, 1995). The mere presence of *S. thermophilum* is however quite essential. Ross and Harris (1983) suggested that visible but dormant biomass of this species in composts fills an otherwise biological vacuum, which in turn allows growth of *A. bisporus* mycelium. Other thermophilic fungal species such as, *Chaetomium thermophilum, Malbranchea sulfurea, Myriococcum thermophilum, Stilbella thermophila, Thielavia terrestris* and two unidentified Basidiomycetes were also found to be promotory for mycelial growth of *A. bisporus* on sterilized compost along with *S. thermophilum* (Straatsma *et al.*, 1994a).

Among various groups actively engaged in studying the ecology of mushroom compost, Straatsma's group in Holland had laid considerable emphasis on population dynamics of *S. thermophilum* for better compost management. It is a natural inhabitant of compost ingredients, and drainage from compost, and has been documented to be present throughout composting. The disappearance of ammonia and selectivity of compost for the growth of *A. bisporus* mycelium that occurs in Phase II at temperature 45-55°C are linked to the presence of *S. thermophilum* (Ross and Harris, 1983). Rajni *et al.* (1998) and Rawat (2004) observed nearly similar microbial distribution pattern in compost as reported by Straatsma *et al.* (1991) with predominance of *S. thermophilum* although inputs in the European and Indian composts are substantially different. They also observed the morphological variability of this fungus in compost with differences in their growth rate and promotion of mycelial growth. In twenty days schedule of compost preparation, *S. thermophilum* was detected from the first turning (*i.e.*, after 5th day of composting) till fifth turning (*i.e.* 17th day) whereas species of *Paecilomyces* were present only during the last turning *i.e.*, 20th day. Two isolates of *Malbranchea cinnamonea* were recorded between the second and fifth turning stage (8th to 17th day). The presence of *Paecilomyces* sp. and *M. cinnamonea*, which are normally slow growers, provided an opportunity to evaluate their influence on *in situ* mycelial extension of *A. bisporus*. At 24th day of composting the population of *S. thermophilum* was 10^8 propagules g^{-1} of compost.

Aspergillus fumigatus and *Humicola grisea* var. *thermoidea* are the dominant member of spent mushroom compost. Others include *Aspergillus flavus, A. nidulans, A. terreus, A. versicolor, Chrysosporium luteum, Mucor* spp., *Nigorospoa* spp., *Oidiodendron* spp., *Paecilomyces* spp., and *Penicillum chromogenum*. The cultivation of mushrooms under controlled conditions in a growth room changes the quality and quantity of mycoflora in the compost. However, Klamer *et al.* (1998) reported *A. fumigatus* and *Rhizomucor pusillus* as predominant species before peak heating and *P. variotii, S. thermophilum* and *Thermomyces lanuginosus* as dominant forms after peak heating. Tewari (2000) reported the presence of *H. lanuginosa*, and *S. thermophilum* during peak heat stage of phase II.

Antagonisms, too plays a significant role in determining the population structure. Volatile compounds and antibiotics are produced by microorganisms *in vivo* and also *in vitro*. The volatile of *Sporotrichum thermophile* and *Chaetomium thermophile* can inhibit conidial germination of *Humicola lanuginosa* by impairing essential metabolic processes. Volatiles of *Chaetomium thermophile* inhibit mycelial growth of *Humicola lanuginosa* and *Torula thermophila*. However, effect of the fungistatic volatile factors in compost ecosystem is only marginal in view of high temperature at which they grow (Johri and Satyanarayana, 1984).

(b) Genetic diversity

The genetic diversity of compost microflora has not been widely studied. It is however attracting attention in order to assign correct taxonomic status based on the traditional and molecular tools. The study of genomic DNA/RNA composition, gene structure mainly with a view to examine intron architecture/ organization and also regulatory sequences and "motifs" related to gene expression, is in progress. The GC content of DNA of all bacterial strains isolated from thermogenic compost at temperature between 65 and 82°C ranges from 60.5- 63.4 mole % (Beffa *et al.*, 1996) and is similiar to that reported for *Thermus* spp. (Hudson *et al.*, 1987). All isolates except one showed high degree of DNA-DNA homology with *T. thermophilus* (65-71%) than with *T. acquaticus* (50-62%) and poor DNA – DNA homology with *T. filiformis* (39-48 %) and *T. ruber* (30-40 %).

During their early studies of *Humicola*, De Bertoldi *et al.* (1972, 1973) observed the mole % GC content of mesophilic and thermophilic species of *Humicola* to be newely in the same range; for the former it ranged between 30.6 and 56.9, and for the latter, 34.8 and 55.5. Although, *H. brevis* and *Torula terresteris* possess almost similar GC content (30.6 and 30.4 %), these two organisms are not related suggesting base sequence dissimilarities (Meyer and Phaff, 1969). Rodrigues *et al.* (1991) suggested that high mutation rates could cause variability and genomic differences in *Humicola* spp., which are further accentuated by polynucleated characteristic of this species.

The most widely applicable and rapid genetic fingerprinting techniques are based on the use of PCR. Genetic fingerprinting includes restriction fragment length polymorphism (RFLP) analysis of major components of the rDNA operon (rrn), namely, 16S, 23S, rDNA intergenic spacer region i.e., IGS, internal transcribed spacer region (ITS – PCR) and random amplified polymorphic DNA (RAPD) analysis (Blanc *et al.*, 1997). A comparison of nucleotide sequence of 5.8 S rRNA was the central focus of early studies on genetic diversity of thermophilic fungi. The nucleotide sequence of *T. lanuginosus* was closely related to other fungi, but exhibited high overall homology with other 5.8 S rRNA (Wildeman and Nazar, 1981).

In hot compost (60 – 80°C) the isolates were found to have similar RFLP profiles as *Thermus thermophilus* HB 8 while strains of *T. thermophilus, T. acquaticus, T. filiformis* and *T. ruber* showed different RFLP profiles (Beffa *et al.*, 1996). A high level of intraspecific polymorphism amongst black *Aspergillus* species, representative of considerable divergence among sexual lines with no, or very little, genetic exchange, was reported based on RAPD analysis (Megnegneau *et al.*, 1993). Muthumeenakshi *et al.* (1994) studied the intraspecific molecular variation among *Trichoderma harzianum* isolates colonizing mushroom compost based on ITS sequences.

The genetic variation exhibited by *Torula-Humicola* complex has drawn wide attention. Azevedo *et al.* (1999) have distinguished homokaryotic strains of *Humicola grisea* var. *thermoidea* in two groups displaying uniform DNA profile reflecting heterogeneity in the wild genome using RAPD analysis. Straatsma and Samson (1993) studied the genetic diversity among *Scytalidium thermophilum* isolates using RAPD analysis. These strains exhibited a distinct pattern of amplified DNA bands. Rajni (1999) studied the genetic diversity between the black and white strains of *S. thermophilum* by RAPD analysis. A distinct band pattern was exhibited by these isolates. The protein profiling of *Torula-Humicola* complex exhibited *S. thermophilum* was more closely related to *H. grisea* var. *thermoidea* while the latter was similar to *H. insolens* than *H. lanuginosa*; *H. insolens* and *H. lanuginosa* were more distantly related.

The RAPD analysis and sequence analysis of ITS region of rDNA exhibited wide genetic variation in *Torula-Humicola* complex (Lyons *et al.*, 2000). RAPD analysis of 34 geographically diverse isolates revealed two distinct groups showing differences in the banding pattern. An examination of the genetic distance matrix indicated differences between isolates belonging to *S. thermophilum* cultural types 1 and 2. The sequence analysis of ITS 1, 5.8 S and ITS 2 region of rDNA suggested high homology between the isolates with minor sequence variation. Genetic distance values, among type 1 and 2 *S. thermophilum* isolates, varied by a value of 0.005%. The RAPD groupings mirror closely the morphological and thermogravimetric data for *S. thermophilum* isolates and provides further evidence of the variation, which exists between the species complex (Straatsma and Samson, 1993; Lyons and Sharma, 1998). Isolates of *S. thermophilum* bear close similarity to those of *H. grisea* var. *thermoidea* and *H. insolens*. Using RAPD analysis (Rawat, 2004) that majority of structurally or functionally dominant fungal species recovered from different stages of composting belonged to *Torula-Humicola* complex.

Beffa *et al.* (1996) reported for the first time the presence of genus *Thermus* in thermogenic (65 to 82°C) composts taken from 2 to 5 week old organic waste samples. Majority of the isolates were probably *Thermus* strains, which had adapted to the conditions prevalent in hot compost ecosystem. The nutritional characteristics, total protein profiles, DNA-DNA hybridization and restriction fragment length polymorphisms (RFLP) profiles of 16S rDNA showed that *Thermus* strains isolated from host composts were closely related to *T. thermophilus* HB8 except strain JT4. All isolates except JT4 showed 65 to 71% DNA-DNA homology with *T. thermophilus* and 50 to 62%, with *T. aquaticus* while JT4 showed 72% homology with *T. aquaticus*. A moderately high number of obiligately autotrophic *Hydrogenobacter* spp. and facultatively autotrophic *Bacillus schlegelii* growing at temperatures above 70^{0}C were isolated from hot composts (Beffa *et al.*, 1996). A number of entrobacteriaceae and members of the genus *Lactobacillus* were observed in the initial stage of composting of organic matter at 41°C. Low G+C gram-positive bacteria constituted a dominant fraction of the bacterial community during hot composting (55-70 ^{0}C) phase (Alfreider *et al.*, 2002).

The genetic profiling techniques which are cultivation independent have great potential in identifying the population structure and community succession. These techniques utilize DNA or RNA directly extracted from environmental samples and amplification of signature genes by PCR or reverse transcription-PCR (RT-PCR) with primers, bind to conserverd regions and produce homologous gene fragments. The products can be subsequently analyzed to determine their nucleotide differences by techniques such as, denaturing gradient gel electrophoresis (DGGE), temperature gradient gel electrophoresis (TGGE), terminal restriction fragment length polymorphism (t-RFLP) and single-stranded conformation polymorphism (SSCP) (Muyzer *et al.*, 1993; Lee *et al.*, 1996; Liu *et al.*, 1997; Schwieger and Tebbe, 1998). SSCP has the potential to be more easily applied in contrast to DGGE and TGGE, as no GC clamps or construction of gradient gel is required (Lee *et al.*, 1996). However, it is quite essential to compare diversity results obtained by both cultivation-dependent and cultivation-independent methods for better understanding and for eliminating the biases associated with genetic profiling.

Rawat (2004) observed that SSCP profile (Fig.1) of fungal community of different stages of composting corroborated well with the culturable fungal diversity; though the total number of fungal species culturable were only 34, a total of 95 distinct bands were obtained. The banding pattern of community underwent a rapid change with the onset of composting; this corroborated well with the reports of culturable fungal diversity by Peters *et al.* (2000). The profile patterns of two samples of zero day compost (PWI and PWII) were quite similar while different stages of phase I composting shared nearly identical pattern profile; pattern of cropping sequences was quite similar to each other. PWI and PWII compost profiles consisted of many faint bands but with the onset of composting these faint bands disappeared; during peak-heat stage only prominent bands were present. This banding pattern corroborated with the observations of Vijay (1996) that fungi encountered on the ingredients play only a minor role in composting and may disappear as soon as composting is started.

The dendrogram (Fig. 2) based on Jaccard's similarity coefficient exhibited a 15-66% similarity amongst the community of different stages of mushroom compost. Community of all the samples could be divided into three main groups. Group I comprised of zero day (PWI & PWII) community exhibiting 26% similarity. Group II could be further subdivided into two groups: Ist subgroup comprised of phase I community. Second (T2) and third (T3) turning community shared maximum similarity at 66% level while first turning (T1) community shared a similarity of 32% with that of second and third community. Fourth turning (T4) and end of phase I (filling) community showed 51% similarity with each other. IInd subgroup comprised of peak-heat, conditioning and end of phase II compost (spawning) community. Peak-heat and conditioning communities shared 38% similarity with each other and 30% similarity with end of phase II compost community. The overall similarity between the two subgroups was 30%. II group had casing soil, cropping, airing and drenching communities.. PWI, PWII, T2, T4, spawning, cropping, airing and drenching community shared a maximum similarity of 66% with each other and 44% similarity with spawn-run community. Drenching community shared 42% similarity with that of spawn-run, cropping and airing community. Casing community showed 22% similarity with cropping, airing and drenching community. The similarity amongst these three subgroups was quite low; 16% between gp I and gp II while they both shared 15% similarity with gp III.

The community profiles of Phase I compost were found to be identical to each other; maximum diversity was observed during end of phase I compost, casing soil and cropping stage (H' = 0.37) and least at peak-heat stage. However, the level of diversity assigned on the basis of SSCP did not match with the culturable fungal diversity; both the approaches however described least diversity value at peak-heat stage as a common point. The diversity indices indicated maximum diversity amongst community of end of phase I compost, casing and cropping community (H'=0.37) while least at peak-heat community (H'=0.22).

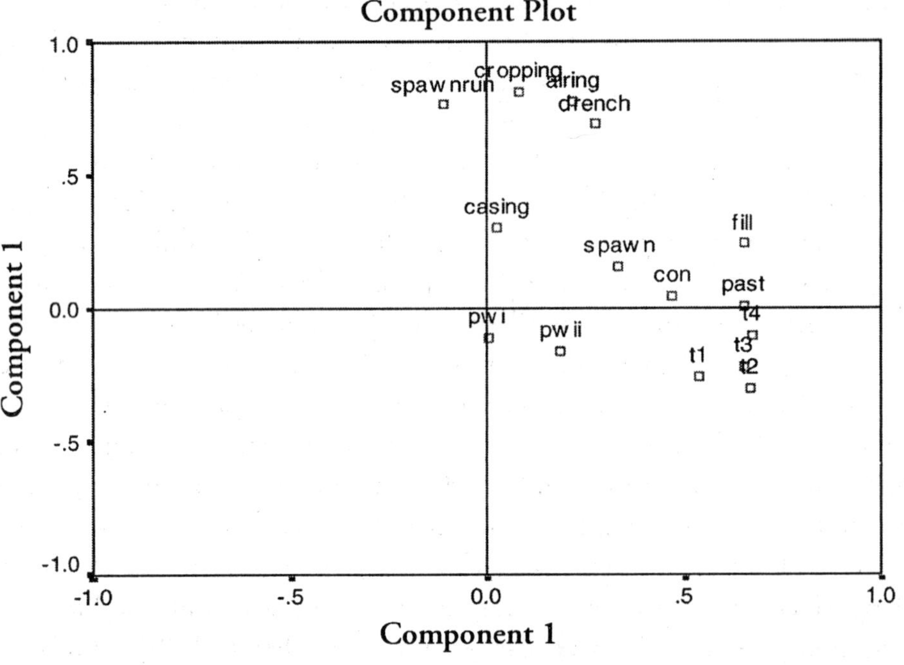

Fig.1: PCA of 18S community of different stages of mushroom compost.

The fungal phylotypes, revealed by cloning and sequencing of fungal internal transcribed spacer (ITS) region of samples of different stages of composting in a full-scale and a pilot-scale composting reactors, could be grouped into those that dominated the mesophilic low pH initial phases (sequences similar to genera *Candida, Dipodascaceae* and *Pichia*) and those found mostly or exclusively in the thermophilic phase (sequences clustering to *Candida, Rhizomucor* and *Thermomyces*), but a few were also present throughout the whole process (Hultman *et al.*, 2010).

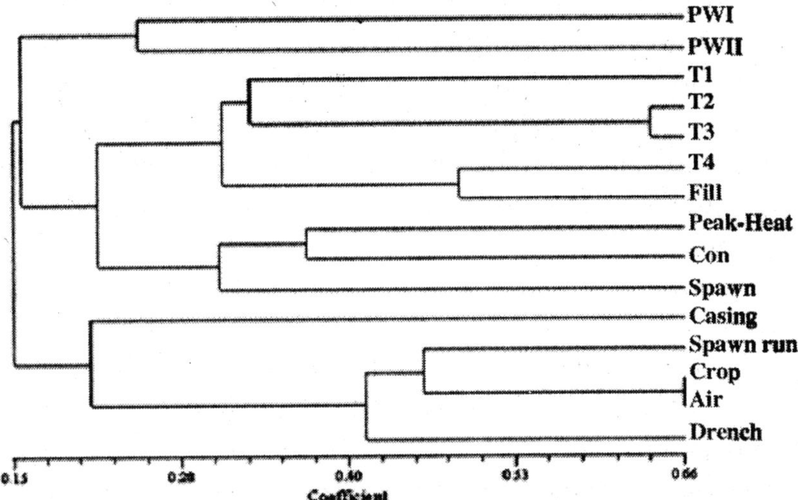

Fig.2: Cluster analysis of 18S community of different stages of mushroom compost based on Jaccard's coefficient.

(c) Functional diversity

Compost is not only structurally diverse but also functionally active. The *in situ* functionality of each microbial component especially the extracellular enzymatic machinery *viz.*, polysaccharases, proteases and lipases, plays an important decisive role in successful colonization and succession in mushroom compost (Rajni *et al.*, 1998; Johri *et al.*, 1999, Rawat and Johri, 2002, Rawat and Johri, 2004). Infact, compost harbours a number of guilds.

The wide exploration of enzymatic machinery of the individual microbial component and their *in situ* enzymatic action in their own niche would help in greater understanding of the relationship between structural and functional diversity. However, the available information appears largely biased towards the functional role of thermophilic mycoflora of compost probably because of they are a dominant component of the functional niche occupied by compost microbiota. The wide enzymatic potential exhibited by these fungi along with enzyme multiplicity with different phsico-chemical characteristics helps in the functioning of each component under different biophysical conditions. The increase in cellulolytic and amylolytic activity during composting is a reflection of change in the population and community structure of the resident microflora. The enzymatic diversity of compost microbiota is known to result in specificity and successional change besides providing a niche to various species to survive in the absence of simple sugars (Rawat and Johri, 2002).

In compost the pioneer microflora in general can utilize simple sugars but such biota disappear soon and only those organisms with wide polysaccharolytic ability persist. The cellulolytic and hemicellulolytic ability of thermotolerant *Aspergillus fumigatus* allows it to persist in the wheat straw compost whereas thermophilic *Mucor pusillus* could not recur even after temperature become suitable for growth due to lack of polysaccharolytic ability. *Humicola lanuginosa* persists throughout composting due to its ability to lead commensal life with other along with its cellulolytic and hemicellulolytic ability. *Chaetomium thermophile, Humicola insolens, Humicola lanuginosa* and *Talaromyces dupontii* develop abundantly in the 'plateau' period and rapidly utilize cellulose and hemicellulose. When compost temperature drops, thermophilic *Sporotrichum thermophile* and mesophilic *Coprinus cinereus* and *Clitopilus pinsitus* appear which can utilize cellulose and hemicellulose in wheat staw at a slower rate (Chang and Hudson, 1967).

The success of microflora in competitive saprophytic colonization (CSC) such as that operative in plant residues and soil depends upon its intrinsic ability to decompose that substrate and the ability to succeed in the competition (Garret, 1963). The strongly cellulolytic and hemicellulolytic *Aspergillus fumigatus, Sporotrichum thermophile* and *Torula thermophila* exhibit greater colonization ability than the weakly cellulolytic, *Humicola lanuginosa* (Johri and Satyanarayana, 1984). The dominance of *S. thermophilum* has been attributed to the presence of complete complement of polysaccharlytic enzyme machinery (Grajek, 1987; Rawat, 1998; Khokar, 1999; Tewari, 2000). The hydrolytic potential of such thermophilic fungi is responsible for solubilization of complex ingredients of compost and making the nutrient availabile to *A. bisporus*. The stimulation of growth of mushroom mycelium by cellulose decomposing mycoflora has been well-documented (Stanek, 1969).

Considering the fact that culturable microbial populations are limited on account of our poor understanding of their nutritional requirements, detailed, *in situ* enzymatic investigations are likely to provide a better understanding of the relationship between structural and functional diversity of thermophilic fungal community. Iiyama *et al.* (1996) observed that the loss of cellulose and lignocellulose and increase in protein content during the composting period was a result of increased polysaccharolytic activity of the fungal biomass; this resulted in increased level of reducing sugars. In mushroom compost the level of enzymes was found to increase from zero day to the end of phase I compost; thereafter it decreased continuously; results were well corroborated with the population structure (Rawat, 2004).

It has been observed that *in situ* changes in lignocellulose, cellulose and loss in weight during composting is corroborated with the activity of thermophilic fungi. The analysis of plant residues after decomposition by pure thermophilic fungal cultures resulted in biochemical changes that were similar to those observed during composting of organic materials by natural mixed microflora (Satyanarayana, 1978). During the 24 day composting sequence of button mushroom, Rajni (1999) reported that the level of organic carbon decreased from 18.12 to 10.57% while that of cellulose and lignocellulose from 32.3-23.0% and 52.4 – 43.1%, respectively. An increase of 77.5% and 86% in protein and reducing sugar level was observed. The level of amylase, endo-cellulase and exo-cellulase also changed. Based on the regression analysis between the population of *Scytalidium thermophilum* and chemical parameters, a 92.5% change in lignocellulose was found as a result of rise in population of *S. thermophilum* in compost. The levels of dehydrogenase activity decreased by 72%; protease by 32%; xylanase and cellulase by 50 % during peak heat while during of phase II and post peak heat stageof mushroom compost, levels again increased by 71%, 23% and 33.3%, respectively. These changes were found to be well corroborated with the change in population structure of compost microflora (Tewari, 2000).

In compost the pioneer microflora, in general, can utilize simple sugars but such biota disappear soon and only those organisms with wide polysaccharolytic ability persist. The cellulolytic and hemicellulolytic ability of thermotolerant *Aspergillus fumigatus* allows it to persist in the wheat straw compost whereas thermophilic *Mucor pusillus* does not recur even after temperature became suitable for growth due to lack of polysaccharolytic ability. *Humicola lanuginosa* persists throughout composting due to its ability to lead commensal life with other along with its cellulolytic and hemicellulolytic ability. *Chaetomium thermophile, Humicola insolens, Humicola lanuginosa* and *Talaromyces dupontii* develop abundantly in the 'plateau' period and rapidly utilize cellulose and hemicellulose. When compost temperature drops, thermophilic *Sporotrichum thermophile* and mesophilic *Coprinus cinereus* and *Clitopilus pinsitus* appear which can utilize cellulose and hemicellulose in wheat straw at a slower rate (Chang and Hudson, 1967).

The success of microflora in competitive saprophytic colonization (CSC) such as that operative in plant residues and soil depends upon its intrinsic ability to decompose that substrate and the ability to succeed in the competition. The strongly cellulolytic and hemicellulolytic *Aspergillus fumigatus, Sporotrichum thermophile* and *Torula thermophila* exhibit greater colonization ability than the weakly cellulolytic, *Humicola lanuginosa* (Johri and Satyanarayana, 1984). The dominance of *S. thermophilum* has been attributed to the presence of complete complement of polysaccharolytic enzyme machinery (Rawat, 1998; Tewari, 2000; Rawat *et al.*, 2005). The hydrolytic potential of such thermophilic fungi is responsible for solubilization of complex ingredients of compost and making the nutrient availabile to *A. bisporus*. The stimulation of growth of mushroom mycelium by cellulose decomposing mycoflora has been well-documented (Stanek, 1969; Straatsma *et al.*, 1994a; Johri and Rajni, 1999).

Mushroom compost that harbours high population of thermophilic flora yields more mushroom produce (Shandilya, 1982; Vijay, 1996). The selectivity of compost is brought about by the static population of thermophilic flora, which becomes inactive at the time of spawning. The thermophilic microbial biomass is a concentrated source of nutrients required for the growth of *A. bisporus*. Thermophilic fungi appear to use up all the readily available nutrients during the process of composting and thus a major portion of the available nutrients is locked up inside their cells (Betterely, 1993). *Agaricus bisporus* posseses the complement of enzymes viz., β-N-acetyl galactosaminidase, laminarinase, protease etc. by which it can degrade thermophilic bacteria, fungal and actinomycete mycelium for its own growth (Fermor and Grant, 1985; Rawat *et al.*, 2005). Sparling *et al.* (1982) reported that microbial biomass contributed less than 10% to mushroom biomass and therefore *A. bisporus* probably obtained bulk of its carbon nutrition from straw. However, the microbial biomass can act as a concentrated source of nitrogen and minerals. The selectivity of compost is lost if dormant thermophilic biomass is destroyed by heat or chemicals (Ross and Harris, 1983), and growth rate of *A. bisporus* mycelium is reduced on sterilized compost (Wood and Matchman, 1980). Rawat (2004) found that dominant functional forms in mushroom compost were representatives of T3 stage.

The role of *Scytalidium thermophilum*, predominant component of compost, in compost management has been well documented by various workers (Ross and Harris, 1983; Straatsma *et al.*, 1989; Johri and Rajni, 1999; Rawat, 2004; Rawat *et al.*, 2005). The disappearance of ammonia and selectivity of compost for the growth of *A. bisporus* mycelium that occurs in Phase II at temperature 45-55°C is linked to the presence of *S. thermophilum* (Ross and Harris, 1983). The density of *S. thermophilum* was found to be positively correlated with mushroom yield (Straatsma *et al.*, 1989). The causal relationship between the presence of *S. thermophilum* and the crop yield of mushroom remains still obscure. Straatsma *et al.* (1991) observed that this fungal species merely affects the radial extension rate rather than having a positive influence on the surface growth rate of *A. bisporus* mycelium. It reduces the growth of pathogenic microorganisms by virtue of inhibitory influence. Respiratory CO_2 of this species may play a stimulatory role (Weigant, 1992) but under different experimental conditions, neither volatiles nor CO_2 were stimulatory (Straatsma *et al.*, 1994a). The mere presence of *S. thermophilum* is however quite essential. Ross and Harris (1983) suggested that visible but dormant biomass of this species in composts fills an otherwise biological vacuum, which in turn allows growth of *A. bisporus* mycelium. The disappearance of ammonia and selectivity of compost for the growth of *A. bisporus* mycelium that occurs in Phase II at temperature 45-55°C are linked to the presence of *S. thermophilum* (Ross and Harris, 1983). Other thermophilic fungal species such as, *Chaetomium thermophilum*, *Malbranchea sulfurea*, *Myriococcum thermophilum*, *Stilbella thermophila*, *Thielavia terrestris* and two unidentified Basidiomycetes were also found to be promotory for mycelial growth of *A. bisporus* on sterilized compost along with *S. thermophilum* (Straatsma *et al.*, 1994a).

Considering the fact that culturable microbial populations are limited on account of our poor understanding of their nutritional requirements, detailed, *in situ* enzymatic investigations are likely to provide a better understanding of the relationship between structural and functional diversity of thermophilic fungal community. Iiyama *et al.* (1996) observed that the loss of cellulose and lignocellulose and increase in protein content during the composting period was a result of increased polysaccharolytic activity of the fungal biomass; this resulted in increased level of reducing sugars. In mushroom compost the level of enzymes was found to increase from zero day to the end of phase I compost; thereafter it decreased continuously and the results were well corroborated with the population structure (Rawat, 2004; Rawat *et al.,* 2005).

It has been observed that *in situ* changes in lignocellulose, cellulose and loss in weight during composting is corroborated with the activity of thermophilic fungi. The analysis of plant residues after decomposition by pure thermophilic fungal cultures resulted in biochemical changes that were similar to those observed during composting of organic materials by natural mixed microflora (Satyanarayana, 1978). During the 24 day composting sequence of button mushroom, Rajni (1999) reported that the level of organic carbon decreased from 18.12 to 10.57% while that of cellulose and lignocellulose from 32.3-23.0% and 52.4 – 43.1%, respectively. An increase of 77.5% and 86% in protein and reducing sugar level was observed. The level of amylase, endo-cellulase and exo-cellulase also changed. Based on the regression analysis between the population of *Scytalidium thermophilum* and chemical parameters, a 92.5% change in lignocellulose was found as a result of rise in population of *S. thermophilum* in compost. The levels of dehydrogenase activity decreased by 72%; protease by 32%; xylanase and cellulase by 50 % during peak heat while during of phase II and post peak heat stage of mushroom compost, levels again increased by 71%, 23% and 33.3%, respectively. These changes were found to be well corroborated with the change in population structure of compost microflora (Tewari, 2000).

Yu *et al.* (2007) reported that hemicellulose and cellulose were partially degraded during initial stage of composting of agricultural wastes and thereafter the degrading ratio was almost unaltered due to high temperature followed by the large decomposition during the temperature falling phase (12-20d) and initial stage of the second fermentation (21-40d of composting). Lignin was slightly decomposed during the initial stage of composting. When the temperature was lower than the maximum value during thermophilic phase, lignin was greatly degraded until the temperature began to fall.

Stanek (1969) observed stimulation of growth of mushroom mycelium by cellulase decomposing microflora mainly, actinomycetes and fungi. It is, however, difficult to draw a conclusive relationship between restricted cellulolysis and growth promotion of *A. bisporus* since species such as *Aspergillus fumigatus* and *Corynascus thermophilum* are cellulolytic but not growth promotry whereas the reverse is true for *Chaetomium thermophilum* and *Sporotrichum thermophile.* Thus, growth promotory species can be cellulolytic but not necessary pioneers colinizers of the compost biota. Such an influence is exerted by the climax species of mushroom compost, *Scytalidium thermophilum*, perhaps due to production of a complete complement of enzyme machinery (Rawat, 1998; Tewari, 2000; Rawat, 2004).

An increase in cellulase activity and decrease in laccase activity was observed after the addition of casing soil to the surface of compost colonized by *Agaricus bisporus* (Gillman *et al.*, 1994). The increase in cellulolytic and amylolytic activity during composting is a reflection of change in the population and community structure of the resident microflora. The enzymatic diversity of compost microbiota is known to result in specificity and successional change besides providing a niche to various species to survive in the absence of simple sugars.

The importance of polysaccharolytic enzymes in mushroom compost has led some to stimulate composting by supplementing the substrate with commercial enzymes. Savoie and Libmond (1994) observed that microbial enzyme activities, number of bacteria, and solubilization of carbon and nitrogen were greater in compost treated with polysaccharidases. However, this had no positive effect on mushroom yield. Libmond *et al.* (1995) observed that supplementation of wheat straw with Express (trade name of poysaccharidase complex) reduced the time of mushroom composting besides, releasing low quantities of readily available sugars, increased enzyme activities and number of microrganisms, particularly aerobic bacterial population in the substrate.

Some workers have exploited the enzymatic potential of thermophilic fungi by preinoculation of compost with thermophilic fungi. Salar and Aneja (2007) observed the growth of *Agaricus bisporus* on sterile compost pre-colonized with four thermophilic fungi *viz.*, *Chaetomium thermophile, Malbranchea sulfurea, Thermomyces lanuginosus* and *Torula thermophila*, either singly or in different combinations. A mixed inoculum of *Malbranchea sulfurea* and *Torula thermophila* was found to be the best amongst the various treatments that promoted the growth of *A. bisporus* to the plateau of 7.7 mm day^{-1} and the yield of the mushroom was almost twice compared to the pasteurized control. The effect of *T. lanuginosus* when inoculated singly or in combination with other thermophilic fungus/ fungi in compost was insignificant resulting in lower growth rates. The study revealed that thermophilic fungi provide for compost selectivity and protection against negative effects of compost bacteria on mycelial growth of *A. bisporus.* Improved growth of *A. bisporus* mycelium in composts treated with *S. thermophilum* has been extensively reported in literature (Ross and Harris, 1983; Weigant, 1992; Straatsma and Samson, 1993; Straatsma *et al.*, 1994 a, b; Rawat, 2004; Salar and Aneja, 2007). Straatsma *et al.* (1994a) reported that nine thermophilic fungi *viz.*, *Chaetomium thermophilum*, an unidentified *Chaetomium* sp., *Malbranchea sulfurea, Myriococcum thermophilum*, *S. thermophilum*, *Stilbella thermophila*, *Thielavia terrestris*, and two unidentified basidiomycetes, promoted mycelial growth of *Agaricus bisporus* on sterilized compost.

The study of physiological diversity of compost by techniques like phospholipid fatty acids analysis (PLFA) and metabolic fingerprinting has been instrumental in tracking changes in microbial communities and thus understanding the *in situ* community structure (Garland and Mills, 1991; Petersen *et al.*, 1991; Kennedy and Busacca, 1995; Inssam *et al.*, 1996; Boggs *et al.*, 1998; Campbell and Cooper, 1999; Cahayani *et al.*, 2002). Neither of the PLFA markers of fungi, i.e., 18:2 ω6C and 18:3 ω6C changed significantly over time, indicating that the proportion of fungi in the community varies little with compost maturation although the type of fungi present or active at any stage could vary significantly (Boggs *et al.*, 1998).

Community analysis of composting of dairy manure and pine shave beddings based on carbon source utilization as tool revealed that microbial utilization of γ-aminobutyric acid was increased over time while histidine was utilized at similar level at all sampling times. The ability to utilize sucrose, galactose and fructose increased while trehalose utilization decreased during composting (Boggs *et al.*, 1998).

Rawat (2004) studied physiological diversity of different stages of mushroom compost by 'metabolic fingerprinting' employing Biolog. Average well colour development (AWCD) was higher for mesophilic community than the thermophiles. Principal component analysis (PCA) analysis revealed that the thermophilic community can be divided into three clusters (Fig. 3): Cluster I comprised of drenching, spawn-run and PWI community; Cluster II comprised of casing, second turning, end of phase I composting, first turning, third turning, peak-heat, conditioning, end of phase II (spawning) community; Cluster III comprised of PWII, cropping, airing and fourth turning community. The physiological distribution of community, however, did not correlate well with the structural distribution. Maximum physiological diversity was exhibited within the mesophiles representing zero day community and least amongst peak-heat community. Among thermophiles, zero day community was maximally diverse and conditioning community of phase II, the least.

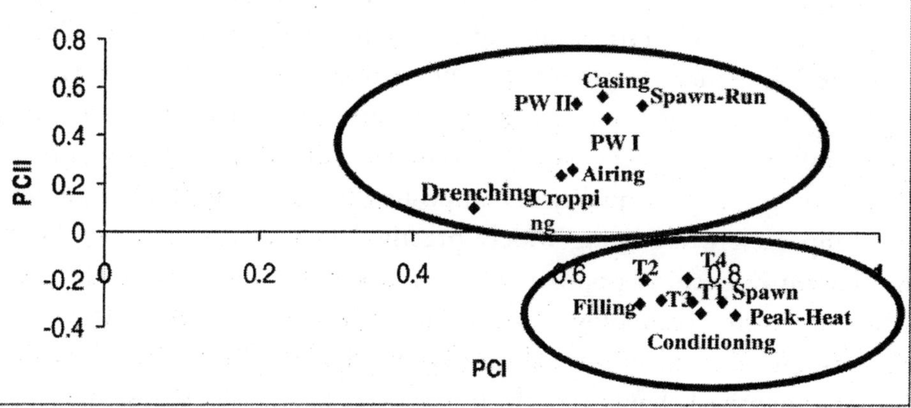

Fig.3: PCA of physiological diversity of thermophilic microflora in mushroom compost based on metabolic fingerprinting.

The wide enzymatic potential exhibited by thermophilic fungi with different physico-chemical characteristics helps in the fuctioning of each component under different biophysical conditions. They also exhibit enzyme multiplicity which helps them function efficiently under different eco-physiological conditions. Multiplicity of hemicellulolytic enzymes has been widely reported in *Chaetomium thermophile* var. *coprophile*, *Humicola grisea* var. *thermoidea*, *Melanocarpus albomyces*, *Talaromyces emersonii*, *Thermoascus aurantiacus* and *Scytalidium thermophilum* (Thakur *et al.*, 1992; Tuohy *et al.*, 1993; Johri and Rajni, 1999; Rawat, 2004; Rawat *et al.*, 2005).

Conclusions

Compost represents an interesting example of thermogenic, solid-state fermentation process that results from a succession of microbial communities. While it is a complex ecosystem, study of structural diversity complemented with functional diversity provides a fair picture of the species spectrum and their associated interactions within the substrate. Application of phenotypic and genetic tools with culturable and non-culturable components of the microbial diversity, coupled to utilization of diversity indices and PCA analysis has permitted deeper insights at subtle changes that result in physico-chemical and biological conditioning of compost. The survival strategies and competitive behaviour of microflora in this interesting niche requires to be understood. The *in situ* functionality of thermophilic fungi is still poorly known whereas tapping the biopotentiality of microflora in this niche would help in hastening the composting process besides improving the quality of compost by pre-inoculation of microflora. This approach while being widely practised in mushroom compost and has currently become popular in other composting systems as Effective Microorganism Technology but is at its nascent stage. New vertical dimensions will add to our existing knowledge, which would throw light into the adaptive behaviour of the microflora in compost ecosystem besides helping unravel new and novel forms. The present moment of organic agriculture augurs well for use of microbial consortia in ecotechnological endeavours of which composting is an essential component. Use of modern molecular tools will permit better and more defined picture of community level changes as also the role of dominant forms in deriving strategies for more effective degradative processes and conservation tools.

Acknowledgement

Thermophilic fungal research of BNJ cited in this chapter has been supported by INSA, DST, ICAR and CSIR whereas more recent molecular microbial ecology component has been supported by the Deptt. of Biotechnology and Ministry of Environment and Forests.

References

Alfreider, A., Peters, S., Tebbe, C., Rangger, A. and Insam, H. (2002). Microbial community dynamics during composting of organic matter as determined by 16S rDNA analysis. *Sci. Util.,* **10**: 303-312.

Antizar-Ladislao, B., Lopez-Real, J. and Beck, A.J. (2005). Laboratory studies of the remediation of polycyclic aromatic hydrocarbon contaminated soil by in-vessel composting. *Waste Mgmt.*, **25**: 281-289.

Azevedo, M.O., Felipe, M.S.S. and Satyanarayana, T. (1999). Molecular and general genetics. *In*: Thermophilic moulds in biotechnology (eds. B.N. Johri, T. Satyanarayana and J. Olsen), Kluwer Academic Publishers, Netherlands, 317-342.

Beffa, T., Blanc, M. and Aragno, M. (1996). Obligately and facultatively autotrophic, sulfur and hydrogen-oxidizing thermophilic bacteria isolated from hot composts. *Arch. Microbiol.,* **165**: 34-40.

Biddlestone, A.J. and Garay, K.R. (1985). Composting. *In*: Comprehensive biotechnology (ed. M. Moo-Young). Pergamon Press, New York. vol. 4. pp. 1059-1070.

Bilai, V.T. (1984). Thermophilic micromycete species from mushroom composts. *Mikrobiol. Zh. (Kiev).,* **46**: 35-38.

Blanc, M., Marilley, L., Beffa, T. and Aragno, M. (1997). Rapid identification of heterotrophic thermophilic, spore forming bacteria isolated from hot composts. *Int. J. Bacteriol.*, **47**: 1246-1248.

Boggs, L.C., Kennedy, A.C. and Reganold, J.P. (1998). Use of phospholipids fatty acids and carbon source utilization patterns to track microbial community succession in developing compost. *Appl. Environ. Microbiol.*, **64**: 4062-4064.

Brown, B.S., Mills, J. and Hulse, J.M. (1974). Chemical and biological degradation of waste plastics. *Nature*, **250**: 161-163.

Cahayani, V.R., Watanabe, A., Matsuya, K., Asakawa, S. and Kimura, M. (2002). Succession of microbiota estimated by phospholipids fatty acid analysis and changes in organic constituents during the composting process of rice straw. *Soil Sci. Plant Nut.,* **48**: 735-743.

Campbell, C.D. and Cooper, J.N. (1999). Community succession and decomposition of microbial biomass during the composting of pot ale liquor. *In*: Microbial process during composting. Proc. 8th Int. Symp. on Microbial. Ecol. (eds. C.R. Bell, M. Brylinsky and P.G. Johnson), Halifax, Canada.

Chang, Y. and Hudson, H.J. (1967). The fungi of wheat straw compost. I. Ecological studies. *Trans. Br. Mycol. Soc.*, **50**: 649-666.

Cooney, D.C. and Emerson, R. (1964). Thermophilic fungi: An account of their biology, activities and classification (eds. D.C. Cooney and R. Emerson). Freman and Co. San Francisco Pub., 180p.

De Bertoldi, M., Lepidi, A.A. and Nuti, M.P. (1972). Classification of the genus *Humicola* Traeen:I. Preliminary reports and investigations. *Mycopathol. Mycol. Appl.*, **46**: 289-304.

De Bertoldi, M., Lepidi, A.A. and Nuti, M.P. (1973). Significance of DNA base composition in classification of *Humicola* and related genera. *Trans. Br. Mycol. Soc.*, **60**: 77-85.

Eggins, H.O.W. and Mills, J. (1971). *Talaromyces emersonii*, a possible biodeteriogen. *Int. Biodet. Bull.*, 7: 105-108.

Ellis, D.H. (1980a). Thermophilic fungi isolated from a heated acquatic habitat. *Mycologia*, **72**: 1030-1033.

Ellis, D.H. (1980b). Thermophilic fungi isolated from some Antarctic and sub-antarctic soil. *Mycologia*, **72**: 1030-1033.

Eriksson, K.E. and Larson, K. (1975). Fermentation of waste mechanical fibres from newsprint mill by the fungus *Sporotrichum pulverulentum. Biotechnol. Bioeng.*, **17**: 327-348.

Fergus, C.L. (1964). Thermophilic and thermotolerant molds and actinomycetes of mushroom compost during peak heating. *Mycol.*, **56**: 267-284.

Fermor, T.R. and Grant, W.D. (1985). Degradation of fungal and actinomycetes mycelia by *Agaricus* bisporus. *J. Gen. Microbiol.*, **131**: 1729-1734.

Finstein, M.S. and Morris, M.L. (1975). Microbiology of municipal solid composting. *Adv. Appl. Microbiol.*, **19**: 113-151.

Forsyth, W.G. and Webley, D.M. (1948). *Proc. Soc. Appl. Bacteriol.*, 34p.

Galler, W.S. and Davey, C.B. (1971). *In*: Livestock waste management and pollution abatement. Publ. Proc. 271, Amer. Soc. Agr. Eng. St. Joseph, Michigan. pp. 159-162.

Garland, J.L. and Mills, A.L. (1991). Classification and characterization of heterotrophic microbial communities on the basis of community-level sole carbon source utilization. *Appl. Environ. Microbiol.*, **57**: 2351-2354.

Gillman, L., Lebeault, J.M. and Cochet, N. (1994). Influence of casings on the microflora of compost colonized by *Agaricus bisporus. Acta Biol. Technol.*, **14**: 275-282.

Grajek, W. (1987). Production of D-Xylanase by thermophilic fungi using different methods of culture. *Biotechnol. Lett.*, 353-356.

Herrmann, R.F. and Shann, J.F. (1997). Microbial community changes during composting of municipal solid waste. *Microbiol.* Ecol., **33**: 78-85.

Hiraishi, A., Narihiro, T. and Yamanaka, Y. (2003). Microbial community dynamics during start-up operation of flowerpot-using fed-batch reactors for composting of household biowaste. *Environ. Microbiol.*, **5**: 765–776.

Horng, J. M. (2003). Food waste utilize effectively. Taiwan, ROC: Environmental Protection Union of Taiwan.

Hultman, J. (2009). Microbial diversity in the municipal composting process and development of detection methods. *Ph.D. Thesis*, University of Helsinki.

Hultman, J., Vasara, T., Partanen, P., Kurola, J., Kontro, M., Paulin, L., Auvinen, P. and Romantschuk, M. (2010). Determination of fungal succession during municipal solid waste composting using a cloning-based analysis. *J. Appl. Microbiol.*, **108**: 472–487.

Iiyama, K., Stone, B.A. and Macauley, B.J. (1996). Changes in the concentration of soluble anions in compost during composting and mushroom growth. *J. Food Sci. Agric.*, **72**: 243-249.

Inssam, H., Amor, K., Renner, M. and Crepaz, C. (1996). Changes in functional abilities of the microbial community during composting of manure. *Microbiol. Ecol.,* **37**: 77-87.

Johri, B.N. and Rajni. (1999). Mushroom compost: Microbiology and applications. *In*: Modern Approaches and Innovations in Soil Management (eds. D.J. Bagyaraj, A. Verma, K.K. Khanna and H.K. Kheri), Rastogi Publ., Merrut, India, pp. 345-358.

Johri, B.N. and Satyanarayana, T. (1984). Ecology of thermophilic fungi. *In*: Progress in Microbial Ecology (eds. K.G. Mukherji, V.P. Agnihotri, and R.P. Singh), Print House, Lucknow, pp. 349-361.

Johri, B.N., Satyanarayana, T. and Olsen, J. (1999). Thermophilic moulds in biotechnology (eds. B.N. Johri, T. Satyanarayana and J. Olsen), Kluwer Academic Publishers, Netherlands, 354p.

Kane, B.E. and Mullins, J.T. (1973). Thermophilic fungi in a municipal compost system. *Mycologia*, **65**: 1087-1100.

Khiyami, M., Masmali, I. and Abu-khuraiba, M. (2008). Composting a mixture of date palm wastes, date palm pits, shrimp and crab shell wastes in vessel system. *Saudi J. Biol. Sci.*, **15(2)**: 199-205.

Khokar, A. (1999). Production and characterization of lipase from *Scytalidium thermophilum. Ph.D. Thesis*, G.B. Pant University of Agril. & Technol., Pantnagar, 116p.

Klamer, M. and Baath, E. (1998). Microbial community dynamics during composting of straw material studied using phospholipid fatty acid analysis. *FEMS Microbiol. Ecol.,* **27**: 9-20.

Klamer, M., Lind, A.M., Gams, W., Balis, C., Lasaridi, L., Szmidt, R.A.K., Stentiford, E. and Lopez-Real, J. (2001). Fungal succession during composting of miscanthus straw and pig slurry. Proc. Int. Symp. on Composting of organic matter, Macedonia, Greece, 30 Aug-1Sept., *Acta Hortculturae*. 549: 37-546.

Klamer, M., Sochting, U. and Szmidt, R.A.K. (1998). Fungi in a controlled compost system with special emphasis on the thermophilic fungi. *Proc. Int. Symp. On composting and use of Composted materials for horticulture,* Ayr, U.K., 5-11 Apr. 1997, No. 469, pp. 405-412.

Kleyn, J.G. and Wetzler, T.F. (1981). The microbiology of spent mushroom compost and its dust. *Can. J. Microbiol.,* **27**: 748-753.

Kowalchuk, G.A., Naovmenko, Z.S., Derikx, P.J.L., Felske, A., Stephen, J.R. and Arkhipchenko, I.A. (1999). Molecular analysis of ammonia-oxidizing bacteria of the β subdivision of the class proteobacteria in compost and composted materials. *Appl. Environ. Microbiol.*, **65**: 396-403.

Lee, D.H., Zo, Y.G. and Kim, S.J. (1996). Non-radioactive method to study genetic profiles of natural bacterial communities by PCR-single stranded conformation polymorphisms. *Appl. Environ. Microbiol.,* **62**: 3112-3120.

Libmond, S., Savoie, J.M. and Elliott, T.J. (1995). Stimulation of mushroom composting by a polysaccharidase complex induction of cellulase production in *Bacillus subtilis.* Proc. 14th Int. Cong. Sci Cult. Edible fungi, UK, pp. 195-202.

Lindt, W. (1886). Mitteilungen iiber einige neve pathogene Schimmelpilze. *Arch. Exp. Path. Pharmakol.*, **21**: 264-298.

Liu, W.T., March, T.L., Cheng, H. and Forney, L.J. (1997). Characterization of microbial diversity by determining terminal restriction fragment polymorphisms of genes encoding 16S rRNA. *Appl. Environ. Microbiol.,* **63**: 4516-4522.

Lyons, G.A., McKay, G.J. and Sharma, S.H. (2000). Molecular comparison of *Scytalidium thermophilum* isolates using RAPD and ITS nucleotide sequence analysis. *Mycol. Res.,* **104**: 1431-1438.

Lyons, G.A. and Sharma, H.S.S. (1998). Differentiation of *Scytalidium thermophilum* isolates by thermogravimetric analysis of their biomass. *Mycol. Res.*, **102**: 843-849.

Mantel, E.F.K., Agarwala, R.K. and Seth, P.K. (1972). A guide to mushroom cultivation Ministry of Agriculture, Farm Information Unit, Directorate of Extension, New Delhi, Farm Bull, No. 2.

Margesin, R., Cimadom, J. and Schinner, F. (2006). Biological activity during composting of sewage sludge at low temperatures. *Int. Biodeterior. Biodegrad.*, **57**: 88–92.

Megnegneau, B., Debets, F. and Hoekstra, R.F. (1993). Genetic variability and relatedness in the complex group of black *Aspergilli* based on random amplification of polymorphic DNA. *Curr. Gen.*, **23**: 323-329.

Meyer, S.A. and Phaff, H.J. (1969). Deoxyribonucleic acid base composition in yeasts. *J. Bacteriol.*,**97**: 52-56.

Miehei, H. (1905). Uber die Selbsterhitzung des Heus. Arb. Deutsch. Landwritisch-Gesellsch., **111**: 76-91.

Miehei, H. (1907). Die Selbsterhitzung des Heus. Fine biologische Studie. Gustav fischer, Jena. pp. 1-127.

Miller, F.C. (1994). Conventional composting system. *In*: *Agaricus* compost (ed. N.G. Nair), Australian mushroom growers association, Windsor, Australia, pp. 1-18.

Mouchacca, J., Allsopp, D., Colwell, R.R. and Hawksworth, D.L. (1995). Thermophilic fungi in desert soils: a neglected extreme environment. *In*: Microbial diversity and ecosystem function. Proc. of IUBS-IUMS Workshop, U.K., 10-13 Aug. 1993, 265-288.

Muyzer, G., deWaal, E.C. and Uitterlinden, A.G. (1993). Profiling of complex microbial populations by denaturing gradient gel electrophoresis analysis of polymerase chain reaction- amplified genes coding for 16S rRNA. *Appl. Environ. Microbiol.,* **59**: 695-700.

Namkoong, W., Hwang, E.Y., Park, J. S., and Choi, J.Y. (2002). Bioremediation of dieselcontaminated soil with composting. *Environ. Pollut.*, **119**: 23–31.

Opden Camp, H.J.M., Stumm, C.K., Straatsma, G., Derikx, PJ.L. and van Griensven, L.J.L.D. (1990). Hyphal and mycelial interactions between *Agaricus bisporus* and *Scytalidium thermophilum* on agar medium. *Microb. Ecol.*, **19**: 303-309.

Peters, S., Koschinsky, S., Schwieger, F. and Tebbe, C.C. (2000). Succession of microbial communities during hot composting as detected by PCR-Single Strand-Conformation Polymorphism-based genetic profiles of small subunit rRNA gene. *Appl. Environ. Microbiol.*, **66**: 930-936.

Petersen, S.O., Henriksen, K., Blackburn, T.H. and King, G.M. (1991). A comparison of phospholipid and chloroform fumigation analyses for biomass in soil: potentials and limitations. *FEMS Microbiol. Ecol.,* **85**: 257-268.

Rajni. (1999). Molecular ecology of *Scytalidium thermophilum. Ph.D. Thesis,* G.B. Pant Univ. of Agril. and Technol., Pantnagar, India, 88p.

Rajni, Rastogi, S., Johri, B.N. and Singh, R.P. (1998). Microbial dynamics and its influence in cultivation cycle of *Agaricus bisporus. Mushroom Res.,* **7**: 63-70.

Rawat, S. (1998). Production and characterization of amylases from *Scytalidium thermophilum. M.Sc. Thesis,* G.B. Pant Univ. of Agril. and Technol. Pantnagar, India, 86p.

Rawat, S. (2004). Microbial diversity of mushroom compost and xylanase of *Scytalidium thermophilum. Ph.D. Thesis,* G.B. Pant Univ. of Agril. & Technology, Pantnagar, India, 199p.

Rawat, S., Agarwal, P.K., Chaudhary, D.K. and Johri, B.N. (2005). Microbial diversity and community dynamics of compost ecosystem. In: *Microbial Diversity: Current Perspectives and Applications* (eds. T. Satyanarayana and B.N. Johri), I.K. International Pvt. Ltd., New Delhi, pp. 181-206.

Rawat, S. and Johri, B.N. (2002). Ecological and functional diversity of thermophilic fungi. *In*: Frontiers of Fungal Diversity in India (Prof. Kamal Festchrift) (eds. G.P.Rao, C. Manoharachari, D.J. Bhat, R.C. Rajak, T.N.Lakhanpal). Int. Book Distributing Co. Lucknow, India, pp. 205-232.

Rawat, S. and Johri, B.N. (2004). Xylanases of thermophilic moulds and their application potential. *In*: Handbook of Fungal Biotechnology (ed. D.K. Arora), Marcel Dekker, USA, vol.20: pp. 299-314.

Rodrigues, E.C., Pizzirani-Kleiner, A.A., Tanaka, Y. and Jorge, J.A. (1991). Cytogenetic and biochemical aspects of the cellulolytic fungus *Humicola* sp. *Mycol. Res.,* **95**: 169-177.

Ross, R.C. and Harris, P.J. (1983). An investigation into the selective nature of mushroom compost preparation. *Scientia Horti.,* **20**: 61-70.

Ryckeboer, J., Mergaert, J., Coosemans, J., Deprins, K. and Swings, J. (2003). Microbiological aspects of biowaste during composting in a monitored compost. *J. Appl. Microbiol.,* **94**: 127-137.

Salar, R.K. and Aneja, K.R. (2007). Significance of thermophilic fungi in mushroom compost preparation: effect on growth and yield of *Agaricus bisporus* (Lange) Sing. *J. Agril. Technol.*, **3(2)**: 241-253.

Satyanarayana, T. (1978). Thermophilic microorganism and their role in composting process. *Ph.D. Thesis.* Sagar University, Sagar, 212 p.

Satyanarayana, T. and Grajek, W. (1999). Composting and solid state fermentation. *In*: Thermophilic moulds in biotechnology (eds. B.N. Johri, T. Satyanarayana and J. Olsen), Kluwer Academic Publishers, Netherlands, pp. 265-288.

Satyanarayana, T. and Johri, B.N. (1981). Volatile sporostatic factors of thermophilic fungal strains of paddy straw compost. *Curr. Sci.*, **50**:763-768.

Satyanarayana, T. and Johri, B.N. (1984). Ecology of thermophilic fungi. *In*: Prog. Microbial Ecol. (eds. Mukherji *et al.*). Print House Publications, Lucknow, India, 349-361.

Satyanarayana, T., Johri, B.N. and Klein, H. (1992). Biotechnological potential of thermophilic fungi. In: *Handbook of Applied Mycology* (eds. D.K. Arora, R.P. Flander and K.G. Mukherjee), Marcel Dekker Inc., New York, pp. 729-761.

Savoie, J.M. and Libmond, S. (1994). Stimulation of environmentally controlled mushroom composting by polysaccharides. *World J. Microbiol. Biotechnol.,* **10**: 313-311.

Schwieger, F. and Tebbe, C.C. (1998). A new approach to utilize PCR-single strand conformation polymorphisms for 16S rRNA gene-based microbial community analysis. *Appl. Enviorn. Microbiol.,* **64**: 4870-4876.

Sen, T.L., Abraham, T.K. and Chakrabarthy, S.L. (1980). Utilization of cellulolytic wastes by thermophilic fungi. *In*: Advances in Biotechnology (ed. M.M. Young). Pregman Press Publ. Proc. 6th Int. Ferm. Symp. 633-638.

Shandilya, T.R. (1982). Composting and casing research at mushroom research centre, during past seven years. *Ind. J. Mushroom.* **8**: 5-13.

Sinden, J.W. and Hauser, E. (1950). The short method of composting during fermentation. *Mushroom Sci.*, **1**: 52-59.

Sparling, G.P., Fermor, T.R. and Wood, D.A. (1982). Measurement of the microbial biomass in composted wheat straw and the possible contribution of the biomass to the nutrition of *Agaricus* bisporus. *Soil. Biol. Biochem.*, **14**: 609-611.

Stanek, M. (1969). Diewitkung der zelulose setzenden mikroorganismen auf das waehstun des champignons. *Mushroom Sci.*, **7**: 161-172.

Straatsma, G., Gerrits, J.P.G., Augustijn, M.P.A.M., OpDenCamp, H.J.M., Vogels, G.D. and Van Griensven, L.J.L.D. (1989). Population dynamics of *Scytalidium thermophilum* in mushroom compost and stimulatory effects on growth and yield of *Agaricus bisporus. J. Gen. Microbiol.,* **135**: 751-789.

Straatsma, G., Gerrits, J.P.G., Gerrits, T.M., Op Dencamp, H.J.M. and van Griensven, L.J.L.D. (1991). Growth kinetics of *Agaricus bisporus* mycelium on solid substrate (mushroom compost). *J. Gen. Microbiol.*, **137**: 1471-1477.

Straatsma, G., Olijnsma, T.W., Gerrits, J.P.G., Amsing, J.G.M., OpDen Camp, H.J.M. and Van Griensven, L.J.L.D. (1994b). Inoculation of *Scytalidium thermophilum* in button mushroom compost and its effect on yield. *Appl. Environ. Microbiol.*, **60**: 3049-3054.

Straatsma, G. and Samson, R.A. (1993). Taxonomy of *Scytalidium thermophilum,* an important thermophilic fungus in mushroom compost. *Mycol.* Res., **97**: 321-328.

Straatsma, G., Samson, R.A., Olijnsma, T.W., Opden Camp, H.J.M., Gerrits, J.P.G., Griensven, L.J.L.D. Van and van Griensven, L.J.L.D. (1994a). Ecology of thermophilic fungi in mushroom compost with emphasis on *Scytalidium thermophilum* and growth stimulation of *Agaricus bisporus* mycelium. *Appl. Environ. Microbiol.*, **60**: 454-458.

Straatsma, G., Samson, R.A., Olijnsma, T.W., Gerrits, J.P.G., OpDen Camp, H.J.M. and van Griensven, L.J.L.D. (1995). Bioconversion of cereal straw into mushroom compost. *Can. J. Bot.*, **73**: 1019-1024.

Stutzenberger, F.J., Kaufmar, A.J. and Lossins, R.D. (1970). Cellulolytic activity in municipal solid waste composting. *Can. J. Microbiol.*, **16**: 553-560.

Subrahmanyam, A. (1980). A new thermophilic variety of *Humicola grisea* var. *indica. Curr. Sci.*, **49**: 30-31.

Subrahmanyam, A., Mehrotra, B.S. and Thirumalacher, M.J. (1977). *Thermomucor*, a new genus of Mucorales. *Geor. J. Sci.*, **35**: 1-6.

Tewari, P. (2000). Thermophilic microorganisms from mushroom compost and their polysaccharolytic activities. *M.Sc. Thesis*, G.B.Pant Univ. of Agril. and Technol., Pantnagar, India, 99p.

Thakur, I. S., Rana, B. K. and Johri, B. N. (1992). Multiplicity of xylanase in *Humicola grisea* var. *thermoidea. In*: Xylan and Xylanases (eds. J.Visser, M.A. Beldman, Kusters-van Someren and A.G.J. Voragen), Elsevier, Amsterdam, pp. 511-514.

Tiquia, S.M., Wan, J.H.C. and Tam, N.F.Y. (2002). Microbial population dynamics and enzyme activities during composting. *Compost Sci. Util.*, **10**: 150–161.

Troussellier, M. and Legendre, P. (1981). A functional evenness index. *Microb. Ecol.*, 7: 283-296.

Tuohy, M.G., Puls, J., Claeyssens, M., Vrsanska, M. and Coughlan, M.P. (1993). The xylan degrading enzyme system of *Talaromyces emersonii* novel enzymes with activity against beta –D- xylosides and unsubstituted xylans. *Biochem. J.*, **290**: 515-523.

Vijay, B. (1996,. Investigations on compost mycoflora and crop improvement in *Agaricus bisporus* (Lange). *Ph.D. Thesis*, H.P. Univ., Shimla.

Weigant, W.M. (1992). Growth characteristics of thermophilic fungus *Scytalidium thermophilum* in relation to production of mushroom compost. *Appl. Environ. Microbiol.*, **58**: 1301-1307.

White, D.C. and Findlay, R.H. (1988). Biochemical markers for measurement of predation effects on the biomass community structure, nutritional status and metabolic activity of microbial biofilms. *Hydrobiol.*, **159**: 119-132.

Wildeman, A.G. and Nazar, R.N. (1981). Studies on the secondary structure of 5.8S rRNA from a thermophile *Thermomyces lanuginosus. J. Biol. Chem.*, **256**: 5675-5682.

Wood, D.A. and Matchman, S.E. (1980). Growth of *Agaricus bisporus* in composted straw. Evidence for microbial interaction. Second International Symposium on Microbial Ecology, Warwick, 233p.

Yu, H., Zeng, G., Huang, H., Xi, X., Wang, R., Huang, D., Huang, G. and Li, J. (2007). Microbial community succession and lignocellulose degradation during agricultural waste composting. *Biodegrad.*, **18**: 793-802.

Biodiversity Conservation and Envir. Management (2012)
Editors: D.R. Khanna et al.
Pub. by Biotech Books. *ISBN: 978-81-7622-262-4*

Pages: 341-352

NATIVE FLORA OF VARIOUS COMPARTMENTS OF GOREWADA FOREST, NAGPUR DISTRICT, NAGPUR (M.S.), INDIA

Kirti V. Dubey[✉]**, Sulbha V. Kulkarni and Pravin Charde**
Sevadal Mahila Mahavidyalaya, Sakkardara Square
Umrer Road, Nagpur, Maharashtra (INDIA)
✉E-mail: kirtivijay_dubey@yahoomail.com

Gorewada forest is one of the tropical forest located in Nagpur district (M.S.), India. It is a rich source of biodiversity which represents valuable global resources such as food, fodder, fruits, fuel, gums and medicines that are beneficial to mankind. Among various resources of forests, legumes and non-legumes represents a very large and diverse group of plant kingdom, ranging from herbs, shrubs, trees,

✉ Corresponding author

climbers and grasses. In the present study, variety of legume and non-legume plants from seven different compartments of Gorewada forest *viz.* 790-796 of Gorewada forest of Nagpur district were surveyed and collected for taxonomic identification of plants and their distribution pattern *i.e.* density to find out the dominant flora. Most of the plants in different compartments are common which includes the leguminous plants which were in large number as compared to non-legumes. Leguminous plants found are *Abrus precatorius*, *Acacia* sp., *Albizia lebbeck*, *Butea monosperma*, *Bauhinia* sp., *Caesalpinia pulcherrima*, *Cassia fistula*, *Cassia tora*, *Leucaena leucocephala*, *Mimosa hamata*, *Mucuna pruriens*, *Pithecellobium dulce*, *Pongamia pinnata*, *Tamarindus indica*, *Tephrosia purpurea*, *Tephrosia hamiltonii etc.* and were as non-legumes are *Blumea* sp., *Tectona grandis*, *Santalabum album*, *Zizipus* sp., *Celosia argentea*, *Cleistanthus collinus*, *Mytragyna perbifolia*, *Partheneium* sp., *Semicarpus anacardium*, *Sida* sp., *vernonia cinerea etc.* This study will be useful in determining the species diversity to characterize the structure of the community available in and around Gorewada International Zoo. The plant diversity study of the Gorewada forest is the first of its kind and the diversity indices calculated was *viz* Simpson's index (D=0.343), Simpson's index of diversity (1-D = 0.657), Simpson's reciprocal index (1/D = 2.9154), Shannon-wiener index (H= 1.204) and Evenness index (E= 0.743)

Keywords: Gorewada forest, Biodiversity, Simpson's index, Shannon-wiener index, Evenness index

Introduction

The forest cover in India is 675,538 sq km and constitutes 20.55 percent of its geographical area. Of this, dense and open forest cover constitutes as 12.68 and 7.87 percent, respectively. However, the area of non-forest in the country is 2,611,725 sq km and constitutes 79.45 percent of the geographical areas (Katiyar, 2005). Development of richness in the vegetation wealth and biodiversity of soil is of prime importance in the forest cover as well as on the non-forest area to conserve the bioresource and tree wealth.

Forests of the tropical zones constitutes about half of the world's forest and mostly occur in developing countries. In recent years, tropical forests have received much attention because of their species richness, high standing biomass and global net primary productivity. Tropical forests consist of world's largest biodiversity and play an important role in the global terrestrial carbon budget. The structure, composition and functioning of these forests are undergoing rapid changes because of anthropogenic activities, biotic pressure and widespread economic growth is also altering the natural vegetal cover and putting tremendous pressure on the sustenance of the few left over tropical forest covers in India. As a result, there is lot of spatial and temporal variations in the reported values of species richness, composition and productivity. There is a pressing need to monitor the rate and extent of changes in the tropical forest covers in countries like India for efficient planning and management leading to sustainable development. The loss of natural habitat and species extinctions around the world have served to focus intense international attention on the issue of biodiversity and it is widely recognized that the path to sustainable development requires the conservation and wise use of the world's biodiversity. Indeed, measuring and monitoring biodiversity must be the first step towards effective conservation and sustainable development.

This chapter highlights the studies on survey of native plant species in and around each compartment of Gorewada forest which is proposed site for the development of International Zoo, in order to find out the native dominant species of each compartment. Species diversity of the plants available in Gorewada forest was estimated through mathematical measures to characterize the structure of a community and their relative abundance.

Aims & Objectives: Aims of the present study is to characterize the structure of a community of the dominant plant species and their relative abundance in and around each compartment of Gorewada forest, Nagpur (M.S.), India.

Materials and Method

Site survey

To achieve the proposed objectives intensive and extensive site surveys were conducted at Gorewada forest during December, 2007 to March, 2008 by the college team of microbiologist, botanist and soil scientist to cover the total area of Gorewada forest area *i.e.* 1881.66 ha.

Survey of native plant species in and around Gorewada forest

The tour in and around different compartments of Gorewada forest was made in phased manner so to cover most of the parts of the Gorewada forest for the survey of native plant species. Specimens were brought to the college laboratory and all plants were identified by using reference floras (Hooker, 1872-1897; Ugemuge, 1986; Almeida, 1990; Joshi, 2000; Singh *et al.*, 2001). For collection and preservation of plant samples, standard procedures were generally followed (Jain and Rao, 1977; Balgooy, 1987). The trips were arranged in such a way so as cover all the compartments of Gorewada forest and collect most of the plants in flowering or fruiting stages. All the specimens collected were serially numbered. The field notes were taken regularly, included habitat, colour of the flowers, association and other pertinent features. Efforts were made to identify the plants from the fresh material; those that could not be satisfactorily identified in the field were brought to the laboratory and identified by checking it with monographs, herbarium specimens and other available literature.

Results and Discussion

Description of Gorewada forest site

Intensive and extensive site surveys were conducted at Gorewada forest during December-2007 to March-2008. During survey it was found that seven different compartments has characteristic features with respect to plants species available during winter season. Total area of Gorewada forest area is 1881.66 ha and the forest area is unequally divided into two parts by Nagpur-Kalmeshwar highway. Among the two, part on left side of the Nagpur-Kalmeshwar highway is divided into three compartments allotted with numbers 790, 791 and 792 and second part, to the right side of the Nagpur-Kalmeshwar highway, has four compartments allotted with No. 793, 794, 795 and 796 along with a water tank of 57.93 ha.

Description of Flora available in different Compartments of Gorewada Forest

Compartment No. 790:

This site has almost a barren look with very few plant species towards its entrance with some of the pits dug for plantation on barren land. However, the density of plants improved towards the center. Types of plant species comprising of trees, shrubs, herbs, climbers and grasses present in this compartment is presented in Table 1. Among the plant species available, *Alternanthera pungens* and *Cassia tora* (herbs) and trees such as *Acacia catechu, Butea monosperma, Cassia siamea, Diospyros melanoxylon, Maytenus emarginata, Mitragyna perbifolia, Ziziphus mauritiana* are the predominant species followed by *Acacia catechu, Albizia odoratissima, Azadirachta indica* and *Dulbergia sisso.* Along the extremity of the compartment *i.e.* towards Nagpur-Kalmeshwar highway monoseries plantation of *Dendrocalamus strictus* was found which was planted to provide protective barriers from invaders. The forest present in 790 compartment is of dry deciduous type.

Compartment No. 791:

This site also has almost a barren and dry deciduous look like that found in compartment No. 790. with comparatively a very few plant species. However, the density of plants improved towards the center. Types of plant species comprising of trees, shrubs, herbs, climbers and grasses present in this compartment is presented in Table 1. Among the plant species available, *Alternanthera pungens, Cassia tora* (herbs) and *Acacia catechu, Butea monosperma, Mitragyna perbifolia, Cassia siamea, Maytenus emarginata, Ziziphus mauritiana, Diospyros melanoxylon* (trees) are the predominant species followed by *Dulbergia sisso, Albizia odoratissima, Acacia catechu* and *Azadirachta indica.*

Compartment No. 792:

In compartment number 792 of Gorewada forest about 42 different plant species classified as trees, shrubs, herbs, climbers and grasses (Table 1). Among the tree species dominant types were *Acacia catechu, A. leucophloea, A. nilotica, Albizia odoratissima, Butea monosperma, Cassia siamea, Mitragyna perbifolia* and *Ziziphus mauritiana.*

Table 1: List of Various Plant Species Present in different Compartments of Gorewada Forest.

S. No	*Various compartments with different plant species*						
	No.: 790	*No.: 791*	*No.: 792*	*No.: 793*	*No.: 794*	*No.: 795*	*No.: 796*
	Trees	Trees	Trees	Trees	Trees	Trees	Trees
1	*Acacia catechu*	*Acacia catechu*	*Acacia catechu*	*Acacia arabica*	*Acacia catechu*	*Acacia catechu*	*Acacia catechu*
2	*Acacia leucophloea*	*Acacia leucophloea*	*Acacia leucophloea*	*Acacia catechu*	*Acacia leucophloea*	*Acacia leucophloea*	*Acacia leucophloea*
3	*Acacia nilotica*	*Acacia nilotica*	*Acacia nilotica*	*Acacia leucophloea*	*Acacia nilotica*	*Azadirachta indica*	*Acacia nilotica*
4	*Adina cordifolia*	*Aegel marmalosa*	*Albizia odoratissima*	*Acacia nilotica*	*Ailanthus excels*	*Butea monosperma*	*Azadirachta indica*
5	*Ailanthus excelsa*	*Ailanthus excelsa*	*Andropogon pumilus*	*Albizia lebbeck*	*Albizia lebbeck*	*Cleistanthus collinus*	*Bauhinia purpuria*
6	*Albizia odoratissima*	*Albizia odoratissima*	*Azadirachta indica*	*Albizia odoratissima*	*Azadirachta indica*	*Emblica officinalis*	*Bauhinia racemosa*
7	*Anogeissus latifolia*	*Azadirachta indica*	*Bauhinia racemosa*	*Annona squamosa*	*Bauhinia racemosa*	*Leucaenea leucocephala*	*Bombaxi cieba*
8	*Azadirachta indica*	*Bauhinia racemosa*	*Butea monosperma*	*Azadirachta indica*	*Butea monosperma*	*Mimosa hamata*	*Butea monosperma*
9	*Bauhinia racemosa*	*Butea monosperma*	*Cassia siamea*	*Bombaxi cieba*	*Cassia fistula*	*Phoenix sylvestris*	*Dalbergia sissoo*
10	*Bombaxi cieba*	*Cassia siamea*	*Dalbergia sissoo*	*Butea monosperma*	*Dalbargia sissoo*	*Pongamia pinnata*	*Dandrocalamus strictus*
11	*Butea monosperma*	*Dalbergia sissoo*	*Dandrocalamus strictus*	*Cleistanthus collinus*	*Dandrocalamus strictus*	*Santalum album*	*Delonix regia*
12	*Cassia angustifolia*	*Dandrocalamus strictus*	*Dolichandrone falcata*	*Dalbergia sissoo*	*Leucaena leucocephala*	*Tamarindus indica*	*Diospyros melanoxylon*
13	*Cassia siamea*	*Delonix regia*	*Maytenus emarginata*	*Eucalyptus* sp.	*Maytenus emarginata*	*Tectona grandis*	*Grewia tiliifolia*
14	*Dalbargia sissoo*	*Diospyros melanoxylon*	*Mimosa hamata*	*Gemelina arborea*	*Mimosa hamata*	*Ziziphus mauritiana*	*Leucaena leucocephala*
15	*Dandrocalamus strictus*	*Leucaena leucocephala*	*Mitragyna perbifolia*	*Hardwickia binata*	*Mitragyna perbifolia*	**Shrubs**	*Maytenus emarginata*
16	*Diospyros melanoxylon*	*Maytenus emarginata*	*Phoenix sylvestris*	*Leucaenea leucocephala*	*Phoenix sylvestris*	*Lantana camera*	*Mimosa hamata*
17	*Dolichandrone falcata*	*Mimosa hamata*	*Semicarpus anacardium*	*Merremia emarginata*	*Pithecellobium dulce*	*Xanthium strumarium*	*Mitragyna perbifolia*
18	*Emblica officinalis*	*Mitragyna perbifolia*	*Tamarindus indica*	*Mimosa hamata*	*Pongamea pinnata*	**Climbers**	*Phoenix sylvestris*
19	*Grewia tiliifolia*	*Phoenix sylvestris*	*Tectona grandis*	*Phoenix sylvestris*	*Tamarindus indica*	*Abrus precatorius*	*Pithecellobium dulce*
20	*Leucaena leucocephala*	*Pongamia pinnata*	*Terminalia bellirica*	*Santalum album*	*Tectona grandis*	*Ipomea aquatic*	*Santalum album*

contd...

S. No	*Various compartments with different plant species*						
	No.: 790	*No.: 791*	*No.: 792*	*No.: 793*	*No.: 794*	*No.: 795*	*No.: 796*
	Trees	Trees	Trees	Trees	Trees	Trees	Trees
21	*Maytenus emarginata*	*Semicarpus anacardium*	*Ziziphus glaberrima*	*Soymida febrifuga*	*Woodfordia fruticosa*	*Ipomea quamoclit*	*Semicarpus anacardium*
22	*Mimosa hamata*	*Tamarindus indica*	**Shrubs**	*Ziziphus glaberrima*	*Xylia xylocarpa*	**Herbs**	*Sesamum orientale*
23	*Mitragyna perbifolia*	*Terminalia bellirica*	*Lantana camera*	*Ziziphus mauritiana*	*Ziziphus mauritiana*	*Ageratum conyzoides*	*Tamarindus indica*
24	*Phoenix sylvestris*	*Ziziphus glaberrima*	*Xanthium strumarium*	**Shrubs**	**Shrubs**	*Alternanthera pungens*	*Tectona grandis*
25	*Pithecellobium dulce*	Ziziphus mauritiana	**Climbers**	*Goniocalon glabrum*	*Ipomea aquatic*	*Blumea eriantha*	*Terminalia alata*
26	*Semicarpus anacardium*	**Shrubs**	*Cocculus hirsutus*	*Lantana camera*	*Lantana camera*	*Cassia tora*	*Woodfordia fruticosa*
27	*Soymida febrifuga*	*Lantana camera*	*Coix lacryma-jobi*	*Xanthium strumarium*	*Xanthium strumarium*	*Goniocalon glabrum*	*Ziziphus glaberrima*
28	*Tamarindus indica*	*Xanthium strumarium*	*Hemidesmus indicus*	*Ziziphus oenoplia*	*Ziziphus oenoplia*	*Hyptis suaveolens*	*Ziziphus mauritiana*
29	*Tectona grandis*	**Climbers**	*Ipomea quamoclit*	*Ziziphus xylopyrus*	**Climbers**	*Ocimum sanctum*	**Shrubs**
30	*Terminali bellirica*	*Cocculus hirsutus*	**Herbs**	**Climbers**	*Combretum ovalifolium*	*Parthenium*	*Cappris zeylanica*
31	*Terminalia alata*	*Hemidesmus indicus*	*Ageratum conyzoides*	*Hemidesmus indicus*	*Cuscuta reflexa*	*Sida acuta*	*Lantana camera*
32	*Ziziphus glaberrima*	*Ipomea quamoclit*	*Alternanthera pungens*	*Ipomea quamoclit*	*Hemidesmus indicus*	*Sida cordata*	*Malachra capitata*
33	*Ziziphus glaberrima*		*Blumea eriantha*	*Mucuna pruriens*	*Mucuna pruriens*	*Tridax procumbens*	*Climbers*
34	*Ziziphus mauritiana*		*Cassia tora*	**Herbs**	**Herbs**	*Vernonia cinerea*	*Cocculus hirsutus*
	Shrubs		*Hyptis suaveolens*	*Ageratum conyzoides*	*Alternanthera pungens*	*Vicoa indica*	*Coix lacryma-jobi*
35	*Lantana camera*		*Parthenium*	*Alternanthera pungens*	*Blumea eriantha*	*Vinca rosea*	*Combretum ovalifolium*
36	*Xanthium strumarium*		*Sida acuta*	*Blumea erintha*	*Cassia tora*	**Grasses**	*Hemidesmus indicus*
	Climbers		*Sida cordata*	*Cassia tora*	*Celosia argentea*	*Cynodon dactylon*	*Ipomoea aquatic*
37	*Cocculus hirsutus*		*Tridax procumbens*	*Celosia argentea or C. Cristata*	*Cyathocline purpuria*	*Hackelochloa granularis*	*Mucuna pruriens*
38	*Combretum ovalifolium*		*Vernonia cinerea*	*Hyptis suaveolens*	*Hyptis suaveolens*		**Herbs**
39	*Hemidesmus indicus*		*Vicoa indica*	*Parthenium hysterophorus*	*Malachra capitata*		*Alternanthera pungens*

contd...

S. No	Various compartments with different plant species						
	No.: 790	No.: 791	No.: 792	No.: 793	No.: 794	No.: 795	No.: 796
	Trees	Trees	Trees	Trees	Trees	Trees	Trees
40	*Mucuna pruriens*		**Grasses**	*Sida acuta*	*Parthenium hysterophorus*		*Blumea eriantha*
	Herbs		*Apluda mutica*	*Sida cordata*	*Sida acuta*		*Cassia tora*
41	*Alternanthera pungens*		*Hackelochloa granularis*	*Tridax procumbens*	*Sida cordata*		*Celosia argentea*
42	*Blumea eriantha*		*Iseilema laxum*	*Vernonia cinerea*	*Solanum xanthocarpum*		*Cyathocline purpuria*
43	*Cassia tora*			*Vicoa indica*	*Tephrosia hamiltoni*		*Hyptis suaveolens*
44	*Celosia argentea*				*Tephrosia purpureai*		*Parthenium hysterophorus*
45	*Cyathocline purpuria*				*Tridex procumbens*		*Pennisetum hohenackeri*
46	*Hyptis suaveolens*				*Vernonia cinerea*		*Sida acuta*
47	*Malachra capitata*				**Grasses**		*Sida cordata*
48	*Parthenium hysterophorus*				*Andropogon pumilus*		*Solanum xanthocarpum*
49	*Sida acuta*				*Apluda mutica*		*Tephrosia purpureai*
50	*Sida cordata*				*Coix lacryma-jobi*		*Tridex procumbens*
51	*Solanum xanthocarpum*				*Eragrostiella bifaria*		*Typha angustata*
52	*Tephrosia hamiltoni*				*Iseilema laxum*		*Vernonia cinerea*
53	*Tephrosia purpureai*				*Pennisetum hohenackeri*		*Xanthium strumarium*
54	*Tridex procumbens*						**Grasses**
55	*Vernonia cinerea*						*Andropogon pumilus*
	Grasses						*Apluda mutica*
56	*Andropogon pumilus*						*Iseilema laxum*
57	*Apluda mutica*						
58	*Coix lacryma-jobi*						
59	*Eragrostiella bifaria*						
60	*Iseilema laxum*						
61	*Pennisetum hohenackeri*						

Compartment No. 793:

This compartment is located at West side of Gorewada Catchments area. Multi-species plantation was found which comprised of predominant species *viz. Acacia catechu, Hardwickia binata, Butea monosperma, Cleistanthus collinus, Dalbergia sisso* followed by *Santallum album, Acacia arabica, Bombox cieba, Acacia leucophloea, Soymida febrifuga, Mimosa hamata* and *Ziziphus mauritiana* (Table 1). Different types of herbs and shrubs were also found. Fruit bearing plants of *Ziziphus oenoplia* and *Abrus precatorius* were also found. Plantation of *Santalum album* was also observed

Compartment No.794:

This compartment is situated on the right side of Nagpur - Kalmeshwar highway. Major part of this compartment is densely populated with multiple plant species (Table 1). However, successful monospecies plantation of *Xylia xylocarpa* is found in small area of this compartment with few plants of *Butea monosperma* and *Azadirachta indica.* Perennial plant species of this compartment are *Maytenus emarginata, Butea monosperma, Acacia catechu, Acacia nelotica, Dalbergia sisso, Cassia fistula, Ailanthus excels, Acacia leucophloea, Albizia lebbeck, Pithecellobium dulci, Pongamia pinnata, Gmelina arborea, Phoenix sylvestris* and *Dendrocalamus strictus* along with herbs, shrubs and climbers. Different types of thorny shrubs *Ziziphus rotandifolia* bearing fruits are found in this dry deciduous compartment. Termite infestation was found in *Butea monosperma* and *Acacia leucophloea.* An example of climber *viz. Macuna pruriens* was present in compartment number 794. Barren patch of land is also found in compartment number 794.

Compartment No.795:

This compartment has a water body measuring 57.93 ha. and multispecies plantation and list of plants found during survey is presented in Table 1. Some of the predominant types of plant species available in compartment number 795 are *Acacia catechu, Acacia leucophloea, Butea monosperma, Cleistanthus collinus, Mimosa hamata, Tectona grandis* and *Ziziphus mauritiana.*

Compartment No. 796:

This compartment is adjacent to Compartment number 794. It is characterized to have natural hilly area. In the centre of this compartment almost barren rocky area with very few plant species *viz., Tectona grandis, Ziziphus mauritiana* and *Ziziphus oenoplia* were found giving dry deciduous look to the compartment. The barren area of this compartment needs plantation. However, on the lower side of hilly area there is Gorewada reservoir and on top of the hilly area dense forest is present. Top of the hilly area is densely covered with green canopy and the dominant plant species available are *Maytenus emarginata, Ziziphus oenoplia, Butea monosperma, Acacia catechu, Acacia nelotica, Dalbergia sisso, Cassia fistula, Ailanthus excelsa, Acacia leucophloea, Albizia lebbeck, Pithecellobium dulci, Pongamia pinnata, Gmelina arborea, Phoenix sylvestris, Dendrocalamus strictus, Tectona grandis, Semicarpus anacardium, Mimosa hamata, Grewia tiliifolia, Hardwickia binata* and *Terminalia arjuna* (Table 1). In this compartment there was mother plant of *Terminalia arjuna.* In this compartment, barren patch of land is present which is almost devoid of vegetation due to inhabitation of Sita Gondi people (the local tribal community). However, *Ziziphus* sp. were found sparsely. Amanalla is passing through this compartment that is a centre of Gorewada catchment area. Few plants of *Tectona grandis* and *Acacia leucophloea* were found to be infested by termites.

Diversity Indices of Plants available at Gorewada forest

It is a measured by a number of diversity indices. There has been a lot of development and proliferation of diversity indices over the years. These indices are mathematical expressions that combine three components of community structure *i.e.* richness (number of species present), evenness (the distribution of individuals among species), and abundance (total number of plants present). The plant diversity studies of the Gorewada forest is the first of its kind carried out and results are presented in Table 2. The diversity indices, Simpson's index (D=0.343): Simpson's index is a dominance measure and a probability measure to study forest plant biodiversity (Magurran, 1988; Sai and Mishra, 1986). The higher value at Gorewada forest could mean the high dominance of the occurring species. There were trees higher in number followed by herbs, shrubs, climbers and grasses. Index of similarity (0.657) indicates high diversity. Reciprocal index was 2.9154. Shannon-wiener index was 1.204 indicating quite low diversity, which is lower than the reported range of 1.5 to 3.5 (Margalef, 1972). Finally, Evenness index was 0.743 which indicates that abundance of different species is similar in proportion to all species.

Table 2: Plant Diversity Indices for Gorewada Forest.

Measure	*Value*
Simpson's index (D)	0.343
Simpson's index of diversity (1 – D)	0.657
Simpson's reciprocal index (1 / D)	0.29154
Shannon-wiener index (H)	1.204
Evenness index (E)	0.743

Conclusion

Present study was carried out to find native as well as dominant plant species in and around each compartment of Gorewada forest, a proposed site for the development of International Zoo. Species diversity of the plants available in Gorewada forest was estimated through mathematical measures to characterize the structure of a community and their relative abundance. Variety of legume and non-legume plants from seven different compartments of Gorewada forest *viz.* 790-796 of Gorewada forest of Nagpur district were surveyed and collected for taxonomic identification of plants and their distribution pattern *i.e.* density to find out the dominant flora. Most of the plants in the different compartments are common which includes the leguminous plants which were in large number as compared to non-legumes. The plant diversity study of the Gorewada forest is the first of its kind and the diversity indices calculated were *viz.* Simpson's index (D=0.343), Simpson's index of diversity (1-D= 0.657), Simpson's reciprocal index (1/D = 2.9154), Shannon-wiener index (H= 1.204) and Evenness index (E= 0.743). This study will be useful in determining the species diversity to characterize the structure of the community available in and around Gorewada International Zoo.

Acknowledgement

Authors acknowledge Director, Gorewada forest, Department of Forest, Government of Maharashtra, Nagpur, India for providing the financial support to carry out the work presented in this manuscript.

References

Almeida, S. M. The flora of Savantwadi. Vol. I–II. Scientific Publishers, Jodhpur - India., (1990).

Balgooy. M. M. J. Van. A plant geographical analysis of Sulawesi. In: TC Whitmore (ed.). Biogeographical evolution of the Malay Archipelago. pp. 94-10, (1987).

Hooker, K. K. The Flora of British India. Vol. I – VII London, (1872-1897)

Jain & Rao A Handbook of field and Herbarium methods. Today & Tomorrow's printers & Publishers, New Delh, (1977).

Joshi, S. J Medicinal plants. Oxford and IBH publishing company Pvt. Ltd., (2000).

Katiyar, S. R. Sustainable management of forest resources and tree wealth in India. In: Biodiversity and conservation. Prof. Arvind Kumar edited, p p. 81- 91, (2005)

Magurran E. Anne. Ecological diversity and its Measurements. Princton University Press, 41 Williams street, Princeton, New Jersey, (1988).

Margalef R. Homage to Evelyn Hutchinson or why is there an upper limit to diversity. Trans. Connect. Acad. Arst. Sci. 44: 211-235, (1972).

Sai V. S. and Mishra M. Comparison of some indices of species diversity in the estimation of the actual diversity in a tropical forest: A case study. Trop. Ecol. 27: 195-201, (1986).

Singh N. P., Narsighan L. Kartikeyan S. and Prasanna P.V. Flora of Maharashtra. BSI Publications, Kolkata, (2001).

Ugemuge, N. R. Flora of Nagpur district. Shree Prakashan, Nagpur, (1986).

Biodiversity Conservation and Envir. Management (2012)
Editors: D.R. Khanna et al.
Pub. by Biotech Books. *ISBN: 978-81-7622-262-4*

Pages: 353-358

DIVERSITY IN STRUCTURE OF HAIR CUTICLE, MEDULLA AND CORTEX OF *HOMO SAPIENS* AND SOME OTHER MAMMALS

Sunil D. Patil✉ and Tahseen Anjum
P.G. Department of Zoology, M.G.Vidyamandir's M.S.G. Arts Science & Commerce College, Malegaon-Camp, Malegaon Dist. Nashik Maharashtra (INDIA)

Hair can be examined under a light microscope and compared to reference samples. The physical appearance of the medulla has many variations. Moreover, scale pattern, whether regular wave or irregular, scale margin are varied from animal to animal. Medulla is amorphous in *Rattus rattus*, *Ovis aries* and *Canis lupus* while lattice in *Homo sapien* and *Bos taurus*. Medullary index is highest in *Bos taurus* and lowest in *Homo sapien*.

Keywords: Hair, Medulla, Cortex, Scales, *Homo sapien*.

✉ Corresponding author

Introduction

Hair is unique characteristic of mammals. No other creature possesses true hair and at least some hair is found on all mammals at some time during their lives. A hair shaft has three layers: the cuticle, the cortex and the medulla. The tough outer layer called cuticle is comprised of overlapping scales that protect the inner layers. Examination of these scales under a microscope gives a quick indicator of whether the hair came from an animal or human-the scales look very different in animal hair. The innermost layer is a hollow tube called the medulla. The physical appearance of the medulla has many variations; it can be continuous, fragmented or even absent from a hair sample. Human hair usually has a fragmented or absent medulla except for Asian people whose hair has a continuous medulla. The medulla can look different in two hairs taken from the same person, even if it came from the same area of the body. Animal hair generally has thicker medulla and cortex layers than human hair because it is an important means of insulation and protection for the animal. Microscopic forensic analysis of hair focuses on hair structure. If a hair sample includes the root structure, its examination can indicate if the hair fell out naturally or was forcefully removed as might happen during a struggle.

Hair can be examined under a light microscope and compared to reference samples or samples from a suspect. For this type of comparison, the hair is typically dry mounted on a glass microscope slide and viewed under a comparison microscope. Hairs can provide investigators with valuable information for potential leads. Identification of hair has practical application in forensic medicine, taxonomy, paleontology, zooarchaeology, anthropology and ecology. The present keys are not sufficient to identify at species level (Mayer *et al.*, 2002). In the present investigation an attempt has been made to compare human hair with few domestic animals.

Materials and Method

For scale casting a clear fingernail polish was used as the medium for impression. After doing a thin coating of the polish on the slide, the hair was placed into the polish before drying. After drying the polish completely the free edge of the hair was pulled to remove it from the polish. The hair was discarded, then the slide was placed onto the microscope and attempt was made to observe the impression that the hair made in the polish. For whole mount of hair, bleaching of hair was done before mounting in DPX. Dimensions of hair cortex and medulla were measured with the aid of Camera Lucida & Occulometry. Human hairs were collected from male and female head while in case of other four pairs of animals collected from trunk region. Animals including *Rattus rattus, Ovis aries, Canis lupus, Bos taurus* and the *Homo sapiens* (Fig. 1).

Table 1: Showing scale margin and scale pattern.

S. No.	*Animal*	*Scale margin*	*Scale pattern*
1	*Homo sapien*	Rough Distant	Regular & irregular wave
2	*Canis lupus*	Smooth Distant	Irregular wave
3	*Rattus rattus*	Smooth Distant	Irregular wave
4	*Bos taurus*	Smooth Distant	Regular & irregular wave
5	*Ovis aries*	Rough	Irregular wave

Table 2: Showing Medullary index in micrometer (μ).

Sr. No	*Animal*	*Medulla Dimension*		*Cortex Dimension*		*Medullary Index*	
		Male	*Female*	*Male*	*Female*	*Male*	*Female*
1	*Homo sapien*	17.5	52.1	20.3	36.4	0.32	0.08
2	*Canis lupus*	50.4	28.7	20.3	17.8	0.56	0.37
3	*Rattus rattus*	16.8	14.7	13.3	14	0.53	0.51
4	*Bos taurus*	27.3	28.7	9.8	13.3	0.74	0.68
5	*Ovis aries*	50.4	32.2	16.1	50.4	0.67	0.66

Results and Discussion

Scale margin and scale pattern observed during study is given in Table 1 while medullary index is given in Table 2.

Hair dimensions cortex & medulla were identified using different key (Marinis and Asprea, 2006). Hair dimension particularly cortex and medulla of animal hair shows variation and these were then compared. Moreover scale pattern, whether regular wave or irregular, scale margin varied from animal to animal. Nomenclature was used to describe hair profile and microstructure followed by Kennedy and Carbyn, 1961; Teerink 1991. Two different medullary microscopic structures *i.e.* Amorphous and Lattice were identified during course of study. The diameter of the cortex was found highest in *Ovis aries* and lowest in *Bos taurus* where as the diameter of the medulla was observed highest in man.Medulla is amorphous in *Rattus rattus*, *Ovis aries* and *Canis lupus* while lattice in *Homo sapien* and *Bos taurus.* Scale pattern was regular and irregular wave in man. Medullary index was found highest in *Bos taurus* and lowest in *Homo sapien.*

Acknowledgement

Authors are thankful to the Principal of M.S.G. College, Malegaon and Head, P. G. Department of Zoology, M.S.G. College, Malegaon for allowing the laboratory facilities and kind cooperation during this work.

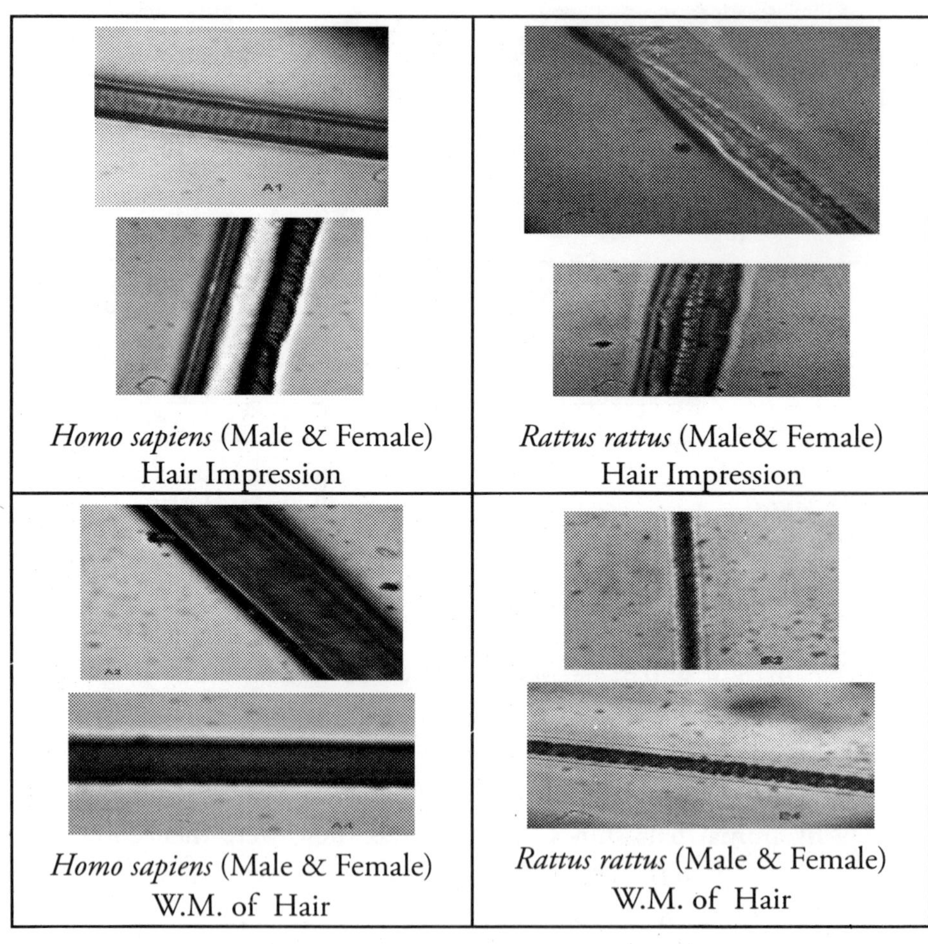

Homo sapiens (Male & Female) Hair Impression

Rattus rattus (Male& Female) Hair Impression

Homo sapiens (Male & Female) W.M. of Hair

Rattus rattus (Male & Female) W.M. of Hair

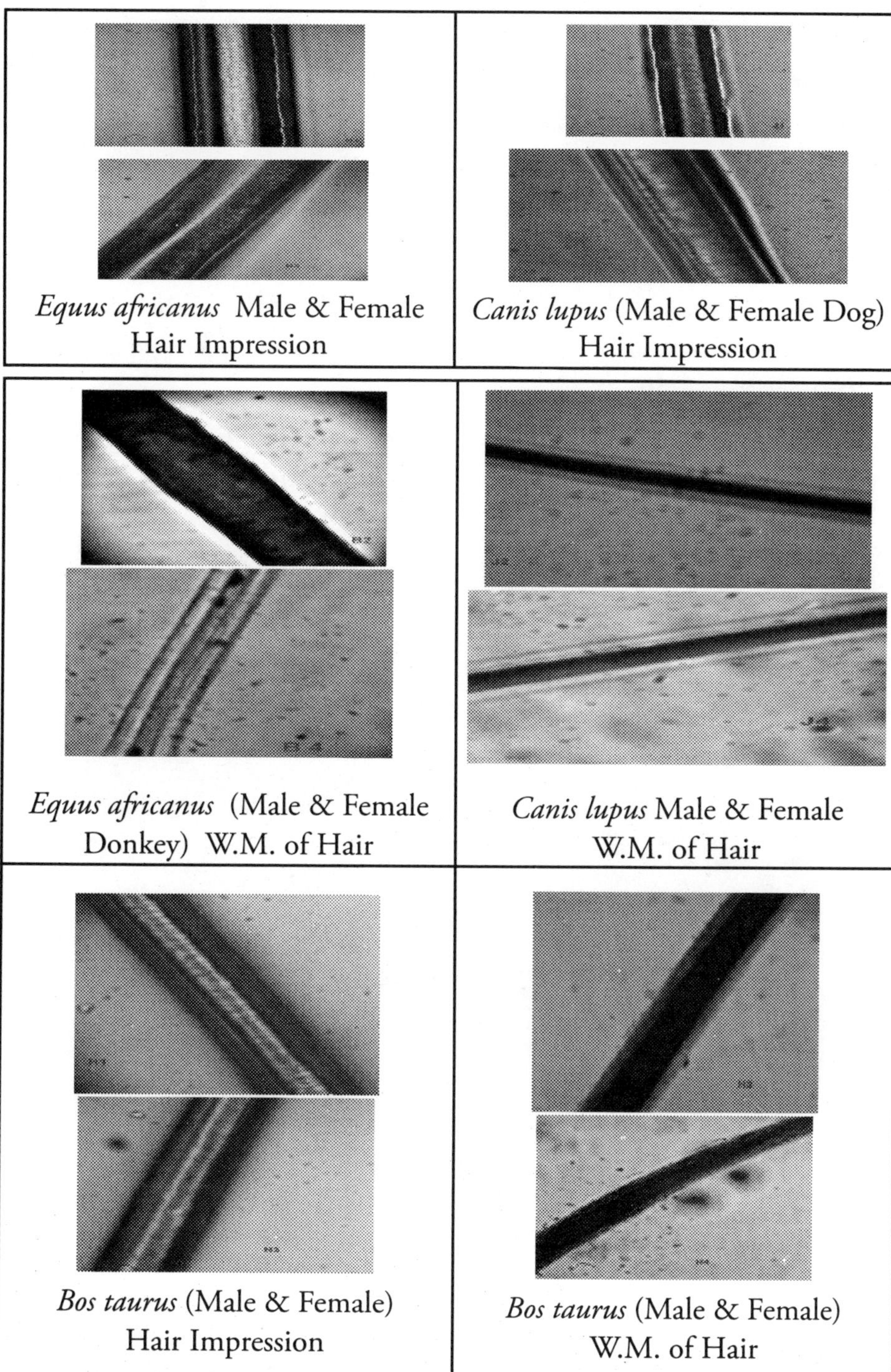
Equus africanus Male & Female Hair Impression
Canis lupus (Male & Female Dog) Hair Impression
Equus africanus (Male & Female Donkey) W.M. of Hair
Canis lupus Male & Female W.M. of Hair
Bos taurus (Male & Female) Hair Impression
Bos taurus (Male & Female) W.M. of Hair

Fig. 1

References

De Marinis, A.M. and Asprea A. 2006: Hair identification key of wild & domestic ungulates from southern Europe. *Wildl. Biol.* 12: 305-320.

Kennedy, A.J. and Carbyn, L.N. 1961: Identification of wolf prey using hair and feather remains with special reference to western Canadian National Parks. *Canadian Wildlife Service, Edmonton, Alberta,* 65 pp.

Mayer, W., Hulmann, G. & Seger, H. 2002: SEM- Atlas of the hair cuticle structure of central European mammals. *Verlag M. & H. Sharper Alfeld Hannover,* 248 pp

Mayer, W. V. 1952. The hair of California mammals with keys to the dorsal guard hairs of California mammals. *American Midland Naturalist,* 48:480-512.

Teerink, B. J. 1991. *Hair of West-European Mammals.* Cambridge Univ. Press, Cambridge. Vii + 224 pp.

Biodiversity Conservation and Envir. Management (2012)
Editors: D.R. Khanna et al.
Pub. by Biotech Books. *ISBN: 978-81-7622-262-4*

Pages: 359-393

BIODIVERSITY OF UTTARAKHAND AND ITS CONSERVATION

Seema Rawat, Amir Khan✉ **and Arun Kumar**
Dolphin (PG) Institute of Biomedical and Natural Sciences, Dehradun, Uttarakhand (INDIA)

Uttarakhand is the 27th state of India which is divided into Kumaun and Garhwal divisions. The state has 13 districts, 49 tehsils and 95 community development blocks. The climate varies from sub-tropical to temperate. The major land cover of the state is under forests followed by agriculture. The state has rich and diverse biological resources. A total of 1618 species of microbes, 6808 species of flora and 4900 faunal species have been reported from the state. Uttarakhand has been traditionally known as 'gold mine of medicinal plants'. 2364 plant species with medicinal uses and 318 species of plants with various ethnobotanical

✉ Corresponding author

uses and 1107 species of agricultural crops have been listed from the state. A total of 259 fossil forms of microbes, flora and fauna have been recorded from the state. However, the rich biodiversity of state is under tremendous pressure due to rapid urbanization, overgrazing, over exploitation of herbs, medicinal plants, poaching and illegal trade. It is imperative to develop a comprehensive plan for the state for the conservation and sustainable use of biodiversity.

Keywords: Uttarakhand, Kumaun, Garhwal, Biological resources, Medicinal plants, Fossil forms, Conservation.

Introduction

Uttarakhand, the 27th state of Indian Republic, was carved out by separating the hill region, with a geographical area of 51,125 sq km of the western Himalaya from the state of Uttar Pradesh (FSI, 2000). It lies between 28° 44' and 31° 28' N Latitude and 77^{0} 35' and 81° 01' E Longitude. Uttarakhand represents a unique geographical area where altitude ranges from 200 m to a magnificent series of snow-clad peaks of more than 2400 m above sea level. Broadly, there are three sub divisions: Himadri or greater Himalaya, Himachal or lower Himalaya and Siwalik or sub Himalaya. Out of its total area, 47,325 sq km is covered by mountains, while 3,800 sq km is Terai plains.

Uttarakhand has been traditionally divided in two parts, namely, Kumaun and Garhwal. The state has 13 districts subdivided into 49 tehsils and 95 community development blocks. The bifurcation of 13 districts in two divisions is: Garhwal Division (Dehra Dun, Haridwar, Chamoli, Rudraprayag, Tehri Garhwal, Uttarkashi and Pauri Garhwal); and Kumaun Division (Almora, Nainital, Pithoragarh, Udham Singh Nagar, Bageshwar and Champawat). These blocks are further divided into 673 Nyay Panchayats comprising 15,669 villages (Kumar *et al.*, 2002).

The state has two distinct climatic regions: the predominant hilly terrain and the small plain region. The climatic condition of the plains is very similar to its counterpart in the Gangetic plain, *i.e.*, tropical. The climate varies from sub-tropical to temperate. The annual rainfall ranges from 1200-2500 mm and temperature varies from less than freezing point in higher hills to more than 40°C in the plains (Kumar *et al.*, 2002). The state's climate is, moderate and tropical, characterized by a hot and dry summer, humid monsoon or rainy season, short pleasant post-monsoon, and cool and dry winter. The climate of the area is conditioned to some extent by the proximity of the Himalayas in the north. Average summer temperature ranges from 12.6 °C to 18.7 °C, whereas winter temperature ranges from 2.9 °C to 11.5 °C. Rainfall peaks during July (15.5 mm average) followed by August (14.1 mm average). High relative humidity occurs during the monsoon season and the maximum relative humidity observed during July and August (95.5–99.2 %) can make the surrounding area rather uncomfortable. A lot of mist, fog, and dew appear at the mountainous areas during the monsoon (Kandari and Gusain, 2001).

The state has a long network of perennial and seasonal streams. It is well drained by numerous rivers and rivulets (locally known as Gad, Gadhera and Naula). There are three main river systems: (i) Bhagirathi-Alaknanda basin–Ganges basin, (ii) Yamuna–Tons basin and (iii) Kali system. The Ganges system drains the major part of the region covering the whole of the Garhwal (except the western part of Uttarkashi district) and the western part of Garhwal Himalaya from an altitude of 7,138 m with confluence at Devprayag and flows as the Ganges thereafter. The main stream is the Bhagirathi. The main tributaries to the Alaknanda and Bhagirathi or to one another ultimately contributing to the waters of the Ganges are the Alaknanda, Saraswati, Dhauli Ganga, Berahi Ganga, Nandakini, Mandakini, Madhu Ganga, Pindar, Atagad, Bhilangana, Jad Ganga, Kaldi Gad and Haipur. The Nayar, which drains more than half the area of Garhwal district, is an important tributary of the Ganges. The Yamuna–Tons system is also located in the Garhwal region. The Yamuna River rises at Yamunotri and is joined by important tributaries such as (i) the Giri and (ii) more importantly the Tons—its biggest tributary (Uniyal *et al.*, 2007).

The major soil type in the state is Brown hill soil (which is found in Uttarkashi, Rudraprayag, Chamoli, Pithoragarh, Bageshwar, Parts of

Almora, Champavat, Pauri Garhwal and Dehra Dun). Other soil types are: Bhabar soil (parts of Dehra Dun, Pauri Garhwal, Nainital and Champavat), Terai soil (parts of Haridwar, Pauri Garhwal, Nainital, Udham Singh Nagar) and Alluvial soil (parts of Haridwar) (Kumar *et al.*, 2002).

The major land cover of the state is under forests followed by agriculture. The forest area is reported to be 3,466 thousand hectares and accounts for around 62.27% of the geographical area of Uttarakhand (FSI, 2000). The Uttarakhand Himalaya is very rich in forest resources and diversity. The plant diversity is found extremely rich from the valley regions to the highly elevated alpine meadows, locally known as kharaks or bugyals (Kumar *et al.*, 2002; Sethi *et al.*, 2002).

The alpine and tropical rainforests that cover most parts of the state make natural habitats of some of the best-known wildlife species. The Corbett Tiger Reserve is home to Royal Bengal Tiger. Another rainforest in the region is Rajaji National Park, famous for its large number of pachyderms. Alpine forests in the region include the Valley of Flowers National Park (known for its amazing variety of flowers), Nanda Devi Biosphere Reserve, Govind National Park and Gangotri National Park. The forest management of the state falls into 3 categories: reserve forests constituting 70% of the area under forests; civil and soyam forests (about 22% of the area) and Van Panchayat's accounting for about 8% of the area under forests. Thus, conservation based forestry that meets local people's livelihood needs is the most crucial aspect of forest management in the state (Sati, 2005).

Agriculture is the mainstay of the people of Uttarakhand. Of the total population, more than 75% people are engaged either with the main occupation of agriculture or its allied practices. In the mainland of Uttarakhand, traditional subsistence agriculture is dominant in farming system, but their viability in terms of sustainable livelihood is insufficient. Therefore, the rate of out-migration from the region is high. Amongst the principal crops, rice, wheat, millets, barley, pulses and oil-seeds are grown in the entire state. The ratio of pulses and oil seeds is comparatively low. Fortunately, the state has high diversity in food grains, vegetables, fruits, oil-seeds and pulses in all the altitudinal climatic zones, such as tropical, temperate and cold. The feasibility of climatic conditions and diversity in crops may help for sustainable farming. The percentage of cropped area is highest for wheat (33.54%) followed by rice (23.51%). Cultivation of vegetables is extending and now accounts for 12.45% of the total cropped area of the state (Sati, 2005).

Though the feasibility of climatic conditions and diversity in crops may help for sustainable farming in the state yet their viability in terms of sustainable livelihood is insufficient. Therefore, the rate of out-migration from the region is high which has necessitates better planning for sustainable agriculture in the mountainous mainland as well as for the foothill plains. The economic viability of off-season vegetables, pulses and oil-seeds in the farming system of the state is noteworthy while the cropped area with these crops is very small. The area with off-season vegetables can be raised for sustainable livelihood since the environmental conditions are very suitable for their production.

The different climatic regions in the state have proved to be suitable for growing different types of temperate, sub-tropical and tropical fruits. Also, the state has wide scope for growing different kinds of flowers, ornamental plants, mushrooms and medicinal plants in its different climatic zones. The land under fruit cultivation is 190,192 ha and under vegetable 80,332 ha, accounting for 29 % and 15 % of the total area sown, respectively. Apple is the most important crop among the various fruits grown in the region and its cultivated land is 54,000 ha. Among vegetables, potato is the most important crop. In the state, fruits are grown in the different altitudinal zones and among them apple, peach, citrus, mango, plum, apricot, walnut, litchi and other fruits are important (Sati, 2005).

Uttarakhand has been traditionally known as 'gold mine' of medicinal plants. Uttarakhand has declared itself as an 'Herbal State' in 2003 and has plans to develop medicinal plant sector as a priority area. Herbs are naturally grown in the meadows (Kharaks). Fortunately, the state has extensive meadows along the Great Himalayan ranges. These herbs are locally utilized for medicinal purposes with the positive results (Kumar *et al.*, 2002; Sethi *et al.*, 2002).

Livestock farming, along with agriculture, is the main occupation of the people of the state. The climatic conditions are very conducive to rearing of animals. Sheep and goats are reared in high altitudinal mountain regions. They are reared in the forest-cleared grasslands. These pasturelands (Kharaks) are usually above 2000 m. Cows are reared in the middle and low altitudinal regions and they are the major source of milk. A very small proportion of crossbreed animals are also reared in the state. Goats, sheep, horses and ponies are used for transportation of goods. Cattle including cows and buffaloes occupy 67.78 % of the total livestock farming. Cows rank first (41.79%) followed by buffaloes (23.72%) and goats stand third (23.52%) (Pers. Comm. State Board for Livestock, Dehra Dun, 2008).

The major fishery in Uttarakhand is in the fast flowing rivers and tributaries, natural and man made lakes, ponds and reservoirs. The available total length of principal rivers and tributaries is reported to be 2,138 km. The natural lakes of the state occupy an area of about 262 ha and the ponds in the plains of Dehra Dun and Udham Singh Nagar cover about 344 ha. Additionally five man made reservoirs of 1205 ha area are a flourishing centre for fish development in the state under the aegis of Uttarakhand Matsya Vikas Nigam Limited (Kumar *et al.*, 2002; Sethi *et al.*, 2002).

The state's rich and varied bioresources have widely attracted the naturalists, wild lifers, taxonomists and conservationists since time immemorial and have been explored extensively. Comprehensive information on the floral and faunal diversity of the state is recorded in 'Flora and Fauna of India' and subsequently in a series of publications by various Government and non Government agencies including Botanical Survey of India, Zoological Survey of India, Forest Research Institute, Wildlife Institute of India, Navdanya, VPKAS and many colleges and universities in the state (Kumar *et al.*, 2002).

Biodiversity Profile of State

The state has a rich and diverse floral, faunal and microbial wealth including rare and threatened species of plants and animals. The state's biodiversity is bliss as it helps the province to maintain its vitality and therefore its conservation is essential as well as necessary. Uttarakhand is also bestowed with a rich biodiversity owing to the close relationship between the religious, socio-cultural beliefs and conventions. The sacred groves of the state are held in high cultural and religious values by the local folks.

The diverse forest types of the state, grasslands, different types of traditional land use patterns, traditional agricultural practices contribute to varied biodiversity of the state. The traditionally valued natural systems at various levels should be integrated with legal and policy-level interventions for better conservation and also for effective management of biodiversity outside the protected areas. The state has wide spectrum of habitats including two kinds of grassland (i) in plains (*chaur*) with dominant species of Bassi, Ulla, Jarakush, Sindhur and Tinna; (ii) alpine grassland or *bugyals* consisting of many medicinal and insectivorous plants.

The rich floral diversity of the state comprises 5080 species of Angiosperms and Gymnosperms. Uttarakhand is bestowed with large varieties of rice, other cereals like mandua, amaranth, pulses and vegetables. The state has fabulous wealth of medicinal plants, wild edibles, timber yielding and fiber yielding plants. The rural community is largely dependent on them for their livelihood.

Uttarakhand is a home for many species of birds, mammals, and reptiles. A total of 4894 faunal species including mammals, birds, reptiles, etc. have been reported from the state (Kumar *et al.*, 2002). The State tree of Uttarakhnad is Burans (*Rhododendron arboretum*), State flower is Brahm Kamal (*Saussurea obvallata*), State animal is Musk deer (*Moschus chrysogaster*) and State bird is Monal Pheasant (*Lophophorus impejanus*).

(a) Microbial diversity

The state has a rich microbial diversity and is endowed with approx. 1618 species of various microbes as summarized in Table 1. These include bacteria, fungi, diatoms and protozoa.

A total of 93 bacterial species belonging to 17 families have been listed from the state. Cyanobacteriacea is the most dominant family comprising of 53 species (Verma, 2005; Selvakumar *et al.*, 2007). Other dominant families are Enterobacteriaceae, Pseudomonadaceae, Bacillaceae and Rhizobiaceae. The other families reported from the state are Acetobacteriaceae, Brucellaceae, Clostridiaceae, Desulfovibrionaceae, Lactobacillaceae, Micrococcaceae, Mycobacteriaceae, Nitrosomonadaceae, Sphingomonadaceae, Staphylococcaceae, Streptococcaceae and Streptomycetaceae. The dominant genera reported from the state are *Bacillus, Gloeocapsa, Microcotens, Nostoc, Phormidium, Pseudomonas, Oscillatoria* and *Sinorhizobium*.

The dominant families of fungi (1226 species) recorded from the state are, Aspergillaceae, Mycosphaerellaceae, Pucciniaceae, Polyporaceae, Hymenochaetaceae, Glomeraceae, Dematiceae, Saprolegniaceae and Pythiaceae (Bisht and Srivastava, 1983; Khulbe *et al.*, 1983; Uniyal, 2001; Das and Sharma, 2006). The dominant genera reported from the state are *Achyla*, *Alternaria*, *Aspergillus*, *Clavulina*, *Fomes*, *Fusarium*, *Glomus*, *Penicillium*, *Phellinus*, *Phoma*, *Physarum*, *Polystictus*, *Puccinia*, *Pythium*, *Ramaria*, *Rhizopus* and *Ustilago*.

A total of 264 diatom species, belonging to orders Achnanthales, Bacillariales, Cymbellales, Eunotiales, Fragilariales, Naviculales, Pennales, Surirellales, Thalassiophysales and Thalassiosirales, have been reported from Uttarakhand. The dominant orders are Cymbellales (65 species), Naviculales (54 species), Bacillarales (50 species) and Achnanthales (39 species). Dominant genera reported from the state are: *Achnanthes*, *Cymbella*, *Diatoma*, *Gomphonema*, *Navicula*, *Nitzschia* and *Synedra* (Nautiyal *et al.*, 1996; Nautiyal and Nautiyal, 1999; Nautiyal *et al.*, 2004). 35 species of parasitic protozoa have been recorded from Uttarakhand are listed under family Eimeriidae under three genera, namely *Isosporia*, *Eimeria* and *Wenyonella* (Nandi, 2010).

Table 1: Summary of bioresources of Uttarakhand.

S. No.	*Group*	*Number of species*	
Microbes			**1618**
1.	Bacteria	93	
2.	Fungi	1226	
3.	Diatoms	264	
4.	Protozoa (Coccidea)	35	
Flora			**6808**
9.	Algae	346	
10.	Bryophytes	478	
11.	Lichens	539	
12.	Pteridophytes	365	
13.	Gymnosperms	35	
14.	Monocots (excluding Orchids)	1021	
15.	Orchids	238	
16.	Dicots	3786	
Fauna			**4900**
Invertebrates			**3941**
17.	Trematode	48	
18.	Cestodes	36	
19.	Nematodes	196	
20.	Mollusca	128	

contd...

S. No.	*Group*	*Number of species*	
21.	Annelida	69	
22.	Crustacea (Decapoda)	20	
23.	Chilopoda	32	
24.	Arachnida	249	
Insecta			
25.	Collembola	31	
26.	Thysanura	04	
27.	Odonata	163	
28.	Plecoptera	20	
29.	Orthoptera	116	
30.	Dermaptera	43	
31.	Isoptera	52	
32.	Hemiptera	479	
33.	Coleoptera	416	
34.	Lepidoptera	567	
35.	Trichoptera	60	
36.	Hymenoptera	234	
37.	Ichneumonidae	302	
38.	Diptera	676	
Vertebrates			**959**
39.	Pisces	142	
40.	Amphibia	20	
41.	Reptiles	75	
42.	Birds	622	
43.	Mammals	100	
Fossils			**263**
1.	Microbes	75	
2.	Flora	34	
3.	Fauna (Invertebrates)	129	
4.	Fauna (Vertebrates)	25	

Sources: Bisht and Srivastava (1983), Nautiyal *et al.* (1996), Sati (1997), Nautiyal and Nautiyal (1999), Uniyal (2001), Srivastava and Singh (2005), Selvakumar *et al.* (2007), Gaur (1999), Nautiyal *et al.* (2004), Gupta (2005), Uniyal *et al.* (2007), Jalal *et al.* (2008), Punetha and Kholia (2010), Santanu *et al.* (2010), Sharma *et al.* (2010), Upreti *et al.* (2010), Arora and Kumar (1995), Khanna (2003), Mohan and Sinha (2003), Rizvi (2007, 2008), Kumar and Rawat (2010), ZSI (2010), Diener (1908), Gupta (1971), Lakhan Pal *et al.* (1976), Prakash (1988), Phadtare (1989), Tiwari (1999).

(b) Floral diversity

Floristically, Uttarakhand falls under the West Himalayan Biogeographic zone. Being situated in the lap of the Himalaya the region is rich in forest wealth that ranges from subtropical to Alpine types. The forest area of the state is *ca* 3,466 thousand hectares constituting 64.8% of the state forest geographic area. The flora and vegetation of the state ranges from tropical deciduous to alpine vegetation; broadly catagorised into three major types, namely subtropical, temperate and sub alpine and alpine. The studies on the floristic status of Uttarakhand reveals that it comprises *ca* 6808 species belonging to 8 groups, namely algae, bryophytes, pteridophytes, lichens, gymnosperms, orchids and angiosperms (Table 1).

Gupta (2005) listed 346 species of algae from the state. A total of 478 bryophytes species belonging to 54 families have been reported from the state. Amongst liverworts, the dominant families reported from the state are, Aytoniaceae, Ricciaceae, Geocalycaceae, Plagiotheciaceae and Porellaceae. Dicranaceae, Polytrichaceae, Pottiaceae, Batramiaceae, Bryaceae, Miniaceae, Brachytheciae, Entodontaceae and Hypnaceae are the dominant families of mosses reported from the state. Amongst all families, Pottiaceae is the most dominant one comprising of 58 species. The dominant genera reported are, *Barbula, Plagiomnium, Fissidens, Plagiochyla, Astrella* and *Ricia* (Srivastava and Singh, 2005; Shantanu and Uniyal, 2010).

The state of Uttarakhand comprises of 539 species of lichens belonging to 41 families (Srivastava and Singh, 2005; Upreti *et al.*, 2010). Cladomaceae, Lecanoraceae, Pyrenulaceae, Parmeliaceae, Physciaceae, Usneaceae and Collemataceae are the dominant families reported from the state (Annexure 11). Parmeliaceae is the most dominant family reported from the state comprising of 77 species while Phyciaceae comprise of 72 species. *Cladomia, Pertusaria, Heterodermaria, Usnea, Lecanora* and *Collema* are the dominant genera known from Uttarakhand State.

A total of 365 pteridophytes species have been reported from the state (Srivastava and Singh, 2005) belonging to 38 families. Polypodiaceae is the most dominant family reported from the state comprising of 69 species. The other dominant families are, Aspediaceae (56 species), Athyriaceae (47 species), Thelpyteridaceae (31) and Aspleniaceae (29). *Asplenium, Tectaria, Dryopteris, Polystichum* and *Athyrium* are the dominant genera of pteridophytes known from the state. A total of 35 gymnosperms species, belonging to families, Cupressaceae, Ephedraceae, Pinaceae, Cycadaceae, Ginkgoaceae, Podocarpaceae, Araucariaceae and Taxodiaceae, have been reported from the state (Srivastava and Singh, 2005; Sharma *et al.*, 2010). Cupressaceae is the dominant family having 12 species. Among these maximum diversity is seen in genera like *Juniperus, Cupressus, Pinus* and *Ephedera.* Another remarkable feature is the occurrence of both native and exotic species of gymnosperms in the state.

A total of 5045 angiosperms have been described from the state. Of these, a total of 1259 monocots, including 238 species of Orchids have been recorded from the state. The dominant families of Angiosperms in the state are: Asteraceae, Fabaceae, Poaceae, Rosaceae, Scrophulariaceae, Cyperaceae, Lamiaceae, Ranunculaceae and Brassicaceae. 16 species of medicinal orchids have also been reported from the state (Gaur, 1999; Uniyal *et al.*, 2007; Jalal *et al.*, 2008; Joshi *et al.*, 2009).

Medicinal and Aromatic plants

Uttarakhand has been traditionally known as gold mine of medicinal plants. The main three agro ecological zones, *viz.*, (a) The Alpine Zone; (b) The Temperate Zone; and, (c) The Sub-Tropical Zone have significant physiographic differences mainly in terms of altitude, aspects, hill slope and soils, which have helped to develop riches for specific medicinal plants.

Uttarakhand has declared itself as an 'Herbal State', in 2003 and has plans to develop medicinal plant sector as a priority area. The Department of Horticulture and its Herbal Research and Development Institute (HRDI) at Gopeshwar in Chamoli district has been identified as the nodal agency for the sector in the state.

The alpine flora, an integral part of the fragile ecosystem of the mighty Himalayas has been a matter of concern to all people involved in environment protection activities, medicinal plant's trade and conservation. Exposed to the extreme environmental conditions and being buried under the snow cover for a considerable period, these slow growing plants hold vast potential to serve as remedy for various diseases. Ever since the medicinal properties of such plants were known to the local people and the outside world, these species have faced tremendous pressure due to heavy exploitation for commercial gain which has consequently resulted into the dwindling of populations of these medicinal plants. The plants mentioned in the traditional systems of medicines like Ayurveda and Amchi which were once very common, are rarely seen in alpine meadows now-a-days. Species of Aconitum, Dactylorhiza, Fritillaria, Microstylis, Polygonatum, Podophyllum and Picrorrhiza have huge market potential but are now restricted to either inaccessible areas or to the protected habitats due to overexploitation and many other anthropogenic and environmental factors. Efforts have been made to assess the threat to these species as most of these are critically endangered (Table 2).

Research institutions like High Altitude Plant Physiology Research Centre in the state have standardized the cultivation techniques for species of *Aconites*, *Dactylorhiza*, *Picrorrhiza* and *Podophyllum* and are now promoting their cultivation on large scale. *Ex situ* conservation could be one of the strategies of conservation of these valuable species but *in situ* conservation especially the ecosystem conservation approach is far more important. The best way to achieve conservation and promote sustainable utilization of these taxa is to create awareness among local people about their importance in conservation and livelihood support. Capacity building of local community will be the most essential component for long-term survival of these plants and their habitat.

Approximately 2364 plant species with medicinal uses have been reported from Uttarakhand as summarized in Table 2 (Pande *et al.*, 2006).

Table 2: List of medicinal and aromatic plants of Uttarakhand.

S. No.	*Group*	*No. of species*
1.	Plants used in Ayurveda	602
2.	Plants used in human diseases and disorders	1340
3.	Plants used in livestock health care	364
4.	Jari-Booties	58
5.	Important medicinal plants in short supply and prioritized for research and development	49
6.	Medicinal and aromatic plants under cultivation/ cultivation trials	96
7.	Medicinal and aromatic plants cultivated/ suggested for cultivation in different forest nurseries	68

Source: FRLHT; Kala *et al.* (2006); Pande *et al.* (2006).

Ethnobotanical resources

Approximately 318 species of plants with various ethnobotanical uses are listed from the state (Table 3; Kumar and Rawat, In Press). Among these 13 species are used for condiments and spices, 37 species are of oil yielding plants, 20 species of alcoholic and non-alcoholic beverage plants, 17 plants are listed for their use in religion and culture, 20 plants used in art, craft, mattresses and sports goods, 16 species of plants have sacred values and are used in various rituals, festivals or worshipped as sacred tree, while 5 plants are associated with religious festivals in the state. Recently 96 species of plants have been listed from the state for bioprospecting. A total of 34 plant species are recorded for their miscellanous uses like preparation of dyes, insecticides, piscicides, broom preparation, etc. In addition, state has 44 species of wild timber yielding plants and as many as 15 species are used for social forestry in Uttarakhand including as bee forage.

Table 3: List of plants with ethnobotanical uses in Uttarakhand.

S. No.	*Group*	*No. of species*
1.	Spices and condiments	13
2.	Oil yielding wild plants	37
3.	Wild edibles used in beverages	20
4.	Plants used in religion and culture	17
5.	Plants used in arts, crafts and sports goods	20
6.	Sacred plant species	16
7.	Sacred trees associated with religious festivals	5
8.	Plants species useful for bioprospecting	96
9.	Plants with miscellaneous uses	34

Source: Badoni & Badoni (2001); Anthwal *et al.* (2006); Gairola and Biswas (2008); Kumar & Rawat (In Press).

Exotic and cultivable plants

There are a large number (435 species) of cultivated exotic flora grown in gardens, orchards, nurseries, etc. in Uttarakhand, which demonstrate human influence on the floristic composition of the state (Negi and Hazra, 2007). In addition, tissue cultured floriculture has taken roots in the state in a big way in recent years. Altogether, 17 cultivated exotic horticultural species with a large number of varieties have been reported to be tissue cultured in Uttarakhand (Pers. Comm. National Horticulture Board, Dehra Dun, 2009).

Conservation status of flora

The rapid urbanization and changes in land use pattern have adversely affected a large number of plant species. Consequently 38 globally threatened plants, 65 species of threatened medicinal plants, and 312 species of threatened plants (under CAMP criteria) have been listed from Uttarakhand in Table 4. In addition a total of 252 plants species under CITES and 140 species of endemic plants have also been listed from Uttarakhand (Table 4). About 29 invasive alien plant species mostly having their origin in tropical America have also been recorded from the state.

Table 4: Conservation status of flora.

S. No.	*Group*	*No. of species*
1.	Endemic plants from Uttarakhand	140
2.	Threatened (IUCN Red List) plants from Uttarakhand	38
3.	Threatened plants (CAMP Criterion) from Uttarakhand	312
4.	Threatened medicinal plants (CAMP Criterion) from Uttarakhand	65
5.	Plants under CITES from Uttarakhand	252

Source: Nayar and Sastry (1987); Kumar *et al.* (2002); Sethi *et al.* (2002); CITES (2003).

Sacred Groves

In Uttarakhand sacred groves have formed an integral part of hill conservation system. The *dev-van* or *sacred groves* were common particularly in the remote areas of the hills and were associated with religious places. Even today such groves can be found on hilltops in practically every cluster of villages in Garhwal and small shrines dedicated to local deities are found in villages, grazing lands, forest paths and agricultural fields. Most temples in Kumaun and Garhwal are located in or near a grove of trees, *i.e.*, Jageshwar temple is surrounded by a dense deodar grove, which harbour a huge deodar tree reported to be older then the shrine itself. Several forest areas in Almora and Pithoragarh have also been dedicated to local deities. Since such religious practices are slowly perishing due to rapid urbanisation therefore sacred groves are now under threat such as the deodar grove near Jageshwar. A total of 18 sacred groves have been reported from the state (Anthwal *et al.*, 2006). Nearly two dozen sacred groves have been established around Berigad-Chowkori belt of Thal tehsil of Pithoragarh (Nagarwalla and Agrawal, 2010).

Agri-diversity

A total of 1107 species of agricultural crops have been reported from the state in Table 5. These include traditional food crops, paddy, wheat and barley, cultivated and wild fruits, vegetables, grasses, fodder plants, mushrooms, etc.

Table 5: List of agricultural crops in Uttarakhand.

S. No.	*Group*	*No. of species*
1	Traditional food crops	191
2	Paddy, wheat and barley races	230
3	Cultivated wild fruits	133
4	Native vegetable crops	62
5	Wild leafy vegetables with medicinal uses	21
6	Wild edible seed grains	12
7	Commonly used fodder species	22
8	Grasses	167
9	Main indigenous natural fibers	28
10	Mushrooms growing in Uttarakhand	116
11	Edible mushrooms	22
12	Varieties of improved crops released by VPKAS	103

Source: Badoni and Badoni (2001); Shiva and Bhatt (2002); Srivastava and Singh (2005); Kumar and Rawat (In Press).

(c) Faunal diversity

The state is very rich in faunal resources. Though, the diversity of faunal components in Uttarakhand clearly indicates the extent of fauna restricted to the forest zone and above timberline. The richness of fauna is distinctly higher in forest zone especially in the broad-leafed wet forest area than in nival zone. The fauna includes a remarkably high percentage predominance of Indo-Malayan elements, besides the Palaearctic affinities. There are few endemics. A total of 4894 species including 3935 species of invertebrates, and 959 species of vertebrates have been recorded from the state (Table 1).

A total of 48 species of trematodes parasites from various vertebrate hosts have been recorded from Uttarakhand (Chakrabarti and Ghosh, 2010). A total of 36 species of Cestodes under 25 genera have been listed from Uttarakhand. The nematodes are a large and diversified group which is both free living and parasites on plants and animals. Extensive information on plant and soil nematodes of Uttarakhand is not available except a few scattered studies. However, recently Zoological Survey of India, Dehra Dun has published results of systematic surveys of these animals in Uttarakhand. 196 species of nematodes, mostly plant parasites and from soil have been listed from Uttarakhand (Rizvi, 2007; Rizvi, 2008).

A total of 128 species of freshwater and land mollusca, including a few subspecific categories have been recorded from the state (Dey and Mitra, 2002). Of the 37 species of land mollusca, as many as 25 are restricted-endemic to western Himalaya, seven species are restricted to Himalaya. The genus *Aanadenus* is restricted to high altitude between 2000-3000m. Some of the species like *Macrochlamys vesiculata, M. glauca,* etc. are recorded well above 3000m. The soft lime rock hill range of Siwalik offers favourable niche for land molluscs and that could be the reason for Dehra Dun recording highest number of species (22).

The annelide fauna, particularly earthworms, have drawn the attention of several taxonomists in various part of world due to their economic importance. Recently 69 species of soil, freshwater and parasitic Nematodes have been recorded from Uttarakhand by various zoologists (Biswas and Mandal, 2010; Halder and Mandal, 2010). The crustaceans include the popularly known animals such as crabs, shrimp, lobsters, etc. Another group is barnacles, best known to grow on ship's hulls or floating timbers. The data on crustacean biodiversity is still meager. 20 species of decapodes have been reported from Uttarakhand.

The centipede fauna of Uttarakhand is represented by 32 species (Khanna, 2010). 249 species of arachnids belonging to spiders, ixodid ticks and oribatid mites have been listed from Uttarakhand (Kumar and Rawat, In Press). Zoological Survey of India, based on their entomological surveys of the state, listed 31 species under 16 genera and 3 families from Uttarakhand (Hazra and Mandal, 2010). The study of Thysanura fauna has generally been neglected. Hazra and Mandal (2010) listed only four species from Uttarakhand.

Odonata (damselflies and dragonflies) are amphibiotic insects, their eggs and larvae are aquatic, while imagos are aerial. Kumar and Prasad (1981), while dealing with the Odonata of Western Himalaya, have published a detailed account of 162 species from Western Himalaya. Kumar and Rawat (In Press) compiled 20 species of Plecoptera belonging to eleven genera and five families from Uttarakhand.

A number of remarkable and endemic species of acridids have been found to occur in Himalaya. In north-western Himalaya these insects have been recorded up to an altitude of 4875 m. A total of 116 species of this order have been reported from the state recently by Mandal *et al.* (2010). These species belong to 11 families, namely, Acrididae, Pyrgomorphidae, Tetrigidae, Tettigoniidae, Tridactylidae, Gryllidae, Trigonidiidae, Scleropteridae, Mogoplistidae, Eneopteridae and Gryllotalpidae.

A total of 43 species of Dermaptera belonging to 6 families *viz.*, Pygidicranidae, Anisolabididae, Labiduridae, Spongiphoridae, Chelisochellinae and Forficulidae and 18 genera have been reported from the state. Some species belonging to genera *Isolaboidesi* Brindle, *Aneclura* Scudder, *Allodahilia* Verhoeff and *Forficula* L. show Palaerctic affinities and occur generally occur around 5000 m (Srivastava, 1995). A total of 52 species of termites belonging to 5 families viz., Termopsidae, Kalotermitidae, Rhinotermitidae, Stylotermitidae and Termitidae have been reported from the state (Saha *et al.*, 2010). Das *et al.* (2010) reported a total of 479 species of hemiptera from Uttarakhand.

Against an estimated total of 179 families of Coleoptera, about 103 families are known from India; of the 3,50,000 described species from all over the world, 1,50,000 species under 2000 genera are known from India. 416 species of these insects have been listed from the state (Chakraborty,

2010; Chatterjee, 2010). It is expected that many more are yet to be listed from the state, especially from the forest ecosystem of the state.

The lepidopterous fauna, particularly the Rhopalocera, or the butterflies, attracts attention of both the naturalist and the zoologist for their aesthetic as well as scientific values. During the eighteenth century, a large number of butterfly species from the Indian region was named and described including those from Western Himalayan region by Linnaeus and Fabricius. The family Nymphalidae may be regarded as the largest family of butterflies considering the number of species and subspecies. By the conservative estimate, nearly 1450 species of butterflies are known to occur in India. 567 species of butterflies have been reported from the state (Kumar, 2010; Majumdar, 2010; Majumdar and Kumar, 2010). Arora *et al.* (1995) reported 91 species from district Dehra Dun.

A total of 7000 species of Trichoptera are known from the whole world and about 812 species known from India. 60 species have been listed from Uttarakhand belonging to 13 families (Ghosh and Chaudhary, 1995). Extensive studies on the Hymenoptera fauna of western Himalaya were started in early sixties by a number of entomologists resulting in publication of a series of monographs and scientific papers on Ichneumonidae and Braconidae. Scientists from Zoological Survey of India field surveyed the western Himalaya (U.P.) and published an exhaustive list of these insects. A total of 234 species of these insects have been recorded from Uttarakhand (Kundu *et al.*, 2010; Parui and Mitra, 2010; Sheela, 2010).

The family Ichneumonidae of the order Hymenoptera is one of the largest of all animal groups. Of all the species of insect known from the world, 5-8% belongs to family Ichneumonidae. There are about 1200 species known to occur in India. A checklist of 302 species belonging to 18 subfamilies and one family Ichneumonidae has been reported from the state (Jonathan, 1995; Kumar and Rawat, In Press).

Indeed, no other insects present so far have such a great diversity of habits and habitats as the Diptera, and this is the reason why general biologists, parasitologists, medical and veterinary doctors, and others pay serious attention to these insects. India has about 1075 species of these insects recorded. 676 species of Diptera have been recorded from Uttarakhand (Mitra and Bhattacharya, 2010; Mitra *et al.*, 2010).

A total of 959 vertebrates have ben recorded from the state (Kumar and Rawat, In Press). Fishes constitute almost half of the total number of vertebrates. Out of 39,900 vertebrate species recognized the world over 21,723 are living species of fishes of which 8411 are of freshwater (excluding commonly diadromous that may have landlocked populations). In the Indian region alone, 2500 species were reported out of which 930 are freshwater inhabitants and 1570 are marine. Uttarakhand is rich in terms of fish diversity due to the two important perennial rivers of India *i.e.* River Ganga and Yamuna supported by many other tributaries. This new site has wide potential for fisheries especially in hill streams/rivers and their tributaries. Fishes can be propagated easily and can enhance the economy of this area. The state Government is taking ambitious steps to establish fish culture successfully in this region. It is essential to know the status of fish taxonomy, ecology and biology for further policy planning, management and conservation. Dehradun district remain an important research center since past due to its geographically setup and availability of diversified fish fauna i.e. hill stream as well as warm water fishes. A total of 142 species of fishes are recorded from different drainage systems of Uttarakhand (Kumar and Rawat, In Press).

The amphibians form an important link in the evolutionary history of vertebrates. In India, this group is represented by all the three orders *viz.*, Gymnophiona/Apoda/Caecilians, Caudata/Urodela and Salientia/Anura. The amphibian listed from Uttarakhand incorporate 20 species belonging to 11 genera and 5 families under order Anura (Bahuguna and Bhutia, 2010). Reptiles are an interesting group of animals. Their study has always been a challenge to the herpetologists. The reptilian fauna (Crocodiles, Testudines, Lizards and Snakes) has attracted the attention of various workers during the past. 75 species belonging to 43 genera, 3 orders and 14 families have been recorded from Uttarakhand (Bahuguna and Padmanaban, 2007; Bahuguna, 2008). Many of these species are listed under Wildlife (P) Act, 1972 and CITES, besides being threatened.

Uttarakhand is one of the richest bird areas of the country. More than 600 bird species (which is almost 50% of the bird diversity of India) have been reported from this region which has an area of just 1.63% of the country. Besides the species diversity, another intresting aspect is that a very healty bird population exists in the state. Birds can be found anywhere, including the vast network of PA's, in other forests and natural habitats, agricultural fields, etc. Another positive aspect is that its wetlands attaract a seizable population of waterfowl, both resident and migratory. Some of the prominent bird species of Uttarakhand are Himalayan monal, Cheer pheasant, Western tragopan, Satyr's tragopan *etc.* Himalayan monal is the state bird of Uttarakhand. In all, 64 out of the 76 bird families known from India are represented in this region. 622 species of birds have been listed from the state (Mohan and Sinha, 2003; Kumar *et al.*, 2005).

Mammals are the highest evolved group of animals and have always been a special attraction to the mankind. The mammalian fauna of India, particularly of Himalayan ecosystem, is most fascinating and diverse. A total of 100 species belonging to 9 Orders and 26 families have been recorded from Uttarakhand (Sati and Tak, 2010).

Ethnozoological species

Unlike the plants which have tremendous ethnobotanical applications, the animals have limited scope in this important aspect. Intrestingly, there is a record of traditional uses of animal and animal products of 38 animal species in medicine and rituals by tribals in Uttarakhand (Negi and Palyal, 2007). On the other hand 12 trouts, snow trouts, mahaseer and carps and 11 livestock species have been reported from the state (Gusain and Gusain, 2001).

Conservation status of fauna

The trade and utilization of faunal resources from the state constitutes a serious threat to its biodiversity. Among the major species, musk from the Himalayan Musk Deer *(Moschus chrysogaster),* bile from Himalayan Black Bear (*Ursus arctos*), mammalian furs, wool and butterfly forms the 'backbone of the species in trade' from the state (Kumar *et al.*, 2002; Sethi *et al.*, 2002). The updated list of the globally threatened Indian fauna, including the CAMP list and species listed under IW(P) 1972 and CITES from Uttarakhand is provided in Table 6.

(d)Fossil diversity

Uttarakhand has hilly and mountainous terrains that cover approximately 90% of the geographical area. The montane zone consists of sub-Himalaya, mid-Himalaya, and greater Himalaya (virtually covered with snow year-round) throughout the state. The structural and lithological characteristics along with the climatic variations have given rise to multiple types of landscapes in whole of the Uttarakhand Himalaya. The region presents a variety of complex or compound landscapes including glacial, peri-glacial, glacio-fluvial, fluvial, etc. Therefore, geological factors have been more responsible for giving different shapes to the landscapes.

Table 6: List of threatened fauna and conservation dependent species in Uttarakhand.

S. No.	*Group*	*No. of species*
1.	Threatened fauna (IUCN Red List)	96
2.	Fauna under CITES	97
3.	Fauna under Schedule I of Wildlife (Protection) Act, 1972	97
4.	Medicinally used wild animals by tribes	38
5.	Threatened fauna (CAMP Criterion)	141
6.	Restricted range bird species	7

Source: Walker and Molur (2000); Gusain and Gusain (2001); Kumar *et al.* (2002); Sethi *et al.,* (2002); Kumar & Khanna (2006); Negi and Palyal (2007); Kumar and Rawat (2010).

The Himalayan tract of the Kumaon–Garhwal region exposes a wide variety of rocks ranging in age from Himalayan pre-Cambrian to Quaternary. Schists, schistose phyllites, granulites, migmatites, limestone, quartzite *etc.* are the major rocks in region. The area is classified as high seismic zone V. The region experiences very severe soil erosion. Soils of the Uttarakhand Himalaya in general are quite shallow, gravelly, and impregnated with unweathered fragments of parent rocks.

Fossils of various types are found in the foot hills rocks (sub Himalaya), the lesser and Tethys Himalaya. Broadly they can be classified into micro fossils represented by Fungi, bacteria, invertebrates, vertebrates and plant fossils. They range from Paleozoic to proper Cenozoic.

A total of 259 fossil forms of microbes, flora and fauna have been recorded from the state (Table 1). 75 species of fossil microbes have been listed from the state (Tiwari, 2008; Tiwari and Pant, 2009). Some of the genera reported from the state are, *Asterocapsoides*, *Trachyhystrichosphaera*,

Eoentophysalis, Eomycetopsis, Gloeodiniopsis, Gunflinitia minuta, Micrhystridium, Obruchevella, Oscillatoriopsis, Siphonophycus, Callimothallus and *Trichothyrites.*

Fossil flora is widely known both from the foot hills as well as the interior regions of Kumaun. A total of 34 species have been reported from four families, namely, Dipterocarpaceae (4 species), Anonaceae (1 species), Leguminosae (2 species) and Ebenaceae (26 species). The reported genera are, *Anisopteroxylon, Dipterocarpoxylon, Polyalthioxylon, Cynometroxylan, Cassinium, Ebennoxylon, Abies, Albizia, Alisporites, Alnus, Apiculatispris, Betula, Calamospora, Carpinus, Cedrus, Corylus, Dipterocarpoxylon, Ephedre, Eretomnia, Gremia, Itex, Jurvsamia, Laevigatosparites, Morus, Pinus, Pityosporite, Atysacus, Striatites, Striatopodocarpita* and *Trilete* (Gupta, 1971; Lakhan Pal *et al.,* 1976).

Invertebrate fossils (129 forms) represented by orders Brachiopoda, Mollusca, Trilobites, Gastropoda and Bryozoans are known from various localities of Kumaun region (Diener, 1908; Gupta, 1971). The various genera viz., *Anatomites, Arcestes, Arietites, Cladiscites, Griesbachites, Jovites, Juavites, Orthis, Phylloceras* and *Torpites* have been reported from the state.

Vertebrate fossils (25 forms) are confined mainly to two regions, namely, Dehra Dun-Haridwar and Kalagarh regions. Despite being first reported by in 1837, these are not abundant compared to invertebrate and plant fossils occurring in Kumaun region (Falconer, 1837; Nanda and Shukla, 2001). A total of 16 fossil species, belonging to Artiodactyla (9 species), Carnivora (1 species), Perrisodactyla (2 species), Primates (2 species), Probiscidea (1 species) and Rodentia (1 species), have been reported from lower and middle Shiwalik. 4 fossile species of Mammalia and 4 species of Reptilia have been reported from Shiwalik range eastwards of Haridwar while 1 species of Artiodactyla have been reported from middle Shiwalik group near Dehra Dun.

Threats to Biodiversity of State

The past decades have seen an increase in pressures on the state's natural ecosystems. The entire Siwalik ecosystem of Uttarakhand has been virtually degraded of their forest cover, the high altitude grasslands (*Bugyals*) are under enormous pressure from local as well as migrant grazier communities; the glaciers are receding, agri-diversity is shrinking because of introduction of exotic varieties and monoculture practices. Livestock diversity has already lost many of the native varieties (like *Kali gai* etc.) due to introduction of new breeds, poor health management and mismanagement of fodder production and cultivation. Ethno botanical resources that are also one of the prime sources of livelihood in Uttarakhand are also declining. It is therefore, considered timely and necessary to develop a comprehensive plan for the State that would ensure that aspects relating to the conservation and sustainable use of biodiversity.

Major threats to biodiversity in the state are: increase in urbanization, habitat degradation due to illegal, commercial and development activities specially around catchment areas, extraction of dead and damaged timber, deliberate forest fire, overgrazing and trampling of saplings, diversion of forest land for non forest purposes, exotic species and weed infestation, sacrificing temperate biodiversity for horticulture, over exploitation of medicinal plants, faulty agricultural practices, soil degradation, mining, large dams, pollution in water bodies, indiscriminate use of fertilizers, pesticides, poaching and illegal trade in wildlife products (TPCG and Kalpavriksh, 2005). There is a well established network of illegal trade of medicinal plants from wild and also there is a report of overexploitation of valuable medicinal plants like Yartsa Gombu, Kuth and Yew in Uttarakhand.

Some initiatives to protect the state's bioresources are already going on (Kumar and Rawat, In Press). Uttarakhand has already a Protected Area network of National Parks (Corbett, Valley of Flowers, Rajaji, Gangotri and Govind) and Wildlife Sanctuaries, namely, Govind Pashu Vihar, Kedarnath, Askote, Sonanadi and Mussoorie and world heritage sites like Nanda Devi Biosphere Reserve. There are also a very large number of community conserved areas, including *Van Panchayats* and other informally protected forest areas around villages, grasslands such as *bugyals* and wetlands. In many places the traditional practices of common property resource management have helped to conserve biodiversity, both of ecosystems and of species. For *ex situ* conservation, various Institutes have arboretum, populatum and bambusetum for flora, and various research activities are going on to know the status of flora in the State. For *ex situ* conservation of animals only one high altitude zoo at Nainital, two musk deer breeding centres at Kanchula Kharak and Dharamgarh and one mini zoo at Almora are present. Various gaps about the status of micro flora and fauna are still unworked. There is need for improving the management of the PA network and to identify other biodiversity hot spots of conservation value. The soil conservation programme of forest department includes six schemes, viz., Development of civil Soyams in Kumaon and Garhwal, River valley projects in the catchment of Ram Ganga, Integrated wasteland development projects, Integrated watershed management, Reclamation and rehabilitation of abandoned mines of Mussoorie hills and Kheerganga project. For fisheries, five man made reservoirs are a flourishing centre for fish development under the aegis of Uttarakhand Matsya Vikas Nigam Limited, but still there is gap in conservation of Golden mahaseer and the lack in control and regulation of illegal fishing and introduction of exotic fishes (Kumar *et al.,* 2002).

Agri diversity of the State needs revival of traditional practices and seed conservation practices in large scale to give rebirth to native varieties. Livestock diversity needs a proper fodder development programme, policy and implementation strategies to conserve the native varieties. Ethno botanical species are strongly linked to the livelihood of rural community and need focus regarding their habitat protection, research and monitoring and regeneration schemes. People's movements such as the *Beej Bachao Andolan*, and NGO efforts linked to villagers such as by Navdanya, and at places like Munsiari and Nahin Kalan have helped to highlight the decline of agro diversity, and attempted to revive it in many different ways. The importance of biodiversity has been added in the curriculum of lower classes. The forest department has also initiated an award declaration for the forest staff and rural people for conservation efforts (Kumar *et al.*, 2002).

The Union Ministry of Environment and Forests (MoEF), the nodal agency for implementing provisions of CBD in India developed a strategy for biodiversity conservation at macro-level in 1999, as part of environment/development planning process, under project funding by GEF/UNDP. Kalpavriksh, an NGO undertook the technical execution, and the Biotech Consortium India Ltd. coordinated the administrative execution. A comprehensive document on the Biodiversity Strategy & Action Plan (BSAP) was prepared for the country, including Uttarakhand (Kumar and Rawat, In Press).

Conservation Strategies and Action Plan

Considering the need to develop a comprehensive plan for the state for the conservation and sustainable use of biodiversity, Ministry of Environment and Forests, Govt of India, initiated an exercise (1999) as part of environment/development planning process under project funding by GEF/UNDP. Kalpavriksh, an NGO was asked to undertake the technical execution. In the meantime, the development of a State level Biodiversity Strategy and Action Plan was also initiated under the World Bank assisted Uttar Pradesh Forestry Project (2002), which was prepared through a consultative process involving multiple stakeholders by TERI, New Delhi (TERI, 2002). In year 2002 comprehensive biodiversity and action plan was prepared for the state jointly by State Forest Department, UK, TERI and Zoological Survey of India (Kumar *et al.*, 2002).

The issues and objectives relating to the conservation and sustainable use of biodiversity that were identified as being key elements to the development of such a strategy (and which have been addressed in the Strategy and Action Plan) include: identification of flora, fauna, agro-biodiversity and plants of ethno botanical importance and their conservation strategies (TPCG & Kalpavriksh, 2005).

The various goals of conservation strategies are: (a) Conservation of the natural heritage of the state including the unique, biodiversity rich and fragile ecosystems of the state such as forests, grasslands, wetlands and mountain ecosystems and their species including wild and domesticated biodiversity, genetic resources and ecological and environmental processes; (b) primacy to be given to *in situ* conservation of the state's biological and cultural diversity located both within and outside the state's protected areas; (c) the *ex situ* conservation of flora, fauna and floral and faunal genetic resources; (d) to develop strategies and actions for conservation of agriculture, livestock, fodder and ethnobotanical diversity; and (e) sustainable use of biodiversity and natural resources.

The State has taken several steps for the protection, conservation and management of biodiversity including the creation of protected areas, designation of protected trees and the preservation plots.

In-situ Conservation: (a) Protected Area Network (PAN) - The state has a high percentage of land under the protected area network as compared to the national average of 4.75%. There are 6 national parks, 6 wildlife sanctuaries, one biosphere reserve, one World Heritage Site, two elephant ranges and two conservation reserves covering an area of 0.73 million ha, constituting 13.68% of the state's geographic area; b) Protected trees- In Uttarakhand, 27 species of trees numbering around 114 are designated as protected trees for their superior morphological and genetic characteristics. These species are scattered over hills and plains of the state in various forest types; (c) Preservation plots- The representative areas set aside in various forest types for permanent protection are known as the preservation plots. The main objectives of these preservation plots are: (i) preserve examples of existing forests as far as possible in their present form; (ii) protect such forest plots from all forms of injury and to permit progression towards climax form. Most of the preservation plots and protected trees are located in the reserved forest areas. However, some of them are outside protected areas (Kumar *et al.*, 2002; TERI, 2002).

Ex-situ conservation: The Uttarakhand Forest Department is working towards the conservation and rejuvenation of population of endangered and threatened species of fauna through its zoos and conservation breeding centres. Currently the state harbours two musk deer breeding centres, one zoological garden, two botanical gardens (one each in FRI and BSI), three arboretum, one Bambusetum and one medicinal plant nursery. The major institutes working towards conservation of plant diversity are the Forest Research Institute, Wildlife Institute of India, Botanical Survey of India, Dehra Dun and G B Pant Institute of Himalayan Environment and Development, Kosi-katarmal, Almora (GBPIHED) and Centre for Aromatic Plants (CAP), Selaqui. The GBPIHED proposes to undertake a medicinal plant conservation programme with support from Department of Biotechnology, GOI and in collaboration with NBRI and BSI.

In Uttarakhand, arboretums are located at Kalika near Ranikhet, FRI, Dehra Dun and GBPIHED, Kosi-katarmal. The arboretum initiated by GBPIHED at Kosi-katarmal includes a collection of rare and endangered medicinal, edible, multipurpose species, orchids and ferns numbering over 300 species. Lal Kuan Populatum near Haldwani has a collection of more than 400 clones of *Populus* species. Forest Research Institute, Dehra Dun harbours the only bambusetum of the state with a rich collection of bamboo species.

The state has one high altitude zoo at Naini Tal, two musk deer breeding centres at Kanchula Kharak and Dharamgarh and one mini zoo in Almora. The only high altitude ex-situ conservation site of Uttarakhand is located in the city of seven lakes, Naini Tal. The zoo occupies an area of 4.693 ha adjoining the reserved forests of Naini Tal. The zoo is identified by the Central Zoo Authority of India (CZA) as breeding centre for high altitude animals of the state. The Van Chetna Kendra in Almora also harbours a high altitude mini-zoo and was established in 1981. It occupies an area of 32 ha of land. The zoo has collection of about 60 animals. Two musk deer breeding centres were created: one at Kanchula kharak in the

Kedarnath which aimed at breeding and re-stocking; second one in Mahroori village in Dharamgarh taluk of Pithoragarh district with the objective of breeding musk deer in captivity for extraction of musk (Kumar *et al.*, 2002; Sethi *et al.*, 2002).

Acknowledgement

We are thankful to the Uttarakhand Council for Science and Technology (UCOST), Dehra Dun for providing funds to undertake this project. We also extend thanks to Kalpavriksh, Pune, State Forest Department, U.K., ZSI, Dehra Dun, TERI, New Delhi, and others for giving access to State Biodiversity Strategy and Action Plan- Uttarakhand documents.

References

Anthwal, A., Sharma, R.C. and Sharma, A. (2006). Sacred Groves: Traditional way of conserving plant diversity in Garhwal Himalaya, Uttaranchal. *J. Am. Sci.*, **2(2)**:35-43.

Arora G.S., Ghosh S.K., and Chaudhury M. (1995). Lepidoptera. *In*: Fauna of Western Himalaya, U.P., Himalayan Ecosystem Series, 1. *Zool. Surv. India*. pp. 61-74. (Coordinators: G.S. Arora and Arun Kumar).

Arora, G.S. and Kumar, A. (1995). Fauna of Western Himalaya (U.P.). *Zool. Surv. India, Himalayan Ecosystem Series,* Part-1: 1-223.

Badoni, A. and Badoni, K. (2001). Ethnobotanical Heritage. *In*: Garhwal Himalaya Nature, Culture & Society (eds. O.P. Kandari and O.P. Gusain). TransMedia, Srinagar. pp. 125-148.

Bahuguna, A. (2008). Reptilia. *In*: Fauna of Corbett Tiger Reserve, Conservation Area Series. *Zool. Surv. India*. **35**: 143-157.

Bahuguna, A. and Bhutia, P. (2010). Amphibia. *Fauna of Uttarakhand, State Fauna Series*, **18(1)**: 505-532. (Publ. by Director, *Zool. Surv. India*, Kolkata).

Bahuguna, A. and Padmanaban, P. (2007). Reptilia. *In*: Faunal Diversity Western Doon Shiwaliks. *Zool Surv. India*. pp. 65-72.

Bisht, G.S. and Srivastava, R.C. (1983). Fungal diseases of some important crop plants grown in Garhwal Himalaya. *In*: Microbial activity in Himalaya (ed. R.D. Khulbe). pp. 141-153.

Biswas, T. and Mandal, C.K. (2010). Oligocheta. *Fauna of Uttarakhand, State Fauna Series*, **18(3)**: 173-180. (Publ. by Director, *Zool. Surv. India*, Kolkata).

Chakrabarti, S., and Ghosh, A. (2010). Trematode parasites of vertebrates. *Fauna of Uttarakhand, State Fauna Series*, **18(3)**: 15-51. (Publ. by Director, *Zool. Surv. India*, Kolkata).

Chakraborty, S.K. (2010). Insecta: Coleoptera. *Fauna of Uttarakhand, State Fauna Series*, **18(2)**: 283-309. (Publ. by Director, *Zool. Surv. India*, Kolkata).

Chatterjee, S.K. (2010). Insecta: Coleoptera: Scarabaeidae (Cetoniinae, Dynastinae and Rutelinae). *Fauna of Uttarakhand, State Fauna Series*, **18(2)**: 311-321. (Publ. by Director, *Zool. Surv. India*, Kolkata).

CITES (2003). Pictorial Identification Manual of CITES Plants in INDIA. Appendices I, II and III valid from 28 May 2003. CITES-listed species Database hosted by UNEP-WCMC. CITES Appendices.htm.

Das, K. and Sharma, J.R. (2006). Russulaceae of Kumaun Himayala. *Botanical Survey of India*, Dehradun.

Das, S.C. Sengupta, C.K. and Parmanik, N.K. (2010). Insecta: Hemiptera: Aphididae. *Fauna of Uttarakhand, State Fauna Series*, **18(2)**: 133-174. (Publ. by Director, *Zool. Surv. India*, Kolkata).

Dey, A. and Mitra S.C. (2002). Molluscs of Himalaya. *Rec. Zool. Surv. India*. Part II: pp. 5-50.

Diener, C. (1908). Upper-Triassic and Liassic fauna of the exotic blocks of Malla Johar in the Bhot Mahals of Kumaon. *Memoirs of Geological Survey of India, series XV*, **I** (**1**): 1-100.

Falconer, H. (1837). Note on the occurrence of fossil bones in the Siwalik range eastwards of Haridwar. *J. Asiatic Soc. Bengal*, **6(1)**: 233-234.

FRLHT: Encyclopedia on Indian Medicinal Plants. FRLHT ENVIS Centre on Medicinal Plants.

FSI. (2000). State of Forest Report 1999. Supplementary Forest Survey of India, Ministry of Environment and Forests, Dehra Dun.

Gairola, Y. and Biswas, S. (2008). Bioprospecting in Garhwal Himalaya, Uttarakhand. *Curr. Sci.,* **94(9)**: 1139-1143.

Gaur, R.D. (1999). Flora of the District Garhwal North West Himalaya (with ethnobotanical notes). TransMedia, Srinagar. xiv + 811.

Ghosh, S.K. and Chaudhury, M. (1995). Trichoptera. *In*: Fauna of Western Himalaya, U.P., Himalayan Ecosystem Series, 1. *Zool. Surv. India*. pp. 75-79. (Coordinators: G.S. Arora and Arun Kumar).

Gupta, R.K. (2005). *Algal Flora of Dehra Dun District Uttaranchal*: 298p. Botanical Survey of India, Kolkata.

Gupta, V.J. (1971). Indian Paleozoic Fossils. Center of Advanced Study in Geology, Panjab University, Chandhigarh. Pub. No.9.

Gusain, O.P. and Gusain, M.P. (2001). Applied Fisheries: Status and Scope. *In*: Garhwal Himalaya: Nature, Culture & Society (Eds. O.P. Kandari and O.P. Gusain). pp. 199-216.

Halder, K.R. and Mandal, C.K. (2010). Annelid fauna of some selected wetlands. *Fauna of Uttarakhand, State Fauna Series*, **18(3)**: 167-172. (Publ. by Director, *Zool. Surv. India*, Kolkata).

Hazra, A.K. and Mandal, G.P. (2010). Collembola. *Fauna of Uttarakhand, State Fauna Series*, **18(2)**: 1-12. (Publ. by Director, *Zool. Surv. India*, Kolkata).

Jalal, J.S., Kumar, P., Rawat, G.S. and Pangtey, Y.P.S. (2008). List of species, Orchidaceae, Uttarakhand, Western Himalaya, India. Check List, **4 (3)**: 304-320.

Jonathan, J.K. (1995). Hymenoptera: Ichneumonidae *In*: Fauna of Western Himalaya, U.P., Himalayan Ecosystem Series, 1. *Zool. Surv. India.* pp. 91-110. (Coordinators: G.S. Arora and Arun Kumar).

Joshi, G.C., Tewari, L. M., Lohani, N., Upreti, K. Jalal, J. S. and Tewari, G. (2009). Diversity of Orchids in Uttarakhand and their conservation strategy with special reference to their medicinal importance. *Report and Opinion*, **1(3)**: 47-52.

Kandari, O. P. and Gusain, O. P. (eds.) (2001). Garhwal Himalaya: Nature, Culture & Society. TransMedia, Srinagar, 393p.

Khanna, V. (2003). Diversity of Scolopendrid Centripedes (Chelopoda: Scolopendromorpha). *In*: Himalayan Ecosystem and Adjacent Areas-A Review. *Rec. Zool. Surv. India*, **101** (3-4): 207-223.

Khanna, V. (2010). Chilopoda: Scolopendromorpha: Centipedes. *Fauna of Uttarakhand, State Fauna Series*, **18(3)**: 209-241. (Publ. by Director, *Zool. Surv. India*, Kolkata).

Khulbe, R.D., Sati, M.C. and Dhyani, A.P. (1983). Mycoflora associated with the seed of pea (*Pisum sativum* L.) in storage and field condition of garampani area. *In*: Microbial activity in Himalaya (ed. R.D. Khulbe). pp. 121-125.

Kumar, A. and Khanna, V. (2006). Globally Threatened Indian Fauna - Status, Issues and Prospects. 104p. (Publ. by Director, *Zool. Surv. India*, Kolkata).

Kumar, A. and Prasad, M. (1981). Odonata of Western Himalaya, India – an annotated check-list with key and notes on field ecology and zoogeography. *Recd. Zool. Surv. India Occ. Publ.*, **20**: 1-118.

Kumar, A. and Rawat, S. (In Press). Biodiversity Log of Uttarakhand and Its Conservation Strategies. Bishen Singh Mahendra Pal Singh, Dehra Dun. (2010).

Kumar, A., Sati, J.P., Tak, P. and Alfred, J.R.B. (2005). Handbook of Indian wetland birds and their conservation. *Zool. Surv. India*. xxvi+468.

Kumar, P. (2010). Lepidoptera: Rhopalocera. *Fauna of Uttarakhand, State Fauna Series*, **18(2)**: 611-689. (Publ. by Director, *Zool. Surv. India*, Kolkata).

Kumar, A., Bahuguna, A., Rawat, G.S., Mohan, D., Sinha, S., Srivastava, S.K. and Banerjee, A.K. (2002). Biodiversity Strategy and Action plan (BSAP) Uttaranchal. FD Uttaranchal, TERI and ZSI, Dehradun. **I-III**: 515 p.

Kundu, R.G. Kumar, P.G., Kazmi, S.I. and Roychowdhury, S: (2010). Insecta: Hymenoptera: Aculeata (Vespidae and Apidae). *Fauna of Uttarakhand, State Fauna Series*, **18(2)**: 725-748. (Publ. by Director, *Zool. Surv. India*, Kolkata).

Lakhan Pal, R.N., Maheshwari, H.K. and Awasthi, N. (1976). A catalogue of Indian fossil plants, covering all available records from 1821 to 1970. Birbal Sahni Institute of Palaeobiology, Luckhnow.

Majumdar, M. (2010). Insecta: Lepidoptera: Families: Pieridae and Arctiidae. *Fauna of Uttarakhand, State Fauna Series*, **18(2)**: 501-529. (Publ. by Director, *Zool. Surv. India*, Kolkata).

Majumdar, M. and Kumar, J. (2010). Lepidoptera: Sphingidae. *Fauna of Uttarakhand, State Fauna Series*, **18(2)**: 553-568. (Publ. by Director, *Zool. Surv. India*, Kolkata).

Mandal, S.K., Dey, A. and Yadav, K. (2010). Insecta: Orthoptera: Acridoidea. *Fauna of Uttarakhand, State Fauna Series*, **18(2)**: 53-79. (Publ. by Director, *Zool. Surv. India*, Kolkata).

Mitra, B. and Bhattacharya, K. (2010). Insecta: Diptera. *Fauna of Uttarakhand, State Fauna Series*, **18(2)**: 361-411. (Publ. by Director, *Zool. Surv. India*, Kolkata).

Mitra, B., Mirdha, R.S. and Parui, P. (2010). Insecta: Diptera: Calliphoridae. *Fauna of Uttarakhand, State Fauna Series*, **18(2)**: 437-441. (Publ. by Director, *Zool. Surv. India*, Kolkata).

Mohan, D. and Sinha, S. (2003). Birds of Uttaranchal (A checklist). Forest Department, Uttaranchal.

Nagarwalla, D.J. and Agrawal, R. (2010). State Uttarakhand. *In*: Community Conserved Areas in Uttarakhand, A Directory (ed. Neema Pathak). Kalpavriksh, Pune. pp. 707-731.

Nandi, R. (2010). Protozoa: Coccidia. *Fauna of Uttarakhand, State Fauna Series*, **18(3)**: 1-14. (Publ. by Director, *Zool. Surv. India*, Kolkata).

Nanda, A.C. and Shukla, S.D. (2001). Fossil Giraffids from the middle Siwalik subgroup near Dehra Dun. *Him. Geol.*, **22 (2)**: 121-125.

Nautiyal, R., Nautiyal, P. and Singh, H.R. (1996). Pennate Diatom Flora of a Coldwater Mountain River, Alaknanda III. Suborder Biraphideae. *Phykos*, **35 (1 & 2)**: 65-75.

Nautiyal, R. and Nautiyal, P. (1999). Altitudnal Variations in the Pennate Diatom Flora of Alaknanda- Ganga River System in the Himalayan Stretch of GHL Region. *In*: 14th Diatoms Symposium 1996, pp. 85-100.

Nautiyal P., Kala, K. and Nautiyal, R. (2004). A preliminary Study of the diversity of diatoms in the streams of the Mandakini basin (Garhwal Himalaya). Seventeenth International Diatoms Symposium 2002, pp. 235-260.

Nayar, M.P. and Sastry, A.R.K. (1987). Red data book of Indian plants. Botanical Survey of India, Kolkata.

NCBI: Invasive species of India. http://www.ncbi.org.in/invasive/search/index.html.

Negi, C. S. and Palyal, V. S. (2007). Traditional Uses of Animal and Animal Products in Medicine and Rituals by the Shoka Tribes of District Pithoragarh, Uttaranchal, India. *Ethno-Med.*, **1(1)**: 47-54.

Negi, P.S. and Hazra, P.K (2007). Alien flora of Doon valley, Northwest Himalaya. *Curr. Sci.*, **97(2)**: 968-978.

Pande, P.C., Tiwari, L. and Pande, H.C. (2006). Folk-Medicine and Aromatic Plants of Uttaranchal; Bishan Singh Mahendra Pal Singh, Dehra Dun. xviii+ 462 p.

Parui, P. and Mitra, B. (2010). Insecta: Diptera: Asilidae. *Fauna of Uttarakhand, State Fauna Series*, **18(2)**: 413-423. (Publ. by Director, *Zool. Surv. India*, Kolkata).

Phadtare, N.R. (1989). Palaeoecological significance of some fungi from the Miocene of Tanakpur (U.P.) India. *Rev. Palaeobotany & Palynology*, **59**: 127-131.

Prakash, U. (1988). Fossil woods from the lower Siwalik beds of Uttar Pradesh, India. *The Palaeobotanist*, **37**: 376-387.

Punetha, N. and Kholia, B.S. (2010). The floristic diversity of Uttarakhand: Ferns and fern-allies. *In*: The Plant Wealth of Utarakhand (eds. P.L. Uniyal, B.P. Chamola and D.P. Semwal). pp. 293-303.

Rizvi, A. N. (2007). Nematoda. *In*: Fauna of Western Doon Shivaliks. *Zool. Surv. India*. pp. 5-12.

Rizvi, A. N. (2008). Plant and Soil Nematodes. *In*: Fauna of Conservation Areas 35, Fauna of Corbett Tiger Reserve (Selected Groups). *Zool. Surv. India*. pp. 119-121.

Saha, N., Roy, P.H. and Sur, A. (2010). Insecta: Isoptera. *Fauna of Uttarakhand, State Fauna Series*, **18(2)**: 81-104. (Publ. by Director, *Zool. Surv. India*, Kolkata).

Sati, J.P. and Tak, P.C. (2010). Mammalia. *In*: *Fauna of Uttarakhand, State Fauna Series*, **18(1)**: 27-76. (Publ. by Director, *Zool. Surv. India*, Kolkata).

Sati, S.C. (1997). Diversity of aquatic fungi in Kumaun Himalaya: Zoosporic fungi in Himalayan microbial diversity: pp. 1-16.

Sati, V.P. (2005). Natural resources conditions and economic development in the Uttaranchal Himalaya, India. *Jour. Mount. Sci.*, **2(4)**: 336-350.

Srivastava, G.K. (1995). Dermaptera. *In*: Fauna of Western Himalaya, U.P., Himalayan Ecosystem Series, 1. *Zool. Surv. India*. pp. 43-45. (Coordinators: G.S. Arora and Arun Kumar).

Selvakumar, G., Kundu, S., Gupta, A.D., Shouche, Y.S. and Gupta, H.S. (2007). Isolation and Characterization of Nonrhizobial Plant Growth Promoting Bacteria from Nodules of Kudzu (*Pueraria thunbergiana*) and their effect on wheat seedling growth. *Curr. Microbiol.*, **56(2)**: 134-139.

Sethi, P., Bhujang Rao, D.D., Mohan, D., Mohapatra, K.K., Upadhayay, S., Hanfee, F. and Khalid, M.A. (2002). Strategies and Action Plan for Biodiversity Conservation in Uttaranchal. Report No. 1999SF61, TERI, New Delhi.

Shantanu S. S. and Uniyal, P.L. (2010). Bryophytes of Uttarakhand. *In*: The Plant Wealth of Uttarakhand (Eds. P.L. Uniyal, B.P. Chamola and D.P. Semwal). Jagdamba Publishing Company, New Delhi, India, pp. 197-276.

Sharma, P., Semwal, D.P. and Uniyal, P.L. (2010). Gymnosperms of Uttarakhand. *In*: The Plant Wealth of Uttarakhand (eds. P.L. Uniyal, B.P. Chamola and D.P. Semwal). Jagdamba Publishing Company, New Delhi, India, pp. 305-316.

Sheela, S. (2010). Insecta: Hymenoptera: Chalcidoidea. *Fauna of Uttarakhand, State Fauna Series*, **18(2)**: 691-700. (Publ. by Director, *Zool. Surv. India*, Kolkata).

Shiva, V. and Bhatt, V.K. (2002). Nature's harvest: rejuvenating biodiversity. Navdanya, Dehra Dun.

Srivastava, S.K and Singh, D.K. (2005). Glimpses of the Plant Wealth of Uttaranchal, xi+158p. Bishen Singh Mahendra Pal Singh, Dehradun.

The Wildlife (Protection) Act (1972). As amended upto 2003, with rules upto 2003. Wildlife Trust of India, New Delhi (2003).

Tiwari, M. (1999). Organic- walled microfossils from the Chert-phosporite Member, Tal Formation, Precambrian-Cambrian boundary India. *Precambrian Res.*, **97**: 99-113.

Tiwari, M. (2008). Additional Neoproterozoic sponge spicules from Gangolihat Dolomite, Kumaon Lesser Himalaya, India. *Himalayan Geol.*, **29** (**1**):49-55.

Tiwari, M. and Pant, I. (2009). Microfossils from the Neoproterozoic Gangolihat Formation, Kumaon Lesser Himalayas: Their stratigraphic and evolutionary significance. *J. Asian Earth Sci.*, **35(2)**:137-149.

TPCG and Kalpavriksh. (2005). *Securing India's Future: Final Technical Report of the National Biodiversity Strategy and Action Plan.* Prepared by the NBSAP Technical and Policy Core Group, Kalpavriksh, Delhi/ Pune. 62p.

Uniyal, B.P., Sharma, J.R., Choudhary, U. and Singh, D.K. (2007). Flowering Plants of Uttarakhand (A Checklist). Bishen Singh Mahendra Pal Singh, Dehra Dun. vi+ 404p.

Uniyal, K. (2001). Arbuscular Mycorrhizal association of *Populus deltoides. Indian Forester*, **127:** 527-529.

Upreti, D.K., Nayaka, S. and Chatterjee, S. (2010). Lichen diversity of Uttarakhand Himalaya. *In*: The Plant Wealth of Uttarakhand (eds. P.L. Uniyal, B.P. Chamola and D.P. Semwal). Jagdamba Publishing Company, New Delhi, India, pp. 79-196.

Verma, R. (2005). Studies on microbial diversity of effluent of some important industries of Dehra Dun valley. *Ph. D. Thesis*, H.N.B. Garhwal University, Srinagar, Garhwal.

Walker, S. and Molur, S. (2000). Species summaries of Conservation Assessment and Management Plan (CAMP) Workshops, 1995-2000: 1-27.

ZSI. (2010). Fauna of Uttarakhand. *State Fauna Series, Zool. Surv. India,* **18(1-3).**

Biodiversity Conservation and Envir. Management (2012)
Editors: D.R. Khanna et al.
Pub. by Biotech Books. *ISBN: 978-81-7622-262-4*

Pages: 395-400

EVALUATION OF BOTANICALS IN STORAGE AGAINST KHAPRA BEETLE (*TROGODERMA GRANARIUM* EVERTS) AND GRAIN QUALITIES OF WHEAT (*TRITICUM AESTIVIUM* CV. HD2329)

Sanjiv Charjan[1], Parikshit Bokare[2], Amol Patil[2], Raviraj Udasi[2], Ashish Lambat[3]✉, Rajesh Gadewar[3]

[1]College of Agriculture (Dr. PKV's), Nagpur, Maharashtra, (INDIA)
[2]Student, College of Agriculture (Dr. PKV's), Nagpur, Maharashtra
[3]Sevadal Mahila Mahavidyalaya and Research Academy, Nagpur, Maharashtra (INDIA)
E-mail : lambatashish@gmail.com

The khapra beetle (*Trogoderma granarium* Everts) is a serious store grain pest of wheat during storage. In the present investigation, the experiment was conducted to know the effect of different organic grain protectants on infestation percentage of khapra beetle and seed qualities of wheat

✉ Corresponding author

(*Triticum aestivium* cv. HD2329). It was observed that seed treated with sweet flag powder (2.5 %) and custard apple seed powder 2.5 % showed significantly higher 100 seed weight, germination percentage, seedling vigour, field emergence percentage and adult mortality as compared to other seed treatment and controll during storage.

Keywords: Organic grain protectants, Sweet flag, Custard apple, Khapra beetle, Wheat infestation, Storage.

Introduction

The khapra beetle (*Trogoderma granarium* Everts) is a serious pest of stored grain causing considerable damage to almost all cereals in storage. It is largely responsible for damage and frequently harboring stores, mill and warehouses. Khapra beetle is cosmopolitan in nature attributing about 50% loss in seed weight during storage.

The steady rise in the use of pesticides for control of store grain pests can be dangerous to human being and cattle as well due to their residual toxicity. With view to find out safe and organic seed protectants, present investigation was taken up to evaluate the organic grain protectants in seed storage against khapra beetle in wheat.

Materials and Method

Wheat (*Triticum aestivium* cv. HD2329) seed were used in various phases of this study, produced in 2008-2009. The seed were cleaned and dried (moisture content 10.5 %). The wheat seed were treated (May 2009) with six plant product *viz.*, Neem leaf, sweet flag, Tulsi, Custard apple, Turmeric and Pongamia powder is in the proportion of 2.5% by weight of the seeds. The experiment was conducted in glass bottle of 1 lit capacity with seven treatments including untreated control. Each glass bottle was then filled with 500 gm of wheat seeds. 10 pairs of 2-3 days old khapra beetle were released in each glass bottle covered with muslin cloth. The set of experiment was kept in well ventilated wire mesh almirah in masonry building having cemented walls, roof and floor under ambient temperature (24.1-45.8 °C) and relative humidity (20-85 %)from May to July, 2009. After three months the seed from each treatment were keenly observed and those found infested were separated out weighted to determine the infestation percentage on weight basis hundred seed weight and germination were tested in quadruplicate with 100 seed in each replication. The germination percentage was evaluated on the value for normal seedling (Anonymous, 1985). the vigour index were workout following the method of Abdul Baki and Anderson (1973). For field emergence test, sowing of wheat seed was done in randomized block design, with four replication with inter and intra row spacing of one feet and six inches respectively. Observations for field emergence were recorded daily and finally the established seedlings were counted after one month of sowing. The experimental data was statistically scrutinized as per Panse and Sukhatme (1967).

Results and Discussion

The data regarding the effect of the different organic grain protectants on population behaviour (Adult mortality), infestation percentage 100 seed weight, germination percentage, vigour index and field emergence percentage after three months of storage are given in Table 1.

The results indicated variation in number of khapra beetle (adult) in each treatment. The khapra beetle adult mortality was significantly highest in wheat seed treated with sweet flag (100 %) which is closely followed by custard apple (92 %) neem leaf (60 %), pongamia (56 %), tulsi (51 %) and turmeric (47 %). Where as significantly lower mortality was observed in untreated control (5 %). Saxena *et al.* (1976), Tikku *et al.* (1978) and Khan and Borle (1985) found *Acorus calamus* L oil vapour responsible for causing infecundity among the female of a number stored grain pest. Biradar (2000) reported that sweet flag has got insecticidal and ovicidal effect. The significantly wheat seed weight loss was observed in untreated control followed by turmeric, tulsi, pongamia, neem leaf, custard apple and sweet flag treatment during entire period of storage. This might be due to sterilizing effect of sweet flag rhizome powder mixed with wheat seeds. Charjan and Tarar (1994), Deshpande *et al.* (2010), Lambat *et al.* (2011)

and Cherian *et al.* (2011) reprted that the sharp declined in infestation percentage of store grain pest in seed treated with *Acorus calamus* powder.

Table 1: The Effect of different organic protectants on khapra beetle mortality, infestation percentage, 100 seed weight, germination percentage, seedling vigour and field emergence percentage.

Treatments	*Adult Mortality (%)*	*Percent weight loss due to infestation (%)*	*100 seed weight*	*Germination (%)*	*Seedling vigour index (SVI)*	*Field emergence (%)*
Neem leaf (2.5%)	60	5.8	3.02	69	1338	57
Sweet flag (2.5%)	100	0.05	3.32	96	2018	88
Tulsi (2.5%)	51	6.4	2.91	65	1291	55
Custard apple (2.5%)	92	2.4	3.17	93	2006	84
Turmeric (2.5%)	47	6.8	2.85	61	1124	50
Pongamia (2.5%)	56	6.0	2.97	64	1202	53
Untreated control	5	17.14	2.01	35	682	22
SEm(±)	0.35	0.09	0.03	0.12		0.10
CD at 5%	1.06	0.29	0.10	0.36		0.31

The seed quality parameters *viz.*, 100 seed weight, germination percentage and seedling vigour index was highest in seed treated with 2.5% concentration of sweet flag powder followed by custard apple, neem leaf, pongamia, tulsi, turmeric and untreated control. The 100 seed weight, germination percentage and vigour index decreased with increasing infestation of stored grain pest (Howe, 1972); Charjan and Tarar, 1994; Deshpande *et al.*, 2010 and Lambat *et al.*, 2011. Since the stored grain pest have been eaten off major portion of the endosperm which leads to reduction in weight of the wheat seeds and in turn affect the seed germination and vigour index because of lack of stored food and is in conformity with the findings of Narayanaswami (1985). Handerson and Christensen (1961) reported that pulse beetle attack the embryo and germination potential of seed reduced or totally destroyed.

The field emergence percentage of wheat seeds follow the same trends of seed quality parameters. The field emergence percentage was highest in seed treated with 2.5% concentration of sweet flag powder as compared to other treatments and untreated control. This might be due to the least infestation of khapra beetle and higher 100 seed weight, germinability and seedling vigour index. The results are agreement to those reported by Charjan and Tarar (1994), Deshpande *et al.* (2010), Lambat *et al.* (2011) and Cherian *et al.* (2011).

Among the plant products sweet flag powder 2.5% were found to be significantly effective against khapra beetle through out the period of investigation. These findings are in agreement with Deshpande *et al.* (2010), Charjan and Tarar (1994) and Lambat *et al.* (2011). Thus, sweet flag and custard apple naturally occurring botanicals which are not toxic can be used as pre storage seed treatment, dispensing with the use of costly and toxic chemicals to control khapra beetle damage without adversely affecting the germination of wheat seed.

References

Abdul Baki, A.A. and Anderson J.D. 1973. Vigour determination in soybean seed by multiple criteria. Crop sci. 13: 630-633.

Anonymous, 1985. International rules for seed testing. Seed sci. and Technol. 13: 299-513.

Biradar, B.S. 2000. Prevention of cross infestation by Sitophilus oryzae L. and Rhizopertha dominica in stored wheat. M.Sc. (Agri.) Thesis, University of Agricultural Sciences, Dharwd.

Charjan S.K.U. and Tarar, J.L. 1994 The influence of some plant products on seed quality of lobia during storage. Ann. Plant Physiol. 8(2)L: 153-156.

Cherian Konglath, Charjan Sanjeev, Mohod Vandan, Lambat Ashish and Gadewar Rajesh. 2011. Studies on the influence of Acorus calamus L. rhizome Powder seed treatments against stored grain pest of wheat. Proc. of National Seminar on Environmental Management and Biodiversity conservation, held during 26-27 February 2011 at Rishikesh (U.K.) India. Abstract no A127 page 101.

Deshpande, V.K. Deshpande, H.H. and Masuthi, D. 2010. Evaluation of grain protectants in seed storage against Sitophilus oryzae (L.) in sorghum. Green farming 1(5):512-514.

Handerson, L.S. and Christensen, C.M. 1961 Preharvest control of insect and fungi. U.S. Dept. Agri. Ybk, pp. 348-356.

Howe, R.W. 1972 Insect attacking seed during storage. Seed Biology Vol. III (ed. Kozlowski, T.T.) Academic press: New York pp. 247-300.

Khan, M.I. and Borle, M.N. 1985 Efficacy of some safer grain protectants against the pulse beetle, Collosobruchus chenensis L. infecting stored Bengal gram (Cicer arietinum L.). P.K.V. Res. J. 9(1): 53-55.

Lambat Ashish, Gadewar Rajesh, Charjan Sanjeev, Cherian Konglath and Lambat Prachi, 2011. Evaluation of organic grain protectants in seed storage against rice weevil in wheat. Proc. of International conference on Sustainable Environment held during 19-20 February 2011 at Aurangabad, Maharashtra, India. Special issue 6: 122-123.

Narayanaswamy, S 1985. Effect of Pulse beetle damage on seed quality of field bean and Pigeon pea. Seed Res. 13(2): 138-141.

Panse, V.G. and Sukhatme, P.V. 1967. Statistical methods for agricultural workers. I.C.A.R. Pub., New Delhi.

Saxena, B.P., Koul, O. and Tikku, K. 1976. Non toxic protectants against the stored grain insect pest. Bull. Grain Technol. 14(5): 190 193.

Tikku K., Koul, O. and Saxena B.P. 1978. The influence of Acorus calamus L. oil vapour on the histocytological pattern of the ovaries of Trogoderma granarium Evert. Bull. Grain Tehnol. 16 (1): 3-9.

Biodiversity Conservation and Envir. Management (2012)
Editors: D.R. Khanna et al.
Pub. by Biotech Books. *ISBN: 978-81-7622-262-4*

Pages: 401-407

A STUDY OF ETHNOBOTANICAL USES & BIODIVERSITY OF *ACONITUM* (ATISH) IN DAYARA BUGYAL REGION OF DISTRICT UTTARKASHI

Ramdas[1], G.K. Dhingra[1✉], P. Pokhriyal[1], Sanjeev Lal[1] and M. A. Rather[2]
[1]Department of Botany Govt. P. G. College, Uttarkashi, Uttarakhand (INDIA)
[2]Department of Chemistry Govt. P. G. College, Uttarkashi, Uttarakhand (INDIA)
✉E-mail: gulshan_k_dhingra@yahoo.com

Aconitum is the botanical name of the genus of plant commonly known as Atis, monkshood or wolfsbane. The *Aconitum* genus belongs to the Ranunculaceae (commonly called Buttercup) family of flowering plants. There are over

✉ Corresponding author

250 species of *Aconitum*. Different *Aconitum* species (and their varieties) can be found in various parts of the world. Ethnobotany and biodiversity conservation of *Aconitum* sps. of Dayara Bugyal of Uttarkashi was studied in the present research work. This valuable medicinal herb was studied in terms of its distribution pattern, traditional knowledge of this plant by local people and discussion of its medicinal properties with the inhabitants of adjacent regions.

Three species namely *Aconitum heterophyllum*, *A. ferox* and *A. balfourii* were reported from this region. Man has used *Aconitum* as a medicine and poison for thousands of years. Villagers inhabiting this local area use the roots for treating headaches, paralysis (hemiplegia), rheumatism, arthritis, contusions, bruises, broken bones and to control the overheating of the body. *Aconitum ferox* is used in many ways. Very small doses are given to people that lack motivation or suffer from permanent fatigue.

Keywords: Bruises, Broken Bones, Contusions, Endangered Medicinal Plants, Indigenous (Ethnobotanical), Monkshood, Wolfs Bane.

Introduction

Aconitum species belong to the family Ranunculaceae. Genus *Aconitum* consists of herbaceous perennial plants which are chiefly natives of the mountainous parts of the northern hemisphere, growing in moisture retentive but well drained soil on mountain meadows. Their dark green leaves lack stipules. It is non-toxic if used properly. As Ayurvedic medicine, it is used for children suffering from fever and diarrhoea. The root is the main part of this plant that is used. The main place where this herb commonly found is the alpine and sub-alpine belts of the Himalayan region at altitudes between 1,800 and 4,500 m from Indus to Kumaun. The dried roots are used as analgesic, anti-inflammatory, antipyretic, antiperiodic, aphrodisiac, astringent, cholagogue, febrifuge and tonic. It is used in the treatment of liver disorders, dyspepsia, diarrhoea, indigestion, nausea, vomiting, throat pain, anorexia, piles and coughs and also slows the heart rate.

It is also used in Tibetan medicine, where it is said to have a bitter taste and a cooling potency. It is used to treat poisoning from scorpion or snake bites, the fevers of contagious diseases and inflammation of the intestines. The whole plant is highly toxic - simple skin contact has caused numbness in some people. This is a very poisonous plant and should only be used with extreme caution and under the supervision of a qualified practitioner. Seeds are stimulant, aromatic, emmenagougue, stomachic, carminative, anti-pyretic, hyperacidity, useful in hiccup and bad breath. It is regarded as a valuable tonic and digestive. Chemically *Aconitum* contains Diterpenoid alkaloid atisine which is the main constituent of the root. Others include atidine, histisine, hetisine, hetidine, heterophyllisine, heterophylline, heterlophylline, isoatisine, dihydroatisine and hetisinone and benzoyl heteratisine.

Materials and Methods

Before starting the research work on this important *Aconitum* species in the Dayara Bugyal regions of Uttarkashi District, general information about the area was collected from the local peoples. About five villages around the area were visited and surveyed where information from around 100 senior villagers was gathered. This region is above 2700 masl and extends up to 3800 masl. Ethnobotanical information (*Aconitum* plants and their uses) was gathered from each site by using a semi-structured and close ended questionnaire. Plant collection and data recording for traditional/ indigenous uses of these plants in various localities were primarily done by carrying the collected specimens to local peoples. The informants were asked questions in Hindi and Garhwali (local language). Collected plant material has been dried, pressed, preserved, accessioned, identified and finally deposited in the Herbarium of the Department of Botany, R. C. U. Govt. P. G. College, Uttarkashi (Uttarakhand). Identification of the field collected medicinal plants was done by confirming them by the respondents and comparing them with those in the various Herbaria of Uttarakhand. Necessary literature has also been assessed from Departmental and Central library of Govt. P. G. College, Uttarkashi.

Parameters Studied

Ethnobotanical information and plant species found were gathered from each site by using a semi-structured and close ended questionnaire containing questions such as:-

(1) Do you know the *Aconitum* (Atis) plant species in your local area; if yes, please name them.

(2) What is the use of these *Aconitum* plant species? How do you use them (as a medicine) and for which ailment?

(3) Which part of these plant species are used for medicinal purposes?

(4) When do you collect these plants?

(5) Do you collect them for your personal use or for selling them to pharmaceutical companies *etc*?

(6) Traditional uses of plants, their vernacular names, distribution, morphology and economical importance.

(7) Endangered *Aconitum* plant species and its collection *etc*.

Observation

The dried leaves of Himalayan monkshood are burned as incense, the seeds are sometimes used in rituals and the entire plant is considered to be an agent of protection. The seeds crushed in honey are applied locally on throat in tonsillitis. Nasal insufflations of roots are beneficial in headache (especially migraine). The roots of *Aconitum ferox* supply the Nepalese poison called ***bikh***, ***bish*** or ***nabee***. It contains large quantities of the alkaloid pseudaconitine, which is a deadly poison. *Aconitum palmatum* yields another of the bikh poisons.

The action of *Aconitum* on the circulation is due to an initial stimulation of the cardio-inhibitory centre in the medulla oblongata (at the root of the vagus nerves) and later to a directly toxic influence on the nerve-ganglia and muscular fibres of the heart itself. The fall in blood-pressure is not due to any direct influence on the vessels. The respiration becomes slower owing to a paralytic action on the respiratory centre and in warm-blooded animals; death is due to this action, the respiration being arrested before the action of the heart. Poisoning may also occur following picking the leaves without wearing gloves; the aconitine toxin is absorbed easily through the skin. From practical experience, the sap oozing from eleven picked leaves will cause cardiac symptoms for a couple of hours.

The Ativisha (*Aconitum heterophyllum*) has small, yellowish-white, bulbous roots, which shaped like a large bud about four to six times that of a jasmine bud. The root is best harvested in the autumn as soon as the plant dies down and is dried for later use. This plant has many green leaves and grows as a greedy plant, inhibiting the growth of nearby species. The root is best harvested in the autumn as soon as the plant dies down and is dried for later use.

Discussion

The plant parts used for medicinal preparations were bark, flower, fruit, leaf, root, rhizome, tuber, seed, shoot, resin, wood *etc.* In some cases the whole plant was utilized. The most frequently utilized plant part was the root/rhizome/tuber (26.15%), followed by leaf (23.84%) and flower and fruit (21.53%). The high importance of the underground part was attributed to having high concentrations of bioactive compounds. The largest number of remedies (21.70%) were used to treat respiratory tract infections (asthma, cold, cough, fever, headache, pneumonia, sinusitis *etc.*) while 19.92% remedies were taken to cure gastro-intestinal problems (cholera, gastritis, intestinal pain, stomachache *etc.*), 7.82% remedies were used for skeleto-muscular problems (arthritis, fracture, rheumatism, sprain, swelling *etc.*) and dermatological problems (scabies, skin disease *etc.*), 4.27% for ENT (ear, nose and throat) problems and to a lesser extent for cuts and wounds, dental, cardiovascular system, circulatory system *etc.* Plants were also important as tonic, astringent, anthelminthic, insecticide, incense stick, appetite stimulant, antidote *etc.* The preparation methods included decoction, juice, oil, paste, powder, extract, smoke, and raw (unprocessed). The majority of remedies were prepared as juice (29.52%), followed by raw (19.04%), paste (16.19%), extract (13.33%), decoction (11.42%), powder (6.66%) *etc. Jurinea dolomiaea* is used as incense and ethnomedicine for diarrhoea and stomachache in Uttarkashi district. Seed of *Juniperus indica* is eaten in Uttarkashi to get relief from kidney problems, whereas its leaf juice is taken for cough, cold and paralysis. Fruit juice and seed coat of *Juglans regia* are employed to treat wounds, while the paste from bark is applied for arthritis and hair growth.

Fig.1: Flowering Twig of *A. heterophyllum* plant

Fig.2: Dried Roots of *A. heterophyllum*

Fig.3: Young Planted *A. heterophyllum* plant

Fig.4: Mature plants of *A. heterophyllum*

Conclusion

People inhabiting this region have strong belief and faith on traditional herbal medicinal practice for health treatment and therefore, the conservation of medicinal plants is not only vital to their livelihood but also has immense cultural significance to them. Medicinal plants are now found growing sporadically in forests/pastures as well as in village groves. Uses of the plants and their produce/products from nearby forests by rural people are common because there is no alternative method to adopt. Forests have commercially exploitable medicinal species, which if manage properly can serve as a sustainable income sources for local communities. An urgent need, therefore for conservation of medicinal plant species and their habitats and indigenous knowledge is required. Ethnoecological knowledge, plant life forms and growth patterns are imperative to consider for management of Himalayan medicinal herbs.

Acknowledgement

The authors are grateful to Forest Department of Uttarkashi and Zila Beshaj Sangh Ikai, Uttarkashi for their supports during field studies. Thanks are also due to locals who gave invaluable time and cooperation during the work in this area.

References

Aumeerudy, Y. (1996). Ethnobotany, Linkages with Conservation and Development. In: *Proceedings of First Training Workshop on "Ethnobotany and its application to conservation"* NARC, Islamabad, pp. 152-157.

Bukhari, A.H. (1994). *Ethnobotanical survey and vegetation analysis of Machyara National Park Azad Kashmir, Pakistan.* M.Sc. Thesis University of Azad Kashmir. Cunningham, A.B. 1993. *African Medicinal Plants.* People and plants working paper. Division of Ecological Sciences. UNESCO 1: 1-50.

Elisabetsky. (1990). Plants used as analgesics by Amazonian Capbocols. *International Journal of Crude Drug Research*, 28: 309-320.

Harshburger, J.W. (1896). Purpose of Ethnobotany. *Botanical Gazette*, 21: 146-154.

Martin, G.J. (1995). *Ethnobotany: A People and Plants Conservation Manual.* Clapham & Hall, London, New York, Tokyo.

Shenji, P. (1994). *Himalayan Biodiversity Conservation Strategies.* Himavikes Pub. No.3.

Shinwari, M.I. and M.A. Khan. (2000). Folk use of medicinal herbs of Margalla Hills National Park, Islamabad, Pakistan. *Journal of Ethno pharmacology*, 69: 45-56.

Yang, S. (1988). A review on the derivation of Zizang (Tibetan) drugs and the advance of its research. *Acta. Bot. Yunnanic Adit.* 1, China 1988. Bhotasrity, Katmandu Nepal.

Zandial, R. (1994). *Ethnobotanical studies and population analysis of Machyara National Park Azad Kashmir.* M.Sc. Thesis University of Azad Kashmir.

Biodiversity Conservation and Envir. Management (2012)
Editors: D.R. Khanna et al.
Pub. by Biotech Books. *ISBN: 978-81-7622-262-4*

Pages: 409-418

ASSESSMENT OF AMBIENT NOISE LEVEL IN THE CITY OF ALIGARH (U.P.) INDIA

J.P.[1] Singh and Shakun Singh[2]✉
[1]U.P. Pollution Control Board, Aligarh, Uttar Pradesh (INDIA)
[2]Dept. of Applied Science, Babu Mohanlal Arya Smarak Engineering College (SGI Group), Agra, Uttar Pradesh (INDIA)

Aligarh is a city in the North Indian state Uttar Pradesh. It is located 90 miles south-east of New Delhi from where a Grand Trunk road passes through the city and has a population of half a million. It is also famous for lock manufacturing and Aligarh Muslim University. Monitoring was carried out to assess the ambient noise level at different locations in the city of Aligarh. Different locations were categorized as Commercial, Residential, Silence Zone or Sensitive area. Hospitals, Courts, Educational institutions, and eminent religious places are included in the Silence

✉ Corresponding author

Zone. Sources of sound generating equipments/ crackers *etc.* are not allowed in the radius of 100 meter of the Silence Zone. It has been observed that during the noise level monitoring of different categorized area, the noise level average data 48.5-82.5 dB(A) was found beyond the standards limits prescribed for different area as compared during night 38.2-54.4 dB(A) except values 66.3-71.5 dB(A) observed at G.T. Road. Higher noise level values may affect the hearing power; nervous system of human beings, disturbance to animals and may cause uneasy, stress, tension, heart, blood pressure etc to inhabitants of that area. Noise pollution is mainly due to automobiles, D.G. sets, musical instruments during parties, loud speakers, burning of crackers. To control the noise pollution people must be aware about its impact on the health of living beings and creatures. Concerned authorities should take a necessary action against the generators of noise pollution.

Keywords: dB (A), Leq., Ambient, Sensitive/Silence zone, Noise Pollution.

Introduction

Aligarh is located at 27.88°N 78.08°E. It has an average elevation of 178 metres (587 feet).The city is situated in the middle portion of Doab, or the land between the Ganges and Yamuna Rivers. The Grand Trunk Road passes through the city. This city is divided into two areas known as Old Aligarh (City) and New Aligarh (Civil lines). The city also has a few known markets-Railway Road market, Centre Point market, Amir Nisha and Shamshad market. There is a Government District hospital situated in the heart of the city.

Aligarh is particularly famous for locks and brass castings (sculptures) and also known as an educational centre. Aligarh is synonymous with the University that is spread across much of the city civil lines area. Aligarh

Muslim University is a residential academic institution of international repute offering more than 250 courses in traditional and modern branches of education. It is a premier central university with several faculties and maintained institutions and draws students from all corners of the world, especially Africa, West Asia and South East Asia. This university has also a famous health care facility known as J.N.Medical College. There are about two dozen famous schools, colleges and Technical and management institutions.

Grand Trunk Road and Agra-Mathura road are the busiest roads where vehicular movement generally causes Traffic jam every day. As a result of increase in urbanisation and population day by day, number of vehicles are also increasing. Besides this large no. of DG sets are also used mostly in commercial and industrial areas that cause noise pollution.

Keeping above facts in mind, it has been decided to assess the noise level of this city. Five places were selected for ambient noise level monitoring at different categorized areas of the city in the month of January-2011. Number of studies has been conducted on noise pollution in various cities of India (Edison *et al.*, 1999; Singhal, 2000; ETI, 2003; Bhatt *et al.*, 2004 and Khanna *et al.*, 2004).

Materials and Method

Only a sound level monitor was used to observe the values of sound level at different places. Regarding materials no more equipment was required during noise level monitoring. Methodology of sound level monitoring was adopted as per the manual of the instruments and CPCB guidelines. Sound pressure level was taken in "A" weightage. The values of noise level have been recorded in the form of minimum, maximum and average. Sound level monitoring was conducted at different categorized areas.

Monitoring Stations

Five monitoring stations were selected to conduct the ambient noise level monitoring at different categorized areas in the city which are given below:

S.No.	*Name of monitoring station*	*Categorised area*	*Monitoring point*
1.	Sut Mill Chauraha /GT road Crossing	Commercial	A
2.	Ramesh Vihar colony	Residential	B
3.	J.N.Medical college	Silence	C
4.	District Hospital	Silence	D
5.	Dubey Padao Chauraha/ GT road Crossing	Commercial	E

Brief Details about Monitorig Stations:

Sut mill Chauraha/Grand Trunk Road (G.T. Road) Crossing is generally a busy and crowded crossing/chauraha from where generally heavy and light vehicles pass to the city and to bypass roads. It is an outer crossing of the city. Commercial activities are held near this chauraha. This monitoring station is represented as–A.

Ramesh Vihar colony is an absolutely residential area where no commercial and industrial activities are held. It is situated in new Aligarh city (Civil line Area). This monitoring station is represented as–B.

J.N. Medical College is a main medical college of Aligarh Muslim University around which mostly residential areas are situated. This point was selected as a silence zone to study the noise level and it is represented as monitoring station-C.

District Hospital is known as Malkhan Singh Hospital and is situated in the dense city of Aligarh, around which commercial areas are situated. Grand Trunk road passes behind this hospital and Railway station is also situated. This point was taken as a silence zone which is represented as a monitoring station–D.

Dubey Padao Chauraha is situated at G.T.Road about 200 metre far away from an old road ways bus stand known as Gandhi Park bus stand, around which several activities like commercial and hotels are held. It is a busiest and crowded chauraha at G.T. Road due to traffic jam from where a Ramghat road passes over railway track bridge which is known as Meenakshi Bridge. This monitoring station is represented as–E.

Sound level monitoring was conducted for at least 10 minutes at each sampling point during day and night time to assess the ambient noise level at different places in the city. Monitoring places were categorized as Commercial, Residential, Silence Zone area. Silence Zone has been mainly defined as Hospitals, Courts, Educational institutions, historical monuments and eminent religious places where no sources of sound generating equipments, loud speakers, musical instruments on the occasion of functions/ parties, burning of crackers, diesel generators, blowing of horns of automobiles *etc.* are allowed in the radius of 100 meter from the Silence Zone.

Results and Discussion

Values of noise level data are given in Tables-1 and 2. Ambient noise level standards are mentioned in Table 3.

Table 1: Results of sound level data at different monitoring stations of the Aligarh city.

S. No.	*Monitoring place*	*Station Denoted as*	*Date of monitoring*	*Categorised Area*	*Values in dB(A) Leq.*					
					During Day time			*During Night time*		
					Min.	Max.	Avg.	Min.	Max.	Avg.
1	Sut Mill Chauraha	A	8.1.2011	Commercial	66.1	92.7	79.4	55.1	84.3	69.7
2	Ramesh Vihar colony	B	8.1.2011	Residential	46.6	61.6	54.1	26.4	50.5	38.45
3	J.N.Medical college	C	8.1.2011	Silence Zone	48.8	76.4	62.6	23.3	50.1	36.7
4	District Hospital	D	8.1.2011	Silence Zone	61.5	86.6	74.05	35.2	65.2	50.2
5	Dubey Padao Chauraha	E	8.1.2011	Commercial	67.8	95.2	81.5	51.1	76.3	63.7
			Average		48.47	82.5	58.608	38.22	54.4	51.75

Table 2: Results of sound level data at Grand Trunk Road in the city of Aligarh.

7	Sut Mill Chauraha	G.T.Road	8.1.2011	Commercial	During Day time	73.1	During Night time	69.7
8	Dubey Padao Chauraha	G.T.Road	8.1.2011	Commercial	During Day time	69.9	During Night time	62.9
			Average			71.5		66.3

Table 3: Ambient Noise Level Standards.

Sl.No.	*Category of area*	*Limits in dB(A) Leq.*	
		Day time	*Night Time*
1	Industrial Area	75	70
2	Commercial Area	65	55
3	Residential Area	55	45
4	Silence Zone	50	40

Notes:

1. Day time is reckoned in between 6.00 a.m.-10.00 p.m.
2. Night time is reckoned in between 10.00 p.m.-6.00 a.m.
3. Silence Zone is defined as areas upto 100 metres around such premises as hospitals educational, institutions, courts and eminents temples/monuments.
4. Use of vehicular horns, loudspeakers and bursting of crackers shall be banned in those zones (Notification, 2000).

It is evident from the Table 1 that during day time minimum sound level values varied between 46.6dB (A) - 67.8d B(A) and maximum values 61.6 dB (A) to 95.2 dB (A) where as during night time the values of minimum sound level was found between 23.3 dB (A) to 55.1 dB (A) and its maximum values varied from 50.1 dB (A) to 84.3 dB (A).The average values of minimum and maximum of day time monitoring was recorded 48.47 dB (A) and 82.5 dB (A) and during night time monitoring average values of minimum, maximum was observed 38.22 dB (A) to 54.4 dB (A) of all sampling stations. Sound level data of average values of Grand Trunk Road have been computed for day and night time monitoring period in which only two monitoring stations have been reported in Table 2. The average values were found between 66.3 dB (A) to 71.5 dB (A) of noise level at GT Road Crossings.

The values of minimum and maximum were observed 66.1dB(A) and 92.7 dB(A) and their average value observed 79.4 dB(A) during day time while during night time minimum, maximum and its average values were found as 55.1 dB(A), 84.3 dB(A) and 69.7 dB(A) respectively at the monitoring point-**A**. The average values of day and night time both are found beyond the prescribed standards for a commercial area at this point which may due to blowing of horns by heavy vehicular movements and traffic jam at GT Road.

The data of sound pressure level found between 46.6 dB(A) to 61.6 dB(A) and its average value was found 54.1 dB(A) during day time and values during night time were recorded between 26.4 dB(A) to 50.5 dB(A) and their average 38.45 dB(A) at monitoring point-**B**. A maximum value of day and night time found beyond the prescribed standards for a residential area which may due to blowing of horns by vehicular movements while the minimum and the average values of this point found within the prescribed standards limits during day and night time monitoring periods.

The values of sound pressure level at monitoring point-**C** ranges 48.8dB(A) to 76.4 dB(A) and its average 62.6 dB(A) obtained during day time and during night time the values of noise level found between 23.3 dB(A) to 50.1 dB(A) and its average value was observed 36.7 dB(A).The minimum values of day and night time and average of night period both were obtained within the prescribed standards limit for silence zone but maximum values of day and night time along with the average value of day time were observed higher than the prescribed standard limit for silence zone which may be due to random vehicular movement of emergency cases to Health care facilities.

At monitoring station-**D**, the range of sound pressure level observed between 61.5 dB(A) to 86.6 dB(A) and its average value noted 74.05 dB(A) during day time while during night time the range of noise level observed 35.2 dB(A) to 65.2 dB(A) and its average value found 50.2 dB(A).The minimum, maximum and average value of day time along with maximum, and average value of night time found beyond the limit of prescribed standards for silence zone which may be due to movement of light and heavy vehicles on GT road and trains on the railway track.

The minimum, maximum and its average of noise level were observed as 67.8 dB(A), 95.2 dB(A) and 81.5 dB(A) respectively during day time and 51.1 dB(A), 76.3 dB(A) and 63.7 dB(A) respectively during night time at sampling point-**E.** The minimum, maximum and average values of day time along with maximum and average value of night time found beyond the limit of prescribed standards for commercial area which may be due to light and heavy vehicular movements, traffic jam and working of DG sets during day time.

Sut mill chauraha /GT Road crossing is a place from where heavy traffic move to the by pass roads and to the city and it is a general traffic jam area. During night and day time monitoring the values of sound pressure level found 69.7dB (A) to 73.1dB (A) at Sut mill chauraha and values observed 62.9dB (A) to 69.9dB (A) at Dubey Padao Chauraha. All these values even also their aggregate values 71.5dB (A) and 66.3dB (A)are found beyond the prescribed standards for a commercial area which may be due to blowing of horns by vehicles, traffic jam and commercial activities. Bhatt *et al.*, 2004, Khanna *et al.* (2004) reported 65.0 dB(A) to 81.0 dB(A); and Babu (2003) reported noise level 30.0 to 90.0 dB(A) and Ingle *et al.* (2001) by traffic.

Highest average values are found at monitoring point-E and A during day and night time as compared to monitoring point-D,B,C which may be due to heavy traffic, traffic jam, blowing of horns, operation of DG sets during power failure, sirens of ambulance and VIP vehicles at the Grand Trunk Road. Average values of day time monitoring except the value of sampling point-B were not found within the prescribed standard limit of their categorized areas, where as during night time monitoring period the average values of the noise level were observed within the prescribed standard limit at only two monitoring points-B and C and this may be due to negligence of vehicular movements at these points.

Blowing of pressure horns, vehicular movements, bursting of crackers, playing of musical instruments, loudspeakers, operations of DG sets and industrial activities *etc* are responsible for higher level noise pollution and persons exposed to this level of noise pollution for long time may suffer the menace of noise pollution such as hearing loss can be temporary or permanent. Consequently, students in college, office bearers, visitors, tourists, shop owners, patients in Hospitals, birds and other animals etc are exposed to very high noise level pollution.

Higher noise level pollution can disturb our work, study of students, rest, sleep, communications, hearing power, nervous system of human beings, disturbance to animals, and its impact may cause uneasy, stress, tension, headache, heart problem, blood pressure, loss of hearing etc to the inhabitants of that area. Noise pollution is mainly due to automobiles, D.G.sets, musical instruments during the occasion of functions/parties.

loud speakers, burning of crackers, industrial activities *etc.* Residential areas even silence zones are not exceptional from the exposure to high noise levels. Hence, it should be the duty of every citizen to minimize or control the noise pollution and people must be aware about its impact on the health of living beings.

Effects of Noise Pollution:

1. Even short exposure to noise can produce temporary hearing losses.
2. Prolonged exposure to noise can lead to a gradual deterioration of the inner ear and subsequent deafness.
3. Constant noise causes the blood vessels and muscles to contract. This causes a gradual loss of hearing, tension, nervousness and psychiatric illness. High intensity sound emitted by many industries and supersonic aircrafts, when continued for long time not only disturb but also permanently damage hearing.
4. Noise has harmful effects on non living material too. Numerous examples can be cited where old building and even new constructions have developed cracks under the stress of explosive sounds.

Local authorities must take a necessary action to prevent this important city from the menace of noise by following some main steps against the creators of noise pollution. Some suggestions may be adopted for control of noise pollution as given below:

1. Strict enforcement to ban the use of loud speakers, musical instruments and bursting of crackers on the various occasions of functions and parties.
2. DG sets should not be allowed without noise pollution control system like acoustic enclosures/ canopy in the city.
3. Restrict the movement of unmaintained vehicles and blowing of horns.
4. Specific legislation and regulations should be proposed for designing and operation of machines to include vibration control, sound proof cabins and sound -absorbing materials.

References

Ingle, S T; Attarde, S B; Dhake, R.B; and Panchpande, B.G. 2001:Noise Pollution- An insidious Hazardous in Urban Environment-A case study of Jalgoon City-" *Status of Indian Environment*-(Pre conference Proceedings), ASEA, Rishikesh, P-31.

Singhal, S.P.2000: Noise pollution and control in Urban and Industrial environment National Physical Laboratory, New Delhi. *A short course on Ambient Air monitoring and management.*

Bhatt, C.S.; Sikander, Mohd; Singh, J.P.; Khanna, DR; Gautam, Ashutosh; and Singh, Shakun; 2004: Astudy of ambient air quality and noise pollution in Nainital city. *Environment Conservation Journal* 5(1-3)7-13.

Khanna, DR; Singh, J.P.; and Singh, Shakun; 2004: Assessment of noise pollution in the city of Haridwar (Uttaranchal) India. *Environment Conservation Journal* 5 (1-3)61-66,2004.

Babu, S.Sarvana 2003 Noise Pollution health hazard. ***Noise Pollution.*** Environmental Training Institute, DANIDA (Denmark),Tamil Nadu Pollution Control Board Chennai.P.9.

Edison., R. Raja; C. Ravi Chandran and J. Christal Sagila, 1999. A assessment of Noise Pollution due to automobiles in Cuddalore, Tamil Nadu. *India. J. Env. Health.*41(4):312-316.

ETI, 2003 Noise Pollution Environmental Training Institute DANIDA (Denmark), Tamil Nadu Pollution Control Board Chennai.P.I.

Biodiversity Conservation and Envir. Management (2012)
Editors: D.R. Khanna et al.
Pub. by Biotech Books. *ISBN: 978-81-7622-262-4*

Pages: 419-427

A REVIEW OF EXPERIENCE AND EXPERIMENTS ON TRIPHALA (A TRI FRUIT POWDER COMPOUND) IN PRACTICE

Uttam Kumar Sharma
Department of Panchakarma
Rishikul Govt. Ayurvedic College & Hospital, Haridwar

Introduction

Triphala is a Ayurvedic herbal formulation consisting of the dried fruits of three medicinal plants, Termnalia chebula, Terminalia bellirica and Phyllanthus emblica. Triphala means three (tri) fruits (phala). The constituents of Triphala have been considered as and important Rasayana in Ayurvedic Medicine. The durgs of Rasayana group are believed to promote health, immunity and longevity. According to Ayurveda, these drugs nourishes all tissue of the body, prevent aging, promote intellect and prevent disease.

Triphala is used in the treatment of a variety of conditions and also forms part of many other Ayurvedic formulations. Conditions for which triphala is employed include headache, dyspepsia, constipation, conjunctivitis, skin disorders, liver dysfunction, diabetes mellitus *etc.* It is also used as a blood purifier and a purgative and to improve the mental faculties and is reported to possess anti-inflammatory, analgesic, anti-arthritic, hypoglycaemic and anti-aging properties. Triphala and its constituents plants have been reported to possess numerous biological activities *i.e.* immunomodulatory, adaptogenic, anti-bacterial, anti-fungal, anti-viral, cardiotonic, hepatoprotective anti cancer antivity and radioprotective.

In practice this drug has been found very effective and safe, when administered in cases of different disorders. This drugs is successfully used in all age groups from new born to old age persons in traditional Indian medicine.

The most popular herbal remedies in the health food industry are those that promote bowel movement. The reason is quite simple since a very common problem for so many individuals is constipation and bowel irregularity. Consider how tremendously valuable a formula is that not only regulates bowel movement but at the same time does the following:

- Improves digestion,
- Reduces serum cholesterol,
- Improves circulation (potentiates adrenergic function),
- Contains 31% linoleic acid,
- Exerts a marked cardio-protective effect,
- Reduces high blood pressure,
- Improves liver function,
- Has proven anti-inflammatory and anti-viral properties,
- Expectorant, hypotensive

Termnalia chebula

Haritaki is a rejuvenative, laxative (unripe), astringent (ripe), anthelmintic, nervine, expectorant, tonic, *carminative*, and appetite stimulant. It is used in people who have *leprosy* (including skin disorders), *anemia*, *narcosis*, piles, chronic, intermittent *fever*, *heart disease*, *diarrhea*, *anorexia*, cough and excessive secretion of mucus, and a range of other complaints and symptoms. According to the Bhavaprakasha, Haritaki was derived from a drop of nectar from Indra's cup. Haritaki is used to mitigate Vata and eliminate ama (toxins), indicated by constipation, a thick greyish tongue coating, abdominal pain and distension, foul feces and breath, flatulence, weakness, and a slow pulse. The fresh fruit is dipana and the powdered dried fruit made into a paste and taken with jaggery is malashodhana, removing impurities and wastes from the body. Haritaki is an effective purgative when taken as a powder, but when the whole dried fruit is boiled the resulting decoction is grahi, useful in the treatment of diarrhea and dysentery. The fresh or reconstituted fruit taken before meals stimulates digestion, whereas if taken with meals it increases intelligence, nourishes the senses and purifies the digestive and genitourinary tract. Taken after meals Haritaki treats diseases caused by the aggravation of Vayu, Pitta and Kapha as a result of unwholesome food and drinks. Haritaki is a rasayana to Vata, increasing awareness, and has a nourishing, restorative effect on the central nervous system. Haritaki improves digestion, promotes the absorption of nutrients, and regulates colon function.

Phyllanthus emblica

According to Ayurveda, aamla fruit is sour (amla) and astringent (kashaya) in taste (rasa), with sweet (madhura), bitter (tikta) and pungent (katu) secondary tastes (anurasas). Its qualities (gunas) are light (laghu) and dry (ruksha), the postdigestive effect (vipaka) is sweet (madhura), and its energy (virya) is cooling (shita).

According to Ayurveda, aamla balances all three doshas. While aamla is unusual in that it contains five out of the six tastes recognized by Ayurved, it is most important to recognize the effects of the "virya", or potency, and "vipaka", or post-digestive effect. Considered in this light, aamla is particularly helpful in reducing *pitta* due to its cooling energy. and balances both Pitta and *vata* by virtue of its sweet taste. The *kapha* is balanced primarily due to its drying action. It may be used as a *rasayana* (rejuvenative) to promote longevity, and traditionally to enhance digestion (dipanapachana), treat constipation (anuloma), reduce fever (jvaraghna), purify the blood (raktaprasadana), reduce cough (kasahara), alleviate asthma (svasahara), strengthen the heart (hrdaya), benefit the eyes (chakshushya), stimulate hair growth (romasanjana), enliven the body (jivaniya), and enhance intellect (medhya).

In Ayurvedic polyherbal formulations, Indian gooseberry is a common constituent, and most notably is the primary ingredient in an ancient herbal rasayana called *Chyawanprash*. This formula, which contains 43 herbal ingredients as well as clarified butter, sesame oil, sugar cane juice, and honey, was first mentioned in the *Charaka Samhita* as a premier rejuvenative compound.

Emblica officinalis tea may ameliorate *diabetic neuropathy*. In rats it significantly reduced blood glucose, food intake, water intake and urine output in diabetic rats compared with the non – diabetic control group.

Terminalia bellirica

"This tree, in Sanskrit Vibhita and Vibhitaka (fearless), is avoided by the Hindus of Northern India, who will not sit in its shade, as it is supposed to be inhabited by demons. Two varieties of T. belerica are found in India, one with nearly globular fruit, 1/2 to 3/4 inch in diameter, the other with ovate and much larger fruit. The pulp of the fruit (Beleric myrobalan) is considered by Hindu physicians to be astringent and laxative, and is prescribed with salt and long pepper in affections of the throat and chest. As a constituent of the triphala (three fruits), *i.e.*, emblic, beleric and chebulic myrobalans, it is employed in a great number of diseases, and the kernel is sometimes used as an external application to inflamed parts. On account of its medicinal properties the tree bears the Sanskrit synonym of Anila-ghnaka, or "wind-killing." According to the Nighantas the kernels are narcotic."[6] Ayurveda

- Rasa (taste): All but salty, mainly astringent, bitter, hot, sweet
- Virya (energy): Heating
- Vipaka (post-digestive effect): sweet
- Guna (quality): light, dry
- Dosha: VPK=
- Dhatu: All tissues
- Srotas: digestive, excretory, nervous, respiratory, female reproductive

Different Effects and Therapeutic Uses of Triphala

Smooth Laxative Effect

Triphala is considered the best colon cleanser in the world. All three herbs in Triphala are laxative in nature and very beneficial in removing toxins. Colon cleansing is an ancient and proven practice for rejuvenating the body. Colon cleansing helps detoxify and clean decayed food residues and other toxic substances accumulated in the intestines walls. This not only gets rid of constipation, hemorrhoids, parasites, flatulence, acne, bad breath and foul body but also prevents colon cancer. Colon cleansing also helps to relieve headaches, irritability, and depression. Triphala, formulated by Ayurvedic physicians thousands of years ago and used by literally billions of people since, is considered the most effective and safest laxative and colon tonic by most health care practitioners. Where other colon formulas are depleting, Triphala is mild, non-habit forming, and a rejuvenative. It improves the absorption of food in intestines especially in the duodenum where most chemical digestion takes place. Triphala also helps to stimulate various gastric enzymes that help to convert complex food into more easily digestible form. More over Triphala stimulates the peristaltic action of the intestinal lining (due to its anthroquinones and other bitters) that also aids in proper movement of food at various levels of the digestive tract. It is also helpful in making the stool loose there by facilitation the easy evacuation of the bowel. While triphala is somewhat laxative to many people due to the anthoquinones, the balance of tannins naturally present has a toning effect. In vitro studies have also shown that triphala is lethal to a variety of gastrointestinal pathogens, including bacteria such as Salmonella typhii, Shigella, Klebsiella and Pseudomonas, and fungi like Candida albicans. Triphala, as it is called, is the most popular Ayurvedic herbal formula of India, since it is an effective laxative that also supports the body's strength. The constitution of vegetarian Hindus cannot tolerate harsh laxatives anymore than vegetarians in other countries. Because of its high nutritional value, Triphala uniquely cleanses and detoxifies at the deepest organic levels without depleting the body's reserves. This makes it one of the most valuable herbal preparations in the world.

How is Triphala different from other kinds of laxatives? There are two primary types of herbal laxatives. One is called a purgative and includes herbs such as senna, rhubarb, leptandra, buckthorne and cascara. These often contain bitter principles in the form of anthroquinones that work by stimulating the peristaltic action of the intestinal lining, either directly or by promoting the secretion of bile through the liver and gall bladder.

The second type of laxative is a lubricating bulk laxative, including demulcent herbs such as psyllium and flax seed. This is more nutritional and usually does not have any significant direct effect on either the liver or the gall bladder. Instead, these work like a sponge by swelling and absorbing fluid, thus acting as an intestinal broom A comparative study of the efficacy of Emblica officinalis fruit powder with a conventional antacid formulation was performed. In a 4 week study, 38 patients with dyspepsia and with or without stomach ulcer were divided into two groups : one group received the fruit powder and the other, the antacid. Emblica was used at the level of 3 gm per dose, three times a day while gel antacid was used at 30 ml per dose, up to 6 times a day. The improvement in the clinical symptoms score (belching, fullness, heartburn, regurgitation, nausea and vomiting) in ulcer dyspeptics was from the initial 4.2 to 0.4 ($p < 0.01$) post treatment score for antacid group, and 4.6 to 0.6 ($p<0.05$) for the Emblica group. Endoscopic examination showed all ulcers in the antacid group in the process of healing, while all but one patient in Emblica group had completely healed ulcers. In the non-ulcer group both antacid and Emblica produced a significant decrease in clinical symptoms score from 4.4 to 1.53 ($p<0.01$) and 5.0 to 1.61 ($p<0.01$) respectively (22). Triphala's benefits for the digestive system make it a very useful herbal medicine for treating Irritable Bowel Syndrome (IBS) and Ulcerative Colitis.

Anti obese Effect

Severe obesity is usually accompanied by congestion of the internal organs of elimination, including the liver and bowels. One recent molecular finding supports Triphala's traditional use in treating obesity. Researchers at the BRA Centre for Biomedical Research found that active molecules in triphala bind to the cellular receptor for CCK. CCK, or cholecystokinin, is a satiety hormone, released to indicate that you are full, and is especially responsive to fat. Synthetic analogs of CCK are under development by pharmaceutical companies to help people realize that they're full, thereby controlling appetite and supporting healthy weight. Since both overeating and obesity disrupt digestion and overburden the gastrointestinal tract, including the liver and bowel. As a result, digestion is compromised, leading to poor nutrient assimilation, imbalances and overgrowth in the intestinal microflora, and putrefaction of ill-digested food. For such conditions, triphala can be highly effective in removing stagnation of both the liver and intestines.

Anti Hypertensive Effect

Triphala's strong affinity for micro-organisms prevents them from causing damage in the body and renders them inactive. This astringent property of Triphala helps in purification of blood and also helps in maintaining the proper density of the blood. Hence it is very helpful in eradicating all blood borne diseases and also skin related problems. Triphala reduces the serum cholesterol, and high blood pressure. It significantly improves function of the liver as well as blood circulation. It has the ability to exert a remarkable protection against cardiovascular diseases. Triphala reduces the plaque formation in the arteries thereby reducing the risk of heart related problems caused due to arteriosclerosis

Immuno modulating Effect

Triphala has powerful anti oxidant agents that help in regularizing the metabolism of cells and ease their proper functioning. This reduces the risk of production of free radicals that are the main cause of aging. It also stimulates the functioning of cell organelles like mitochondria, golgi bodies and nucleus that play a vital role in proper functioning of the cell. Recent studies have shown Triphala to be an anti-cancer agent. It has the ability to kill tumor cells while sparing normal ones

Anti Diabetic Effect

Triphala is very effective in treating diabetes mellitus. It helps in stimulating the pancreas. Pancreas contains islet of langerhans that secretes insulin. Insulin is responsible for maintaining the proper glucose level in the body. More over due to its bitter taste it is also advisable to be taken in hyperglycemia. In a study conducted by the American Botanical Council, it was shown that Triphala greatly reduced blood glucose levels in diabetic rats. The study concluded that the mechanism of action of these extracts for lowering blood glucose is not known, but they may decrease the effect of inflammatory cytokine release in diabetics, which in turn might reduce insulin resistance. Interestingly, the authors note that traditional medications used to treat diabetes also have significant antioxidant effects.

Eye Protective

Triphala is also effective in all eye diseases including the treatment of conjunctivitis, progressive myopia, the early stages of glaucoma and cataract. Triphala is also widely taken for all eye diseases including the treatment of conjunctivitis, progressive myopia, the early stages of glaucoma and cataracts. For these conditions, it is taken daily both internally as described above, as well as externally as an eye wash. Steep one tablespoonful of the powder or six tablets in an 8 ounce glass of water overnight. In the morning, strain the infusion through a clean cloth. The resultant tea is used to sprinkle over the eyes or used in an eyewash with an eyecup that can be readily purchased at most drug stores. One can drink the remainder in one or two doses, morning and evening. Taken in this way for at least three months, Triphala becomes an herbal eye tonic.

Nutritive Supplement

Triphala combines both nutritional as well as blood and liver cleansing actions. It has little function as a demulcent or lubricating laxative, however. It possesses some anthroquinones that help to stimulate bile flow and peristalsis. The nutritional aspect is in the form of its high vitamin C content, and the presence of linoleic oil and other important nutrients that make it more of a tonic.

Because of its high nutritional content, Ayurvedic doctors generally do not regard Triphala as a mere laxative. Some of the scientific research and practical experience of people who have used it down through the ages has demonstrated that Triphala is an effective blood purifier that stimulates bile secretion as it detoxifies the liver, helps digestion and assimilation, and significantly reduces serum cholesterol and lipid levels throughout the body. As a result, it is regarded as a kind of universal panacea and is the most commonly prescribed herbal formula.

A popular folk saying in India is: "No mother? Do not worry so long as you have Triphala." The reason is that Indian people believe that Triphala is able to care for the internal organs of the body as a mother cares for her children. Each of the three herbal fruits of Triphala takes care of the body by gently promoting internal cleansing of all conditions of stagnation and excess while at the same time it improves digestion and assimilation.

Other Scientific Studies

One Indian study reported by C.P. Thakur, demonstrated the enormous value and effectiveness of Amla, in reducing serum, aortic and hepatic cholesterol in rabbits. In another study, extracts of Amla fruit were found to decrease serum free fatty acids and increase cardiac glycogen. This helps to prevent heart attacks by providing significantly greater protection and nourishment to the heart muscle.

Studies of the fruit of Bihara found that it contains up to 35% oil and 40% protein. The oil is used in soap making and by the poorer classes,as a substitute cooking oil for ghee. The sweet smelling oil is 35% palmitic, 24% oleic and 31% linoleic. Linoleic oil is an essential fatty acid important for increasing HDL cholesterol, associated with a healthy state and reducing LDL cholesterol, considered to indicate a higher-than-average risk for developing coronary-heart disease.

One of numerous studies of Harada demonstrated its anti-vata or anti-spasmodic properties by the reduction of abnormal blood pressure as well as intestinal spasms. This confirms its traditional usefulness for heart conditions, spastic colon and other intestinal disorders.

Conclusion

With all the virtues of the three individual herbs, Triphala has many wide and varied uses as a therapeutic herbal food. Before considering pathological indications for which Triphala would be appropriate, we should never ignore the value of taking it on some regular basis whether once daily or once or twice a week simply for health maintenance. Triphala, having great nutritional properties, will help to prevent sickness.

Biodiversity Conservation and Envir. Management (2012)
Editors: D.R. Khanna et al.
Pub. by Biotech Books. *ISBN: 978-81-7622-262-4*

Pages: 429-435

EFFECT OF ENVIRONMENTAL FACTORS ON PRODUCTION AND QUALITY OF EGGS OF JAPANESE QUAIL, *COTURNIX COTURNIX JAPONICA*

Jyoti Ramteke[1], Pravin Charde[1], Suresh Zade[2]✉ and Rucha Gabhane[1]

[1] Research Academy, Sevadal Mahila Mahavidyalaya, Nagpur-440009 (M.S.)

[2] Post Graduate Teaching Department of Zoology, R.T.M. Nagpur University, Nagpur-440033(M.S.).

E-mail: profsbzade@redifmail.com

The Japanese quail (*Coturnix coturnix japonica*) is a domesticated species, which is highly adaptable to climatic changes. In the present study, the egg mass and egg volume under control conditions were studied. In the summer season, record for 4 months was maintained to study the egg incubation after fumigation. A significant correlationship

✉ Corresponding author

was observed between climatic conditions of central India (such as the temperature, humidity) and egg production with good egg mass and egg quality.

Key-words: Japanese Quail, temperature, humidity, incubation, egg production

Introduction

Quails are partridge like bird popularly known as '*Bater*' belongs to the class Aves and family *Phasianidae. Bater* is a good table bird known for its delicacy since olden days. They are used as food before chicken was domesticated. Quail meat is rich in vitamins, amino acids, unsaturated fatty acid, and phospholipids which are vital for health of human being. Quail meat therefore can be recommended to be included in diet of children, pregnant women and convalescent patient for speedy recovery. The Japanese quail (*Coturnix coturnix japonica*), a truly domesticated species is highly adaptable to variable climatic changes and are resistant to many common poultry diseases which make them suitable for rearing in large scale in relatively small space.

Quail production is popular in Japan, China and Taiwan but comparatively neglected in developing countries including India. In Europe, particularly in Italy, Japanese Quail have been in use very recently as experimental animals for medical research. Due to commercial exploitation and urbanization of forest and waste land, the quail population in India has declined considerably. Although, *bater* feeds on paddy, jawar and wheat, it is not yet known whether these birds are agricultural pestilent and causes considerable damage to crops. Since these animals are important in ecosystem and to human economy, there is an urgent need to protect them; unfortunately Indian quails are greatly neglected in establishment of quaillary. For conservation of wild species of quail, breeding in capacity and introduction of some new species in quaillary of Central India are most essential. Fundamental knowledge of reproductive process and the endocrine mechanism that control this process is therefore inevitable for profitable management and production of these birds in large scale (Charde, P. N., 1998).

The paucity of information on reproductive endocrinology and developmental biology of Indian quail motivated us to undertake this investigation. Living organisms are continuously facing the environmental impacts and hence in turn they develop an influence on the surrounding environment thus creating a biological unity. The study of this animal-environment relationship helps the researchers to study the modern methods of development under different environmental factors (Tarasewicz. Z, *et al.,* 2006).

In this context, the objective of this research is to standardize the different parameters of environment for the better quality of egg production to be used for incubation and subsequently for hatching.

Material And Methods

Domesticated quail do not have the tendency for broodiness and hence eggs must be incubated under a broody hen or by artificial incubation. Japanese quail eggs can be successfully incubated by using almost any type of commercial incubators. Japanese quail eggs are a mottled brown colour and are often covered with a light blue chalky material. Each quail hen appears to lay eggs with a characteristic shell pattern and colour. The average egg weighs about 10 gms (±2 gms) that is about 8% of the body weight of quail hen. Young chicks weigh 6-7 gms when hatched and are brownish with yellow stripes. The shells are very thin and susceptible to shell damage, so it is to be handled with care (M. Randell and G. Bolla, 2007).

The eggs were collected from the local poultry farm. The egg number, egg mass and egg volume was observed for four months, which gives the basic idea about the effect of environmental factors such as temperature and humidity. The eggs were made ready to incubate after fumigation. BOD incubator is thoroughly washed and disinfected with commercial disinfectant before using and is set at 35°C temperature and 60% humidity. Eggs were cleaned and then the eggs were fumigated before placing in the incubator. The eggs were kept under one large wooden box for fumigation (M. Randell and G. Bolla, 2007). The eggs were fumigated by using potassium permanganate (25gms) in earthenware added with formalin.

After fumigation the eggs were kept for incubation. The temperature of the incubator was set at 35.5°C and humidity was set at 60% by keeping the tray half filled with water along with the sponge. During incubation, the tilting movement was to be done manually after every 3 hours. The eggs were marked on the vegetal pole of the egg for easy tilting. The tilting was stopped after 15 days of incubation. During incubation, some eggs were removed and opened to observe if the development of embryo is proper. Some eggs were destroyed which shows the egg mortality.

Results And Discussion

Japanese quail shows variation in egg mass, egg colour and shape. It is seen that the egg mass is influenced by temperature. In the month of March, temperature is warm i.e. around 30°C-35°C. We received 231 eggs (n=231) in the month of March were observed and the average egg mass was maximum as compared to other three months i.e. April, May, June (Du Plessis and Erasmus, 1972; Ayorinde *et al.*, 1988) reported consistent reduction in body mass and attributed to increased use of physiological reserve to meet the demand of egg production (Oke et al., 2004). The egg mass is less but the production is more in the month of April. Production of May, the highest temperature month is nearer to the production of March but the egg mass is less. In the humid weather of June, the production increases but the egg mass remain constant.

Table 1: Monthwise average egg mass and egg volume of eggs collected.

Sr. No.	*Month*	*No. of eggs collected (n)*	*Average Egg mass (gms)*	*Average Egg volume (mm³)*
1	March	231	10.83	10604.18
2	April	324	10.51	10494.25
3	May	235	10.23	21611.93
4	June	266	10.34	10104.82

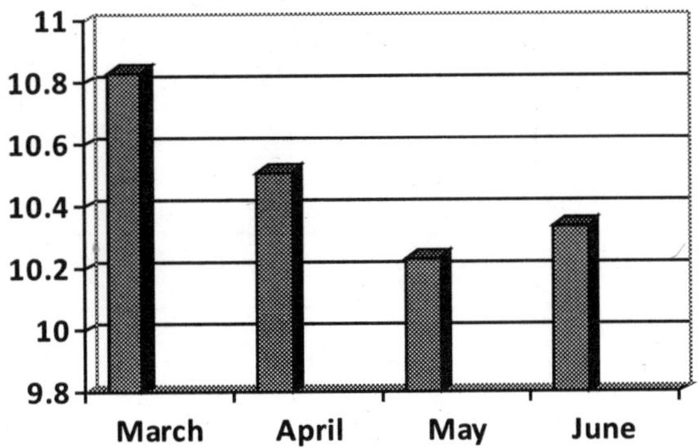

Graph 1: Monthly variation in average egg mass (gms) during summer.

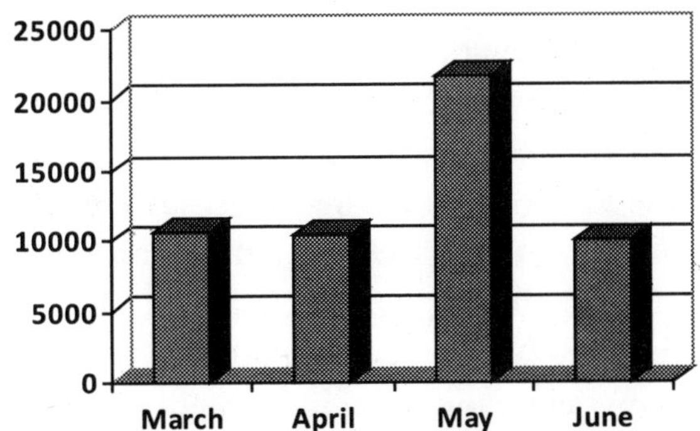

Graph 2: Monthly variation in average Egg (mm^3) volume during summer

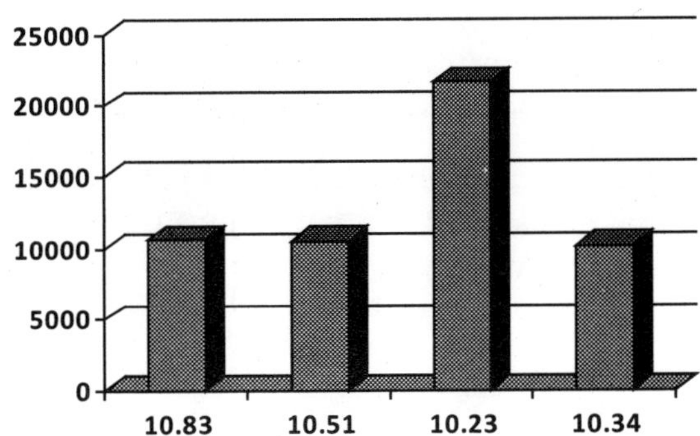

Graph 3: Graph showing correlation between average monthly egg mass (gms) and average monthly egg volume (mm^3) during summer

Table 2: Climatic Conditions of Central India

Sr. No.	*Month*	*Average Temperature (°C)*		*Average Relative Humidity (%)*		*Average Rainfall (mm)*
		Max.	*Min.*	*Max.*	*Min.*	
1	March	39	21	50	29	0.6
2	April	43	26	33	22	0.0
3	May	45	29	36	22	0.1
4	June	39	27	65	45	4.5

Source: Metrological Department

The above results show that the temperature directly affects the egg production as well as the egg mass. Similarly, the humidity was high in the month of June and egg mass as well as egg production is the maximum. This shows that the temperature and humidity are equally responsible for the production of eggs (Ar. A, 1991). As the temperature increases, the egg mass decreases and the egg volume increases. This shows the water consumption of the birds (Ar and Rahn, 1980). In the month of June, when the temperature is moderate and humidity is high, the egg production with normal weight (around 10 to 12 gms.) is increased much. The egg volume is equalized with the egg weight (Martin, P. A. and Arnold T. W., 1991).

Conclusion

A significant correlationship was observed between climatic conditions of central India (such as the temperature, humidity) and egg production with good egg mass and egg quality. Subsequently both can help us to incubate the Japanese quail eggs in climatic conditions of Central India for the economical and nutritional benefit of society.

References

1. Ar, A., and H. Rahn. (1980). Water in the avian egg: overall budget of incubation. Am. Zool. 20:373-384.

2. Ar. A (1991). Egg water movements during incubation. In S. G. Tullet (editor). Avian incubation. London, (Buterworth-Heinemann) 157-173.

3. Ayorinde K L Toye A A and Aruleba, O. A. (1988). Association between body weight and some egg production traits in a strain of commercial layer. Nigeria Journal of Animal Production 15:119-121.

4. Charde, P. N. (1998). Influence of Exogenesis Pharmacological Compounds on Reproductive Biology of *Coturnix coromandelica*. Ph.D. Thesis, Nagpur University, Nagpur.

5. Du Plessis, PHS and Erasmus, J. (1972). The relationship between egg production, egg weight and body weight in laying hens. World Poultry Science Journal 28(2): 73-78.

6. Martin P. A. and Arnold T. W. (1991). Relationship among fresh mass, incubation time, and water loss in Japanese quail eggs. *The Condor*, 93:28-37.

7. Oke U.K., Herbert U., Nwachukwu E.N.(2004). Association between body weight and some egg production traits in the guinea fowl (*Numida meleagris galeata*. Pallas). Livest. Res. Rural Develop. 16, 9, pp. 72.

8. Randall M. (2007). Raising Japanese Quail. NSW Department of Primary Industries, Australia, 1-7.

9. Tarasewicz, Z., Szczerbinska, D., Majeswska, D., Danczak, A., Ligocki and M., Wolska, A. (2006). The effect of magnetic field on hatchability of Japanese quail eggs. Czechkoslavia Journal of Animal Science, 51(8):355-360.